BATTLEGROUND
ENVIRONMENT

BATTLEGROUND

ENVIRONMENT

VOLUME 1 (A–H)

Robert William Collin

GREENWOOD PRESS
Westport, Connecticut • London

Library of Congress Cataloging-in-Publication Data

Collin, Robert W., 1957–
    Battleground : environment / by Robert William Collin.
       p. cm.
    Includes bibliographical references and index.
    ISBN-13: 978–0–313–33865–6 (set : alk. paper)
           978–0–313–33866–3 (v. 1 : alk. paper)
           978–0–313–33867–0 (v. 2 : alk. paper)
1. Environmental sciences.   2. Environmental policy.   3. Environmental degradation.
4. Human beings—Effect of environment on.   I. Title.   II. Title: Environment.
    GE105.C65   2008
    363.7—dc22   2008002114

British Library Cataloguing in Publication Data is available.

Library of Congress Catalog Card Number: 2008002114
    ISBN: 978–0–313–33865–6 (set)
    ISBN: 978–0–313–33866–3 (vol. 1)
    ISBN: 978–0–313–33867–0 (vol. 2)

First published in 2008

Greenwood Press, 88 Post Road West, Westport, CT 06881
An imprint of Greenwood Publishing Group, Inc.
www.greenwood.com

Printed in the United States of America

The paper used in this book complies with the
Permanent Paper Standard issued by the National
Information Standards Organization (Z39.48–1984).

10 9 8 7 6 5 4 3 2 1

These books are dedicated to the spirit and vision
of the late Damu Smith, founder of the National Black Environmental Justice
Network, and former Greenpeace community organizer in
Louisiana and Texas. He recognized and taught that the environmental context
of Truth can bring us together to face the
environmental controversies in our midst.

Battleground· Environment is also dedicated to my two faithful
canine research companions, Ambar and Max.

CONTENTS

Guide to Related Topics   xi

Preface   xv

Introduction   xix

Entries

Acid Rain   1

Air Pollution   13

Animals Used for Testing and Research   19

Arctic Wildlife Refuge and Oil Drilling   28

Automobile Energy Efficiencies   36

Avalanches   46

Big-Box Retail Development   49

Brownfields Development   60

Cancer from Electromagnetic Radiation   69

Carbon Offsets   72

Carbon Taxes   75

Cell Phones and Electromagnetic Radiation   79

Childhood Asthma and the Environment   85

Children and Cancer	90
Citizen Monitoring of Environmental Decisions	95
Climate Change	102
Collaboration in Environmental Decision Making	109
Community-Based Environmental Planning	115
Community-Based Science	124
Community Right-to-Know Laws	129
Conservation in the World	132
Cultural vs. Animal Rights: The Makah Tribe and Whaling	137
Cumulative Emissions, Impacts, and Risks	141
Different Standards of Enforcement of Environmental Law	149
Drought	151
Ecological Risk Management Decisions at Superfund Sites	157
Ecosystem Risk Assessment	162
Ecotourism as a Basis for Protection of Biodiversity	167
Endangered Species	171
Environment and War	176
Environmental Audits and Environmental Audit Privileges	182
Environmental Impact Statements: International	185
Environmental Impact Statements: Tribal	190
Environmental Impact Statements: United States	194
Environmental Justice	198
Environmental Mediation and Alternative Dispute Resolution	210
Environmental Regulation and Housing Affordability	214
Environmental Vulnerability of Urban Areas	218
Evacuation Planning for Natural Disasters	222
Farmworkers and Environmental Justice	231
Federal Environmental Land Use	237
Fire	240
Floods	247
Genetically Modified Food	253
Geothermal Energy Supply	257

Global Warming   260

Good Neighbor Agreements   265

Hemp   269

Hormone Disruptors: Endocrine Disruptors   271

Human Health Risk Assessment   277

Hurricanes   281

Ice   287

Incineration and Resource Recovery   289

Indigenous People and the Environment   294

Industrial Agricultural Practices and the Environment   298

Industrial Feeding Operations for Animals   302

Land Pollution   309

Land-Use Planning in the United States   314

Landslides and Mudslides   318

Lead Exposure   329

Litigation of Environmental Disputes   333

Logging   339

Low-Level Radioactive Waste   344

Mining of Natural Resources   351

Mountain Rescues   354

Multicultural Environmental Education   357

Nanotechnology   363

National Parks and Concessions   366

Nuclear Energy Supply   369

Organic Farming   375

Permitting Industrial Emissions: Air   381

Permitting Industrial Emissions: Water   388

Persistent Organic Pollutants   394

Pesticides   398

Pollution Rights or Emissions Trading   407

Poverty and Environment in the United States   413

Precautionary Principle   417

Preservation: Predator Management in Oregon   423

Public Involvement and Participation in Environmental Decisions   426

Rain Forests   433

Sacred Sites   437

Ski Resort Development and Expansion   440

Solar Energy Supply   448

Sprawl   452

State Environmental Land Use   457

Stock Grazing and the Environment   460

Supplemental Environmental Projects   462

Sustainability   469

"Takings" of Private Property under the U.S. Constitution   475

Total Maximum Daily Loads (TMDL) of Chemicals in Water   481

Toxic Waste and Race   485

Toxics Release Inventory   491

Transportation and the Environment   494

Trichloroethylene (TCE) in Water Supplies   500

True Cost Pricing in Environmental Economics   502

Tsunami Preparation   508

Water Energy Supply   513

Water Pollution   520

Watershed Protection and Soil Conservation   524

Wild Animal Reintroduction   529

Wind Energy Supply   532

Appendices

A. Environmental Database Programs, Applications,
and Portal Web Sites   537

B. Index Chemicals   543

C. Glossary of Environmental Terms   547

General Bibliography   *553*

About the Author and Contributors   *557*

Index   *559*

GUIDE TO RELATED TOPICS

AGRICULTURE AND ENVIRONMENTAL CONTROVERSIES
Genetically Modified Food
Hemp
Industrial Agricultural Practices and the Environment
Industrial Feeding Operations for Animals
Organic Farming

ANIMALS AND ENVIRONMENTAL CONTROVERSIES
Animals Used for Testing and Research
Arctic Wildlife Refuge and Oil Drilling
Cultural vs. Animal Rights: The Makah Tribe and Whaling
Endangered Species
Preservation: Predator Management in Oregon
Wild Animal Reintroduction

CHILDREN AND ENVIRONMENTAL CONTROVERSIES
Cancer from Electromagnetic Radiation
Cell Phones and Electromagnetic Radiation
Childhood Asthma and the Environment
Children and Cancer

CITIZEN ENVIRONMENTAL CONTROVERSIES
Citizen Monitoring of Environmental Decisions
Collaboration in Environmental Decision Making

Community-Based Environmental Planning
Community-Based Science
Community Right-to-Know Laws

EMERGING ENVIRONMENTAL CONTROVERSIES
Multicultural Environmental Education
Sustainability
True Cost Pricing in Environmental Economics

ENERGY
Automobile Energy Efficiencies
Geothermal Energy Supply
Nuclear Energy Supply
Solar Energy Supply
Water Energy Supply
Wind Energy Supply

ENVIRONMENTAL DECISION MAKING
Environmental Mediation and Alternative Dispute Resolution
Litigation of Environmental Disputes
Public Involvement and Participation in Environmental Decisions
Supplemental Environmental Projects

FAIRNESS AND ENVIRONMENTAL CONTROVERSIES
Environmental Audits and Environmental Audit Privileges
Environmental Justice
Different Standards of Enforcement of Environmental Law
Farmworkers and Environmental Justice
Indigenous People and the Environment
Sacred Sites
Toxic Waste and Race

GLOBAL ENVIRONMENTAL CONTROVERSIES
Acid Rain
Air Pollution
Climate Change
Conservation in the World
Ecotourism as a Basis for Protection of Biodiversity
Environment and War
Global Warming
Precautionary Principle
Rain Forests

HUMAN ENVIRONMENTS
Brownfields Development
Environmental Regulation and Housing Affordability
Environmental Vulnerability of Urban Areas
Poverty and Environment in the United States
Transportation and the Environment

IMPACT ASSESSMENT
Environmental Impact Statements: International
Environmental Impact Statements: Tribal
Environmental Impact Statements: United States

INDUSTRY PRACTICES AND ENVIRONMENTAL CONTROVERSIES
Permitting Industrial Emissions: Air
Permitting Industrial Emissions: Water
Persistent Organic Pollutants
Pesticides

LAND USE AND ENVIRONMENTAL CONTROVERSIES
Big-Box Retail Development
Evacuation Planning for Natural Disasters
Federal Environmental Land Use
Good Neighbor Agreements
Land-Use Planning in the United States
Ski Resort Development and Expansion
Sprawl
State Environmental Land Use
"Takings" of Private Property under the U.S. Constitution
Watershed Protection and Soil Conservation

NATURAL DISASTERS
Avalanches
Drought
Floods
Hurricanes
Ice
Landslides and Mudslides
Mountain Rescues
Tsunami Preparation

NATURAL RESOURCES
Logging
Mining of Natural Resources

National Parks and Concessions
Stock Grazing and the Environment

POLLUTION
Incineration and Resource Recovery
Land Pollution
Lead Exposure
Low-Level Radioactive Waste
Pollution Rights or Emissions Trading
Total Maximum Daily Loads (TMDL) of Chemicals in Water
Toxics Release Inventory
Trichloroethylene (TCE) in Water Supplies
Water Pollution

RISKS FROM THE ENVIRONMENT
Cumulative Emissions, Impacts, and Risks
Ecological Risk Management Decisions at Superfund Sites
Ecosystem Risk Assessment
Hormone Disruptors: Endocrine Disruptors
Human Health Risk Assessment

PREFACE

Controversies around environmental issues abound. Human impacts on the environment are larger than ever and expected to grow. Issues of global warming and its human impacts increasingly fill the news. The scientific community is powerful but often feels compromised by politics in environmental conflicts. More laws are passed and court cases litigated. Air and water quality are still eroding in many areas. As human populations swell, conflicts over water quantity and quality will increase. Wars around the globe leave environmental devastation along with land mines. Vulnerability to natural disasters, like hurricanes Katrina and Rita, will increase in coastal communities and significantly impact inland communities in the United States. All these modern social dynamics raise the profile of environmental controversy to a new level that stresses environmental literacy, accurate science, and preventative public policies. This in turn expands the popular mental construct of *environment* to include public health and urban areas for the first time. In this process old controversies of race and class are revived, and new controversies about cumulative impacts and sustainability begin. The discussion of environmental controversies here includes both old and new environmental controversies.

Communities, scientists, government agencies, health care professionals, and industry are all engaged in these conflicts. Teachers, librarians, professors, researchers, parents, and engaged citizens seek to learn more about these environmental issues to keep up with the rapidly expanding environmental literacy of their students and children. Media accounts of these conflicts focus on the adversarial nature of the conflict and overlook other aspects of the conflict, many of which remain unresolved. Judicial decisions alone also suffer from this narrow vision of environmental problem resolution, although these decisions are powerful outposts of policy. The nuances of many environmental controversies lie in the positions and participation of many stakeholders. Each entry identifies the issue involved, various points of view or positions, and approximately where and when the conflict occurred, and explains some of the cultural, social, and political context and dimensions of the conflict. Public policies are described where appropriate. Each entry concludes with a description of the potential for future controversy. Discussions of environmental threat or impairment to future efforts for sustainability are included where relevant. Each entry is followed by a list of cross-references, Web resources, and a short, relevant bibliography suitable for student research.

The individual controversies were chosen based upon their actual and potential environmental impact. Some controversies are smaller parts of bigger controversies. The range of most environmental controversies can be global or local. Controversy can explode around environmental issues at flash points, when powerful stakeholders and values clash over a given environmental decision. Research resources, media coverage, and public documents help document these types of controversies as well as capture the student's imagination. Politics, uncertain science, and public distrust all help push a given issue into the realm of controversy. Environmental issues with serious environmental impacts remain controversial despite politics, science, or trust. They may not go away and may increase exposures for our future generations. Some of these impacts may bioaccumulate or skip generations. Some of these impacts may be perfectly safe and perform socially beneficial and desired functions. There is more than one point of view in every controversy, and they are included. The environmental controversies selected defy easy or quick resolution and have great or potentially great environmental impacts. They are unsettling controversies, with very dynamic information changes. They evoke hard questions, and the entries give a description of the controversy from current stakeholder positions. In describing these complex issues the approach focuses on multiple stakeholders and is factually descriptive and as straightforward and neutral as possible.

This compilation of environmental controversies is written to advance the development of environmental literacy and increase the capacity of the public to understand the overarching nature of environmental and ecological controversies. It is specifically written to be understandable to high school students and all members of the public. Technical, scientific, and legal jargon is reduced as much as possible. However, the Web resources and bibliography will take the interested student into these areas if their interest is stimulated. Each entry has one or two sidebars. These are short descriptions of aspects of the environmental controversy designed to spark interest. They often touch upon especially sensitive parts of a given environmental controversy.

I would like to express my gratitude for all those who helped with this book. Willamette University–College of Law, President Lee Pelton, and the Center for Community Sustainability provided foundational support that greatly facilitated my work. I am also grateful to my editor at Greenwood Press, Kevin Downing, for his support and timely assistance. Contributing entry authors

(Robin Morris Collin on sacred sites and on sustainability, Steven Bonnoris's about supplemental environmental projects, and Cathy Koehn about predator management controversies in Oregon) all made valuable contributions on timely environmental controversies within their expertise and experience. Elizabeth Eames provided timely research assistance for which I am grateful. My spouse, Robin Morris Collin, provided needed moral and intellectual support during the long hours of research and writing. She tirelessly read all the entries and sidebars and constantly acquired new research almost daily. To her I am most appreciative.

I would like to thank all my students from Auckland University, New Zealand; Department of Land Economy, Cambridge University; Department of Urban and Environmental Planning, University of Virginia; Department of Urban and Regional Planning, Jackson State MS; Environmental Studies, University of Oregon; and Willamette and Lewis and Clark Law Schools. Many of the environmental controversies discussed in this book were discussed in my courses. Students' energy and demand for environmental knowledge is a continuing inspiration. So too is the ardent need for environmental information by community groups. Recent advancements in information access have greatly empowered citizen groups. Many of these controversies come from my experience advising the U.S. Environmental Protection Agency under both Clinton and Bush administrations. I am grateful to community groups, such as the Environmental Justice Advisory group in Portland, Oregon, and to federal agencies for their persistent engagement with many environmental controversies.

The environmental controversies that persist are here. The stakeholders in each controversy change in number and power over time. Political platforms that win elections may be popular, changeable, and unreliable. But these outcomes are not always environmentally sound. Economic regimes that view all nature as a tangible commodity for sale and profit may create an artificial view of human control superseding nature. Social and cultural norms of behavior may value conspicuous consumption as a measure of status and increase environmental impacts. Political, economic, and social reasons gird the current ideas about environmental controversies. There are always powerful stakeholders that benefit from the status quo, from keeping things the way they are right now, and from keeping financial and community expectations safe. All-powerful stakeholders seek to control the battleground by containing the environmental controversy to their own choice of venue or forum. For example, the U.S. Environmental Protection Agency is a very powerful stakeholder representing the government. They control the administrative hearings process for most environmental permits to cities and industries. In fact, one must exhaust all administrative remedies for these types of environmental controversies before a federal court will even hear the substance of it.

Through all this, human population increases, human environmental impacts increase faster, and our environmental consciousness and scientific knowledge increase slowly. By increasing the environmental literacy of all stakeholders a common language of the environment can emerge. This is slowly starting to happen at international and domestic levels. As this occurs a new way of thinking about the environment also emerges. This paradigm brings with it new controversies like sustainability and the precautionary principle. Environmental controversies will persist and are signs of a healthy and growing environmental consciousness, and hopefully they are those that future generations will face with much greater knowledge.

INTRODUCTION

Writing about environmental problems is inherently controversial, and most scholars approach these topics cautiously if at all. This partially explains the lack of scholarship and the increasing coverage of environmental controversies in the media. Scholarship and writing about environmental controversies tend to be divided into narrow disciplinary approaches such as economics, law, business, biology, chemistry, and so forth. While these disciplines are narrow in their approach, they can differ widely from one another. The lack of consistent interdisciplinary approaches to environmental studies has also suppressed research and writing on environmental controversies. Environmental studies curriculum is relatively new and also faces definitional challenges in the political economy of disciplines in higher education. Even though there is increasing media coverage and public policy coverage of environmental controversies, academic writing about multidimensional environmental and ecological controversies is sparse. Interdisciplinary writing, research, teaching, and public service are risky in higher education because of the rigidness and narrowness of disciplinary boundaries. Some environmental education scholars have suggested that environmental studies are really a *metadiscipline*. Cross-cultural differences abound in approaches to environmental education. In the early 1990s, all seven New Zealand universities required an environmental studies course, generally part of the philosophy program. All 32 Australian universities had an environmental studies course requirement, generally placed in the engineering program. The contrasting approach to environment is reflected in the curriculum of the universities.

Most of the entries here span several disciplines and professions. Nanotechnology, for example, is an emerging technology with applications to environmental cleanup, computers, war, and medicine. Some have suggested that in the future, nanotechnology could be developed to be self-replicating. It spans the professions and disciplines of engineering, physics, materials science, medicine, and production technology research.

The difficulties of higher education with environmental controversy have not stopped the tremendous increase in environmental literacy in communities and in secondary education. Communities and young people are naturally curious about the environment around them. Controversial issues attract their interest and concern. No other single compilation exists of the most controversial environmental issues facing society today. As these controversies have increased so has the need for a comprehensive reference book. These chosen environmental controversies are contextualized for present and future use. Each controversy discusses environmental threats or impairment to future efforts for sustainability. Each entry also discusses in a nontechnical way the potential for future conflict. These features together with the large number of controversies presented are unique characteristics of this reference set. This type of information is essential for emerging U.S. civic environmentalism, growing environmental education programs, and ultimately sustainability.

These controversies can emerge in different venues and forums. These are the battlegrounds of controversy. These venues and forums generally include some combination of court cases, legislation, agency administrative hearings, political controversies, and community activism. They can be driven by natural disasters, cumulative impacts, population growth, and the rapidly expanding knowledge base on ecological systems. Environmental injustices that cause public health threats can create controversy and remain as powerful grassroots environmental issues. Environmentalists pushing for sustainability on an ecosystem basis often find exploited environments in communities of exploited people. Issues of racism, cancer clusters, and high childhood exposures to toxins exist in degraded landscapes. These conflicts within society at large will shape policy around the world. As the United States unavoidably embarks on a cleanup policy under federal, state, local, and private auspices, communities will seek ways to involve themselves in the environmental decisions of government and industry. Brownfields redevelopment projects, waste siting, and industrial permits are typical environmental decisions that concern residents of a community. There has also been an increasing distrust of government in that communities suspect that it always hides environmental information to protect industry.

Environmental controversies are increasing their political profile and saliene in elective processes. While the congressional Black Caucus has always had one of the strongest environmental voting records in the U.S. Congress, other caucuses are now more aware of environmental controversies such as environmental justice and sustainability. The rise of this civic environmentalism is fueled by the growth, development, and implementation of federal, state, and local right-to-know laws. Continued growth of population, waste, and environmental literacy will increase social concern for the role of the community in environmental decision making. These dynamics unavoidably confront long-held traditional beliefs about private property, an ethos of continued growth, and the role of government. The rise of environmental controversies is directly related to the increase in environmental literacy, resulting in challenges to traditionally held values.

Governmental agencies that help protect our environment, like the U.S. Environmental Protection Agency (EPA), can seem inaccessible because of their complexity. Environmental agencies are recent additions to U.S. government, and their regulatory systems are new, ever-changing, and evolving in ad hoc ways. As environmental results become obvious and inescapable, people become environmentally literate and self-educated. Often they become politically mobilized and seek redress from government. This increased political engagement in environmental issues often leads to conflict that inevitably engages an environmental agency of the government. The battleground for many of these controversies with agencies is often the courtroom. These battlegrounds are described in nontechnical terms where relevant.

The environmental controversies selected are interrelated by 17 or 18 themes. These themes are interrelated with an ecological perspective on the environment. The controversial aspects have pulled them into one battleground or another. For example, the theme of agricultural issues includes topics such as genetically modified food, industrial-feeding operations for animals, and organic farming. The theme of animals and environmental controversies includes entries on animal testing and research, endangered species, and wolf reintroduction. Children and environmental controversies may focus on cancer and asthma in children. Citizen environmental controversies engage the range of controversies around the environment from science, planning, and decision-making perspectives.

Emerging environmental controversies include sustainability, multicultural environmental education, and true cost pricing in environmental economics. Professor of Law Robin Morris Collin succinctly describes the core environmental controversies around sustainability in her entry here. The entries on energy provide a range of perspectives on energy supply and its environmental controversies. Entries on environmental decision making discuss the controversies around current methods of environmental decision making—litigation, mediation and alternative dispute resolution, and public participation/involvement. Here, Steven Bonnoris has written a fascinating entry on supplemental environmental projects that can be done in lieu of a fine. Fairness and environmental controversy describes issues of race, class, and environmental benefit/burden. Global environmental controversies discuss acid rain, ecotourism, global warming, rain forests, and the "precautionary principle," among others. Environmental impact assessment is discussed in three areas of environmental controversy— United States, international, and tribal. There are many points of view on any given environmental controversy. One powerful stakeholder is industry. Industrial permitting processes, pesticides, and persistent organic compounds are some of the entries on this theme. Land use and environmental controversies are an area in which many citizens first engage in an environmental controversy. Big-box retail development, evacuation planning for natural disasters, sprawl, and land-use planning in general are described. Natural resources generate their own set of environmental controversies. Logging, mining, and grazing are covered here.

Natural disasters also generate a unique set of environmental controversies: Mountain rescues, avalanches, landslides, and tsunami preparation are described, among others. Pollution entries cover lead exposure, low-level radioactive waste, emissions trading, and the "toxics release inventory." Last, risks from the environment cover ecosystem and human risk assessment, hormone disruptors, and cumulative impacts.

All environmental controversies are dynamic. Web resources are provided with each entry to update the reader as to the current status of a given issue. There are some environmental controversies that did not make the final list this time. This compilation is not exhaustive, and new controversies around environmental issues are increasing daily. New information, new technologies such as nanotechnology, and increasing environmental literacy all create new controversies, and new battlegrounds for old controversies.

Most U.S. environmental policy has emerged from public controversies. In the late 1960s the Cuyahoga River in Cleveland, Ohio, caught fire twice because of pollution. There were oil spills on the coast of California. One was at San Clemente, home of President Richard Nixon. President Nixon signed the early laws that helped form the U.S. Environmental Protection Agency in 1970, such as the National Environmental Policy Act. Early controversies like acid rain merge into new controversies like climate change. It is the context of expanding human awareness of environmental impacts that creates new paradigms, or ways of thinking, about environmental controversies like sustainability. This relationship is inextricably involved with science. Western science has grown in its understanding of our environment and ecology. As such it is inextricably bound with environmental controversies in their discovery, debate, remediation, and resolution. These days science seems remote and removed from everyday life for most people, but the environment and especially risks from environmental sources command the public's attention. As environmentalism and civic environmentalism rise so will the role of science in everyday life. Science can present some politically inconvenient truths that can cost scientists and others their careers. It can challenge whole economic systems as unsustainable. Private property patterns may have to change to be sustainable. The pursuit of environmental truths in these battlegrounds can be bruising, and they are increasingly unavoidable. The generational transfer of environmental knowledge is especially important for the effective pursuit of sustainable practices. The transfer of knowledge is one of the functions of education in a civil society, which is especially necessary for the leaders of tomorrow. As the role of science increases under civic environmentalism the need for education also increases. In an address called "The Relation of Science to Human Life" Professor Adam Sedgwick discussed the nature of this role.

> Remember the wise, for they have labored and you are entering into their labors. Every lesson which you learned in school, all knowledge... has been made possible to you by the wise. Every doctrine of theology, every maxim of morals, every rule of grammar, every process of mathematics, every law of physical science, every fact of history or geography,

ACID RAIN DYNAMICS

Moisture in the atmosphere forms small droplets of rain that gradually become larger and heavier and sink to the ground as rain. These droplets can form around dust, particulate matter, and each other. There are many sources for the pollution that forms the acid in rain. There is consensus that it is primarily industrial air emissions that contribute to acid rain. Large coal-fired power plants and factories add to this problem. The prevailing air currents carry these emissions all across the United States and the rest of the world. Airborne emissions from other industrialized and newly industrialized nations also travel long distances to other countries. This is the reason the direction of the prevailing winds can determine *the deposition of acid onto architecture and statuary*. Once the acidic gases have been emitted into the atmosphere, they follow prevailing wind circulation patterns. Most industrialized areas of the world are located within the midlatitude westerly belt, and their emissions are carried eastward before being deposited. Acid rain is possible anywhere precipitation occurs. Early scientific controversies were about the number of tree species affected and its long-term ecological impacts. Current controversies are about whether the problem has been solved. It has a greater environmental impact than predicted by early studies. Although sulfur emissions have decreased, other emissions have increased. Some contend that mercury deposition has increased, and others consider the scope of environmental regulation inadequate.

DOES AIR POLLUTION CAUSE ACID RAIN?

Historical weather data come from precipitation records, ice cores, and tree borings. They show an increase in acid rain starting in the late 1930s, 1940s, and 1950s. This was also the same approximate time as a large industrial expansion in the United States, and before the implementation of clean air policy in the early 1970s. Many U.S. cities used coal and natural gas in their everyday activities depending on the dominant industry.

Acidity causes metals, such as aluminum or lead, to become soluble in water. Once in the water acid affects plants and fish and is considered toxic to both. Acid deposition is directly damaging to human health. Its ability to corrode metals, from lead and copper pipes, for example, can be toxic to humans. The sewer and water infrastructures of many older U.S. cities have lead pipes. Increased concentrations of sulfur dioxide and other oxides of nitrogen in the air, common industrial emissions, have been causally related to increased hospital admissions for respiratory illness. In areas that have a large concentration of these airborne industrial emissions, there is an increase in chest colds, asthma, allergies, and coughs in children and in other vulnerable populations. When communities learn that there is acid in the rain that is caused by the polluted air they also breathe, concern increases. There is a strong public health concern with these industrial emissions besides acid rain.

Acid deposition refers to the process of moving acids to the land part of a given ecology. The acids then move through the top surface of the earth, through the soil, vegetation, and surface waters. As metals, such as mercury, aluminum,

ACID RAIN

When pollution enters the atmosphere it can form rain, which can increase the acidity of precipitation, with environmental consequences. Environmentalists and affected communities and countries challenged industry over the environmental consequences in this early battleground.

BACKGROUND

Science, environmentalists, and industry clashed in this important environmental controversy. Issues focused on whether the wind could carry these chemicals that far, and whether they had any significant environmental impact. For a variety of reasons, certain areas and species were affected first and became persuasive environmental examples of the need for a national policy. Fundamental issues of governmental intervention at the federal level to regulate industrial growth in the name of unproven science and environmental advocacy were reflected in Congress and the Clean Air Act of 1970, amended in 1990. Now there is consensus that environmental pollutants can be carried long distances by wind and other media. In the early years of governmental regulation of pollution there was a battle about how far pollution would go in a region. There is also general agreement that the Clean Air Act did reduce some chemical emissions, and needs to do the same with other regulated chemicals. The acceptance of national governmental intervention and some success with an early policy paved the way to bigger battlegrounds of global warming and climate change.

which you are taught, is a voice from beyond the tomb. Either the knowledge itself, or other knowledge which led to it, is an heirloom to you from men whose bodies are now moldering in the dust, but whose spirits live forever and whose works follow them, going on, generation after generation, upon the path which they trod while they were on the earth.... They are the aristocracy of God, into which *not* [emphasis added] many nobles, not many rich, not many mighty are called. Most of them were poor; many all but unknown in their own time; many died and saw no fruit of their labors; some were persecuted, some were slain as heretics, innovators, and corruptors of youth. (Vlahakis et al. 2006, 327)

The environment provides the most honest witness possible regarding humans and our environmental impacts. Scientists, communities, and environmentalists unfolding the mysteries of nature inevitably fall into battlegrounds of environmental controversy. These battlegrounds push research, law, and public policy to evolve further. The men and women who have engaged these controversies are the "the wise" discussed by Professor Sedgwick. By engaging in these controversies it is hoped that future wise men and women can learn about the environmental issues of the past, quickly enough to prevent further environmental degradation.

REFERENCE

Vlahakis, George N., Isabel Maria Malaquais, Nathan M. Brooks, François Regourd, Feza Gunergun, and David Wright. 2006. *Imperialism and Science: Social Impact and Interaction.* Santa Barbara, CA: ABC-CLIO.

and lead, are set free from the increased acidity from the rain, they can have adverse ecological effects. There is also concern that some of the metals may bioaccumulate and intensify as they move up the food chain. The sustainability of an ecosystem depends on how long it takes for the system to recover, in this case from acid in the rain. The ability of some ecosystems to neutralize acid has decreased due to the cumulative impacts of acid rain over time. This slows the recovery of other parts of the ecosystem. This is why environmentalists contend that recent decreases in some industrial emissions are not likely to bring about full ecosystem recovery. The cumulative impacts in spite of partial environmental regulation of acid rain pollutants have damaged sensitive areas of the Northeast, such as the Adirondack State Park, by impairing their ability to recover from acid shock events.

HOW DOES ACID RAIN AFFECT THE ENVIRONMENT?

Acid rain has a range of effects on plant life. Sensitive species go first. Part of the battleground of the acid rain controversy is which species are affected and the scope of the problem. Acid rain goes anywhere it rains or snows. Crops used for food or other purposes can be negatively affected. Acid rain effects on plants depend on the type of soil, the ability of the plant to tolerate acidity, and the actual chemicals in the precipitation. Soils vary greatly from one location to another. Soils with lime are better capable of buffering or neutralizing acids than those that are sandier or weathered acidic bedrock. In other soils increasing acidity causes the leaching of plant nutrients. The heavy metal aluminum causes damage to roots. This can interfere with the plants' ability to absorb nutrients such as calcium and potassium. The loss of these nutrients affects the plants' ability to grow at normal and productive rates, and the intake of metals increases their potential toxicity. For example, acid deposition has increased the concentration of aluminum in soil and water. Aluminum has an adverse ecological effect because it can slow the water uptake by tree roots. This can leave the tree more vulnerable to freezing and disease.

ACID RAIN AND DIRECT GOVERNMENT ACTION

On September 13, 2007, 24 nations signed the Montreal Protocol. It was signed in Canada, a country highly motivated to solve air pollution problems because of acid rain. This landmark environmental treaty required the phaseout of ozone-depleting chemicals and compounds such as chlorofluorocarbons, carbon tetrachloride, and methyl chloroform. There is scientific consensus that these compounds erode the stratospheric ozone layer. Ozone protects the Earth from harmful ultraviolet radiation. This radiation can cause cancer, among other environmental impacts. To date 191 countries have signed the protocol. The United States implemented many parts of it more quickly and at less cost than expected. The thinning of the ozone layer mostly stopped in 1988 and 1989, almost immediately after treaty reductions began to take effect. The U.S. Environmental Protection Agency estimates that 6.3 million U.S. lives will be saved as a direct result of worldwide efforts to implement the Montreal Protocol requirements.

The early successes of the Montreal Protocol laid the groundwork for a national approach to acid rain in the United States.

The 1990 National Acid Precipitation Assessment Program (NAPAP) mandated report to Congress concluded that acid deposition had not caused the decline of trees other than red spruce growing at high elevations. This was a large scientific controversy because it had policy implications for the Clean Air Act Amendments of 1990. Some contended that 20 years of the Clean Air Act was enough and that air pollution was no longer a severe problem. Others strongly disagreed. Recent research shows that acid deposition has contributed to the decline of red spruce trees and other important trees throughout the eastern United States. Other species such as black flies increase under acid rain conditions. Indicator species and mammals at the high end of the food chain show high levels of some of the pollutants in acid rain, which indicates how pervasive the chemical exposure is in the environment. For example, sugar maple trees in central and western Pennsylvania are also now declining. The Clean Air Act Amendments of 1990 included specific provisions about acid rain.

LAKES, RIVERS, AND STREAMS

Acid rain falls to the Earth and gradually drains to the oceans via rivers, lakes, and streams. Acid deposition erodes the water quality of lakes and streams. It causes lakes to prematurely age, speeding up a natural process of eutrophication. It does this in part by reducing the species diversity and aquatic life. Fish are considered an indicator species of ecological health. However, ecosystems are made up of food webs, and fish are just one part of that. Entire food webs are often negatively affected and weakened. Environmentalists and those interested in sustainability are very concerned about ecosystem effects and are very involved in acid rain discussions.

THE CLEAN AIR ACT AND ACID RAIN

Early concerns about acid rain provided a strong impetus to early Clean Air Act legislation. The Clean Air Act was strong, groundbreaking environmental policy. It was never complete in its coverage, nor strong in its enforcement. Legislative and legal exceptions have developed. Emissions and acid deposition remain high compared to background conditions. The Clean Air Act did decrease sulfur dioxide emissions, yet these emissions remain high. Enforcement on nitrogen oxides and ammonia is sporadic. Emissions of these compounds are high and have remained unchanged in recent years. There are other emissions, such as mercury, that have only recently been regulated. The holes in environmental policy often appear as environmental impacts. Is the Clean Air Act enough? This is still a large controversy, going far beyond the debate about acid rain. Some research shows that the Clean Air Act is not sufficient to achieve ecosystem recovery. As the battlefield has moved from acid rain, to clean air environmental policy, to global warming and climate change, this important policy question is certain to be revisited many times.

THE ADIRONDACK PARK AND ACID RAIN

The Adirondack Mountains represent one of the first observable impacts of acid rain in the United States. It was an early battleground, pitting environmental conservationists against Midwestern industrial interests. Scientists also had many controversies here, primarily in the area of environmental impact assessment.

The Adirondack Mountains are located in the Adirondack Park, in upstate New York. New York State was an early center for industry and commerce. Many wealthy families recreated in the Adirondack Mountains. It was considered one of the major sources of water for New York City and other growing metropolitan areas along the Hudson River. The Adirondack Park is one of the largest state parks in the United States. The park is a group of public and private lands protected under state law. About 2.6 million acres within the park are owned and managed by New York State. About 3.5 million acres are owned by private property owners. In general they work in cooperation, but conflicts can arise. Usually private property owners want to develop their land and the Adirondack Park refuses to allow it. The Adirondack Park was established by the New York State Legislature in 1892. New York State gave it more protection in 1894, when these words were added to the New York State Constitution:

> The lands of the state, now owned or hereafter acquired, constituting the forest preserve as now fixed by law, shall be forever kept as wild forest lands. They shall not be leased, sold or exchanged, or be taken by any corporation, public or private, nor shall the timber thereon be sold, removed or destroyed

The term *forever wild* resonated with environmentalists of the day and helped organize a national preservationist movement. This movement was an important part of the national parks movement later. New York State has kept its pledge, although some question if it is still forever wild. Throughout the Adirondack Forest Preserve, New York State now maintains over 2,000 miles of marked trails. The state operates about 42 campgrounds of differing types. There are also many private cabins and campgrounds. The Adirondack Park contains 54 species of mammals and about 200 species of birds. Many of them are in the 17 wilderness areas with a total area of about one million acres. About 1.3 million acres of Forest Preserve land are classified as wild forest. There are about 30,000 miles of rivers and streams and 2,800 lakes and ponds. Many people canoe and kayak these lakes and streams. Wildlife preservation and high levels of human usage make the effects of acid rain all the more observable. One new emerging battleground is the development of a permit system for recreational users. Many private property owners in the park would object to this if it prevented their use of the property, but otherwise support it. Environmental groups are concerned that it may restrict access unevenly but recognize that human overuse is environmentally degrading areas already weakened by acid rain.

The Adirondack Park suffered intensive damage from acid rain. It was literally the bellwether for the political controversy about acid rain that helped open up the first major era of environmental regulation in the United States. Coal-fired energy plants in Ohio, Illinois, Indiana, Pennsylvania, and western New York emit large amounts of sulfur and nitrogen and other chemicals. These emissions are carried Northeast via the jet stream and prevailing wind patterns. The first set of mountains they hit are the Adirondack Mountains. They can form acidic clouds. The clouds are laden with heavy amounts of nitric and sulfuric acid that falls as rain onto the Adirondack Mountains. This precipitation can be more acidic than natural rain, with a pH of 3.3 or less. Now, about 18 percent of the lakes and ponds in the Adirondack Park are too acidic to support plants and aquatic wildlife. This does not include lakes and ponds outside the park's boundaries.

The Adirondack Lakes Survey Corporation determined that 27 percent of all 2,800 lakes and ponds in the park remain at a pH of less than 5.0 all year. This is considered too low in this context. Few species can live in a pH lower than 5.5, including rainbow trout, smallmouth bass, fathead minnow, clam, snail, and mayfly. Each spring, as the previous winter's snow melts, the pH drops to 5.0 causing an *acid shock*. Acid shock can also occur when the precipitation takes the form of fog or snow. The EPA estimates that 43 percent of the park's lakes and ponds will be irreparably damaged by the year 2040 if there are no greater controls on air pollution. This is the subject of intense controversy. EPA scientists say that reduction of both sulfur dioxide and nitrogen oxides to a level 70 to 75 percent below 1990 levels is necessary to prevent this.

Sulfur Dioxide, Nitrogen Oxides, and Now Mercury

The Adirondacks are located downwind of numerous coal-burning power plants. Their emission of mercury contributes significantly to mercury pollution here. The largest single source of mercury emissions in the United States is coal-fired electric utilities. In the last five years there has been an expansion of coal-fired power plants in some regions.

In 1997, the EPA found that 60 percent of the total mercury deposited in the United States came from human, or anthropogenic, air emission sources. Humans have an even greater ecological footprint and impact in certain regions, called hot spots. Atmospheric deposition, or acid rain, is the major contributor to the formation of biological mercury hot spots. Ninety-six percent of the lakes in the Adirondack region exceed the recommended EPA action level for methylmercury in fish. New York has posted 32 mercury health warnings covering 59,228 acres of lakes. More than 50 New York State lakes, rivers, and reservoirs are said to contain fish that are unsafe to eat due to high mercury levels.

Children and fetuses exposed to mercury can suffer from a cluster of behavior problems. Impacts include poor attention span, impaired language development, impaired memory and vision, problems with processing information, and impaired fine motor coordination. Some research indicates that 1 in 12 women of childbearing age has an unsafe blood level of mercury. Three hundred thousand at-risk children are born each year with increased risk for neurological problems and developmental disorders because of mercury exposure. It is estimated that fish in the Northeast had mercury levels 10 times higher than considered acceptable by the EPA human health criteria. This has occurred in areas of the Great Lakes, and other lakes, rivers, and streams near air pollution sources.

How Does Mercury Get into the Air?

When mercury is emitted from a site, it enters the atmosphere. Atmospheric mercury can travel over long distances before it is deposited back on Earth. Bacteria in soils, water, and sediments convert this mercury to methylmercury. In this form, it is consumed by plants and animals. Fish that eat these plants and animals accumulate methylmercury in their bodies, sometimes in the fatty tissues and sometimes in the nervous systems. Cultures that eat fish caught from dangerous sites and eat them whole face greater risk. One of the early battle-grounds was the fish studies and difference, by ethnicity of the angler, in risk from fish consumption. Store bought fish is often only the fillets, which can have lower concentrations of toxic chemicals depending on the fish. Cultures that consume the whole fish are much more exposed to metals that bioaccumulate in the fat or nervous systems of the fish. As fish higher up the food chain eat smaller fish, the methylmercury is concentrated further. This process is called *bioaccumulation*.

As atmospheric mercury enters the soil, biological and chemical processes transform elemental mercury into methylmercury. Methylmercury is the most toxic form of mercury and does not degrade or disappear, similar to other persistent organic pollutants. Bioaccumulation of methylmercury increases as it goes up the food chain. As a result of bioaccumulation, smallmouth bass and common loons can reach methylmercury levels 100 times greater than the levels found in insects. Ultimately humans eat fish with high levels of methylmercury, and it enters the bloodstream, causing severe health effects in our bodies.

One of the most studied species with regard to mercury contamination is the common loon. Mercury is present in two-thirds of Adirondack loons at levels that negatively impact their reproductive capacity. This is a significant risk to their survival. Study of the loon is significant because the common loon is an indicator species of the Northeast's ecology. Loons have high levels of mercury in their bloodstream because they are higher up the food chain. A recent study found that 17 percent of the loons tested in New York State were at risk for harmful effects from mercury contamination. Loons considered to be at highest risk were found in acidic lakes in the Adirondack Park. High levels of mercury are correlated with behavioral changes in common loons. These lead to decreased productivity, decreased survival rates for loons, and possibly increased susceptibility to disease. Twenty-five percent of loons found in the central Adirondacks (New York) have elevated levels of mercury in their blood.

Acid rain issues continue to influence New York environmental policy. The Clean Air Act creates tradable permits for some of the pollutants that cause acid rain. New York State recently passed a law requiring any New York utility that sells its sulfur allowances to an upwind state to pay the state a fee equal to the price of the allowance. This law is currently being challenged in court by the utility companies.

Many important life-forms cannot survive in soils below a pH of about 6.0. The loss of these life-forms slows down normal rates of decomposition, essentially making the soil sterile. When the acid rain is nitrogen based it can have a strong impact on plants. High concentrations of nitric acid can increase the nitrogen load on the plant and displace other nutrients. This condition is called nitrogen saturation. Acid precipitation can cause direct damage to plants' foliage in some instances. Precipitation in the form of fog or cloud vapor is more acidic than rainfall. Other factors such as soil composition, the chemicals in the precipitation, and the plant's tolerance also affect survival. Sensitive ecosystems such as mountain ranges may experience acidic fog and clouds before acid rain forms.

WHAT CAUSES THE RAIN TO BE ACID?

A low pH measure is considered acidic. It is an easy, direct measure of a primary effect of some primary pollutants, and pH is also a standard measure of environmental conditions. Acid rain is caused when gases are emitted into the air by the burning of sulfur and nitrate compounds found in ores or coal. In the United States, a large number of power plants burn coal and are the source of pollutants found in acid rain. Lung ailments such as asthma and emphysema can result from exposure to these airborne pollutants. Pollutant abatement and control technology does exist to remove some of these chemicals. Recycling former waste products into productive products is part of many business concepts of sustainability. There are pollution-control processes available that would clean coal and other ores before they undergo combustion, as they combust, and after combustion. The extraction processes are expensive according to industry. They also yield hazardous waste products in terms of ash that must be considered. Extraction processes are not done voluntarily by most industry. Requiring known but expensive technologies such as scrubbers is a controversial area of environmental policy around the issue of acid rain, as well as climate change and global warming.

ENGINEERING SOLUTIONS TO ACID RAIN

One common concern of industry is that they must meet new and expensive environmental compliance requirements. It is one thing to pass a law, they point out, but more difficult to actually implement it at the point of emission. Pollution and abatement engineering firms faced early challenges with aspects of the Clean Air Act because of the concerns raised about acid rain. The enactment of new rules and regulations for emission controls, such as best available control technologies, have required pollution and abatement control engineers to develop new ways to limit the amounts of sulfur and nitrogen in order to comply with their new permit requirements. How do chemical engineers reduce emissions and abate pollution from sulfur dioxide and nitric oxides? Here are some of the basic methods to date.

Pollution-Control Methods for SO$_2$ (Sulfur Dioxide)

GAS ABSORPTION AND CHEMICAL STRIPPING: This is the standard chemical method for removing a substance from a gas stream. It requires a liquid solvent in which the gaseous component is more soluble than the other parts of the sulfur dioxide gas stream. The sulfur dioxide gas enters the absorber where it then flows up and the liquid stream flows down. Once the gas has been chemically stripped of the sulfur dioxide, it is released into the atmosphere. The toxic ash that remains is shipped to a hazardous waste landfill.

LIMESTONE WET SCRUBBERS: Coal-or oil-burning sources that produce pollutants such as sulfur dioxide use this method. First, the solid ash particulates are removed from the waste stream. They are shipped to a hazardous waste landfill. Then, the remaining sulfuric gas goes to a tower where it travels through a scrubbing slurry. This slurry is made of water and limestone particles that react with the sulfuric acid to neutralize it, producing carbon dioxide and a powdery ash. The legal destination for the final waste stream is a hazardous waste landfill.

DRY SYSTEMS: Some pollution abatement and control approaches are called dry because they do not use a wet slurry. When handling sulfur-based emissions dry systems inject dry alkaline particles into the sulfuric acid gas stream. They neutralize the acid into an ash. The particles are then collected in the particle collection device. Dry systems avoid problems with disposal of wet sludge from wet scrubbers. They increase the amount of dry solids to be disposed of, usually in the form of fly ash. The final destination is again a landfill.

WET/DRY SYSTEMS: These systems are a combination of the wet and dry systems that remove some pollutants, such as sulfur dioxide, from the waste stream. The sulfur-based emissions are essentially watered down and reduced to a powdery ash. Then final destination for this ash is the hazardous waste landfill.

Pollution Abatement and Control Techniques for Nitrogen Emissions

Reducing nitrogen emissions is more challenging in implementation. Basically, there are two ways to reduce NO_x emissions:

- Modifying combustion processes to prevent NO_x formation in the first place, or
- Treating combustion gases after flame to convert NO_x to N_2.

Both methods incur costs and sometimes liabilities for industry. Pollution abatement and control engineers remain in high demand to assist industrial compliance with environmental laws. The next battleground and challenge with acid rain will be removing atmospheric mercury.

CANADA AND THE U.S. ACID RAIN CONTROVERSY

One intrinsic political problem highlighted by acid rain, and part of all air pollution, is that air and water currents do not follow political boundaries. If one country's air pollution goes directly to the neighboring country, little can be done. Canadian concern about the damage from acid rain predates U.S. concern. Canada examines all possible sources for the acid rain problem, including its own contribution. Acid rain resulting from air pollution is severely affecting lakes and damaging forests. Eastern Canada is particularly hard hit because of the prevailing winds. There is the continuing controversy about actual site-specific impacts. Some studies indicate no significant changes in some measures of acid rain. Others find that impacts for a given species are significant. This type of dispute is typical for this controversy. They tend to be ongoing in nature.

U.S. and Canadian relations have overcome some of the initial strain over the issue of acid rain. The U.S. government did not make much progress until U.S. land was at risk. Those efforts, exemplified by the Clean Air Act Amendments of 1990, put into policy proven measures that could reduce emissions of pollutants that cause acid rain. Since then there have been a number of cooperative environmental programs.

ACID RAIN AND ART, ARCHITECTURE, AND ANTIQUITIES

Acid rain occurs all over the world. It is most common in areas of human habitation with a history of industrialization. It has impacts on both the natural environment and the urban environment. The impact of acid rain to antiquity all across the planet is difficult to know. Other parts of the world still burn brown coal, that is, coal containing many impurities such as sulfur. Large industrial processes fueled by coal churn large emissions into the atmosphere with little environmental regard or regulation. Scrubbing the coal is currently expensive but offers a way to remove the sulfur within the coal.

Acid rain has a negative effect on architectural buildings and works of art depending on their materials. Its effects are far ranging, especially over time. Washington, D.C.; Philadelphia; Milan, Italy; Bern, Switzerland; and many other cities feel acid rain impacts. In terms of traditional Western classical works of art Italy may have the highest risk of damage from acid rain. In Italy, many works of art are constructed of calcium carbonate in the form of marble. Similar to most choices of building stone, marble was selected because it was locally available. Calcium carbonate reacts upon contact with acid rain. The marble structures are dissolved. Italy has a severe acid rain problem due to geography, prevailing winds, and their dependence on coal burning as a source of energy. Many classic ancient marble structures and statutes are at risk of being corroded by acid rain in many cities in Italy. Northern Italy has the worst air quality in western Europe. Some of the smaller sculptures have been encased in transparent cases, which are then filled with a preserving atmosphere. Others have continued to corrode. This is one example of the effects of acid rain. There are more.

In the United States limestone is the second most used building stone. It was widely used before Portland cement became available. Limestone was preferred because of its uniform color and texture and because it could be easily carved. Limestone from local sources was commonly used before 1900. Nationwide, marble is used much less often than the other stone types. Granite is primarily composed of silicate minerals that are resistant to acid rain. Sandstone is also composed of silica and is resistant to most types of acid rain. Limestone and marble are primarily composed of calcium carbonate, which dissolves in weak acid. Depending on the building materials, many older U.S. cities suffer acid rain damage.

HOW MUCH ACID IS THERE IN RAIN?

The term *acid deposition* is used to encompass the dry deposition of acidic compounds as well as the wet deposition of acidic compounds in acid precipitation. The most recent term used in place of *acid rain* is *atmospheric deposition,* which includes acidic compounds as well as other airborne pollutants. Atmospheric deposition recognizes that air pollution involves the complex interaction of many compounds in a chemical concoction in the atmosphere.

Unpolluted rain is normally slightly acidic, with a pH of 5.6. Carbon dioxide (CO_2) from the atmosphere dissolves to form carbonic acid, which is why normal rain is slightly acidic. When acidic pollutants combine with the rain the acidity increases greatly. The acidity of rainfall over parts of the United States, Canada, and Europe has increased over the past forty years. This is primarily due to the increased emissions of sulfur and nitrogen oxides that accompany increased industrialization.

The sulfur and nitrogen oxides are the common pollutants from coal-burning activities such as power generation. Many, if not most, of these emissions are legal in that they are within the terms of their permit from the EPA or state environmental agency. Legal or not, these are oxidized in the atmosphere as they are converted to sulfuric and nitric acids. These acids are then absorbed by clouds laden with raindrops. As they become heavier they fall to the earth. This process is called acid deposition. Acidic fog, snow, hail, and dust particles also occur. The acidity of these different forms of precipitation can vary greatly.

SOURCES OF ACID RAIN

Part of the controversy about sources of acid rain revolved around the question of whether environmental policy could really affect the acidity in rain. Scientific debate about natural, human, and industrial causes engulfed much of the political battleground. Although the policy question was answered in that, yes, environmental policy can make the air cleaner, the debate about sources continues.

All forms of precipitation are naturally acidic due to naturally occurring carbon dioxide (CO_2); human activities tend to add to the acidity. Nonpolluted rain is assumed to have a pH value of 5.6. This is the pH of distilled water. Natural sources of these environmentally regulated chemicals may be significant. Man-made emissions tend to be concentrated near historic industrial sites and older population centers. The presence of other naturally occurring substances can produce values ranging from pH 4.9 to 6.5. This scientific dynamic has kept other debates alive regarding whether government has an effective role in environmental policy if the sources are natural. pH factors are one of many measures that are monitored.

The relationship between pollutant emission sources and the acidity of precipitation at affected areas has not yet been determined. More research on tracing the release of pollutants and measuring their deposition rates to evaluate the effects on the environment is under way. This is an area of much scientific and legal controversy. If it were possible to show that a given emission definitely came from a given plant, then government would be able to assign liability to the polluter. Governments would also be able to locate sources of acid rain that comes from other countries.

AIR POLLUTION FROM SHIPPING

In determining emission sources, some sources are easier to find than others. Oceangoing shipping has been easy to locate. They are allowed large industrial emissions. The European Union limits the sulfur emissions in cars to 15 parts per million but allows ships 45,000 parts per million. Shipping is estimated to account for about 10 percent of world sulfur emissions, and a large amount of nitrous oxide and soot. Shipping makes up about 3 percent of the world greenhouse gas emissions. While shipping vessels are easy to locate, they are difficult to prosecute under environmental laws. They are often flagged, or incorporated, in another nation, and jurisdictional disputes are international and generally avoided. It is very expensive to keep an oceangoing vessel moored at harbor, but often that is what a minimal environmental inspection requires. Other environmental controversies such as deep-ocean waste dumping and oceangoing incinerators may increase the environmental impact of shipping. As energy costs of land and air transport increase, water-based transport is receiving stricter environmental scrutiny. In 2005, California forbade cruise ships from dumping their bilges (waste) into the waters around the coastline. Most ports in most of the world currently see very little environmental protection or regulation. Numerous small petrochemical spills over time, illegal dumping, and a terminus for all that the rivers have carried off the land characterize most commercial ports.

As a global industry, shipping is enticed into the market for emissions trading. They are interested in buying and selling their pollution credits. Ocean transport is still a very efficient way to move goods as opposed to the energy costs of land and air transportation. It is estimated that shifting the fuel of oceangoing vessels to a lighter grade would reduce their sulfur dioxide emissions by 60 percent. This fuel may cost more. Some have criticized this plan saying the higher distillation costs and refining offset the greenhouse gases saved from the ships. Currently, shipping firms may have emissions credits to give, which has subjected them to more regulatory scrutiny from governments.

The shipping industry is an important stakeholder in environmental policy implementation. They can serve as monitors of changing ocean conditions. They have a large impact on the environment and remain on the edge of enforceable environmental regulation. Citizen monitoring of environmental conditions, through organizations such as Riverkeepers, has increased public attention to the role of ports and waterways in all areas, including cities. Many ports are so polluted that they need to be dredged of the toxic sludge at the bottom. Large, unknown cleanup costs, rising ocean levels, and expanding dead zones around major ports are environmental concerns of shipping firms.

One problem with acid rain is that the sources of emissions are so diffuse. It is not possible to assign individual liability for a given environmental impact and emission. With industries such as shipping the emissions may be easier to monitor, facilitating environmental policy implementation.

POTENTIAL FOR FUTURE CONTROVERSY

Acid rain as a U.S. controversy has been subsumed by controversies around global warming and climate change. Many of the debates are the same, especially in terms of science and legal issues. Scientific disputes mark the continuing evolution of ways to measure the actual environmental impacts. The controversy around acid rain was an early one, historically documented as a symptom of a larger problem. It is an important historic controversy because it promoted significant, successful policies. It is also a modern controversy because it continues to provide evidence of humans' impact on the environment.

See also Air Pollution; Climate Change; Global Warming; Sustainability; "Takings" of Private Property under the U.S. Constitution

Web Resources

Acid Rain, Atmospheric Deposition, and Precipitation Chemistry. U.S. Geologic Survey. Available at bqs.usgs.gov/acidrain/. Accessed January 20, 2008.
Buildings with Acid Rain Damage. Washington, DC. Available at pubs.usgs.gov/gip/acidrain/fieldguide.html. Accessed January 20, 2008.

Further Reading: Alm, Leslie R. 2000. *Crossing Borders, Crossing Boundaries: The Role of Scientists in the U.S. Acid Rain Debate.* Westport, CT: Praeger; Lehr, J. 1992. *Rational Readings on Environmental Concerns.* New York: John Wiley and Sons; Schmandt, Jurgen, Judith Clarkson, and Hillard Roderick. 1988. *Acid Rain and Friendly Neighbors: The Policy Dispute between Canada and the United States.* Durham, NC: Duke University Press; Wilkening, Kenneth E. 2004. *Acid Rain Science and Politics in Japan: A History of Knowledge and Action toward Sustainability.* Cambridge, MA: MIT Press.

AIR POLLUTION

Smog, acid rain, methane, and other forms of outdoor air pollution, as well as air pollution inside homes and other buildings, can all affect the environment. Cars, trucks, coal burning energy plants, and incinerators all make controllable contributions to air pollution. New environmental air pollution regulations continue to decrease emissions but with industry resistance.

AIR

Air quality has been a driving force for U.S. and global air pollution control. It can be quite different from region to region and over time. Geological features such as deep mountain valleys may facilitate dangerous atmospheric conditions when on the downwind side of industrial emissions, heavy car and truck traffic, and wood and coal stoves. Battlegrounds for air quality are scientific monitoring of air quality conditions, debate over what chemicals to regulate as pollution, and environmentalists' concerns over weak and incomplete enforcement. Each one of these is a controversy itself.

PUBLIC HEALTH

One of the primary criteria for an airborne chemical to be a pollutant is its effect on public health. One of the first areas of public concern about air pollution is breathing.

WHAT OZONE DOES TO LUNGS

The usual regulatory approach to environmental air pollution policy is the application of cost-benefit analysis to human health and environmental conditions. While value neutral, this approach can overlook the actual pain and suffering experienced by people in communities. Many communities experience many contaminant flows and exposures, some over long periods of time. The application of cost-benefit analysis in the development of air pollution policy generally permits a certain risk of death from cancer per population. There are many other risks and costs, many of a currently unknown nature, short of cancer. They can affect both individual health and the health of a community. Particulate matter is associated with early and unnecessary deaths, aggravation of heart and lung diseases, reduction in the ability to breathe, and increases in respiratory illnesses. This in turn can lead to increased school and work absences. Cancer itself can have many causes other than air pollution. Nonetheless, it is known that air pollution can have long-term health effects depending on the type of pollution and the age of the exposed person. The exposure of young people to ozone is particularly controversial because, unless they die from cancer, they do not enter into the cost-benefit analysis. However, by ignoring this cost, society may face even greater costs later.

Ozone causes chronic, pathologic lung damage. Human lungs are like filters, cleaning the air of whatever contaminants are encountered. What they do not remove can enter the bloodstream. At the levels experienced in most U.S. urban areas, ozone irritates cell walls in lungs and airways. This can cause tissues to be inflamed. This cellular fluid seeps into the lungs. Over time, especially if that time includes early childhood, the elasticity of the lungs decreases. Excessive exposure to high levels of air pollution in childhood can impair lung development for life. Susceptibility to bacterial infections can increase. Scars and lesions can develop in the airways of children chronically exposed to ozone. Ozone effects are not limited to vulnerable populations. At ozone levels in most warm-weather U.S. cities, average, healthy, nonsmoking young males who exercise can experience ozone impacts. Ozone exposure can shorten a life and cause difficult breathing.

Hospital admissions and emergency admissions increase as ozone levels increase. School and work absences increase. The level of human concern, from mother to child, increases as concern for our own and our loved ones' health rises. The intangible psychological factors of dread and fear weigh heavily on those who breathe polluted air. Ozone exposure is one of many.

Asthma is becoming more common. This is true even though some air pollutant concentrations have decreased. The increase in asthma is concentrated in people of color and low-income people. The incidence of acute asthma attacks in children doubled in the last 13 years even as very effective medicines were developed. About five million child hospitalizations were children who had asthma attacks. It is the most frequent cause of childhood hospitalization. Deaths of children with asthma rose 78 percent from 1980 to 1993. It is concentrated in high-population urban areas. This one environmental effect of air pollution can spread to inner-ring suburbs then to air regions over time. Asthma is described as breathing through a straw. The serious public health issues around air pollution underscore the intensity of this battleground in air pollution controversy.

Air pollution can have short- and long-term health effects. Asthma from air pollution can have short- and long-term effects. Short-term effects of asthma are irritation to the eyes, nose, and throat. Long-term reactions to air pollution can include upper respiratory infections such as bronchitis and pneumonia. Other symptoms of exposure to air pollution are headaches, nausea, and allergic reactions. Short-term air pollution can aggravate underlying medical conditions of individuals with asthma and emphysema. Long-term health effects are more controversial. Depending on the type of air pollution, there is general consensus that exposure can cause chronic respiratory disease, lung cancer, heart disease, and damage to the brain, nerves, liver, or kidneys. Continual exposure to most kinds of air pollution affects the lungs of growing children by scarring them at early stages of development. Recent studies suggest that the closer one is raised to a freeway in southern California, a notoriously low-quality air region overall, the greater the chance of having one of the listed long-term effects. Cumulative exposure to polluted air does aggravate or complicate medical conditions in the elderly. Some air pollution risk is involuntarily assumed. However, people die prematurely every year in the United States because of smoking cigarettes and voluntarily increasing other risk factors. Members of these communities label this type of risk assessment as blaming the victim. The involuntary assumption of health risks is something most communities strongly object to. With the advent of the Toxics Release Inventory many communities can track airborne industrial emissions. Citizen monitoring of environmental decisions has increased, especially around air quality issues.

STATE OF AIR POLLUTION

The air becomes polluted in different ways. How the air becomes polluted determines the types of problems it causes. Different sources of emissions contain different chemicals. These may interact with other airborne chemicals in unknown ways. As the chemicals mix with moisture in the air they can become rain. The rain can move the chemicals through the ecosystem, including crops and livestock. Mercury, lead, and aluminium all move in this way, with adverse ecological effects. There may be other chemicals with adverse ecological effects that do not last as long as metals do and may therefore be hard to detect while present. Air pollution can expose populations to more than just airborne pollution.

WHAT IS POLLUTION?

The term *pollution* has important legal and environmental meanings. Legally, it means that a person or business is not complying with environmental laws. Many environmentalists do not think this is extensive enough and believe that large environmental impacts can be considered pollution even if they are legal. Many permits do not in fact decrease emissions but permit more emissions.

Many permits have numerous exceptions to emissions. The petrochemical industry is allowed de minimus, fugitive, and emergency emissions beyond the permit, and that industry is leaking a valuable commodity. Industry argues that if it complies with all the environmental laws, then its emissions are not pollution because they are part of the permit issued by the EPA via the respective state environmental regulatory agency. While state and federal environmental agencies argue with the regulated industries, communities, and environmentalists, the actual environmental impact has worsened. While many environmental decisions are made behind closed doors, more and more communities are monitoring the environment themselves.

One type of air pollution is particulate matter. The particles are pieces of matter (usually carbon) measuring about 2.5 microns or about .0001 inches. Sources of particulate matter are the exhaust from burning fuels in automobiles, trucks, airplanes, homes, and industries. This type of air pollution can clog and scar young developing lungs. Some of these particles can contain harmful metals. Another type of air pollution is dangerous gases such as sulfur dioxide, carbon monoxide, nitrogen oxides, and other chemical vapors. Once in the atmosphere they follow the prevailing winds until they condense and fall to the ground as precipitation. This type of pollution can participate in more chemical reactions in the atmosphere, some of which form smog and acid rain. Other atmospheric chemical reactions are the subject of intense scientific controversy and are part of the debates of global warming and climate change.

Most air pollution comes from burning fossil fuels for industrial processes, transportation, and energy use in homes and commercial buildings. Natural processes can emit regulated chemicals at times. It is a subject of continuing scientific debate, both generally and specifically, how much of a given chemical is naturally emitted versus how much of the emission is from human actions.

The Natural Resources Defense Council closely tracks the air emissions of the biggest polluters. They call it their benchmarking project. They are a nonprofit environmental advocacy organization that believes in keeping track of environmental conditions to establish a baseline. Their research is based on publicly available environmental information, much of it available in the Toxics Release Inventory. Key findings of the benchmarking project's 2004 report include:

- Emissions of sulfur dioxide and nitrogen oxides have decreased by 36 percent and 44 percent, respectively, since the stricter pollution-control standards of the 1990 Clean Air Act went into effect.
- Carbon dioxide emissions increased 27 percent over the same period.
- Carbon dioxide emissions are expected to spike in coming years due to a large number of proposed new coal plants.
- Wide disparities in pollution rates persist throughout the electricity industry with a small number of companies producing a relatively large amount of emissions.
- Few power plants use currently available, state-of-the-art emissions control technologies.

- The electric power industry remains a major source of mercury emissions in the United States.

The Natural Resources Defense Council's benchmarking project uses public data to compare the emissions performance of the 100 largest power producers in the United States. They account for 88 percent of reported electricity generation and 89 percent of the industry's reported emissions. Emissions performance is examined with respect to four primary power plant pollutants: sulfur dioxide, nitrogen oxides, mercury, and carbon dioxide. These pollutants cause or contribute to global warming and to environmental and health problems including acid rain, smog, particulate pollution, and mercury deposition.

THE MOST POLLUTED TOWN IN THE UNITED STATES

The town with the most bad air days per year is Arvin, California. It has averaged about 73 bad air days a year since 1974. It is a small town with very little industry or traffic. It is on the valley floor between the Sierra Nevada and Tehachapi Mountains in southern California. Air pollution is coming from the east with the prevailing winds. The air pollution comes from the large industrialized California communities of Fresno, Bakersfield, Stockton, and the San Francisco Bay area. The air pollution problem has been getting worse for this predominantly Hispanic community. The San Joaquin Valley Air Pollution Control District, where Arvin is located, recently passed a controversial cleanup plan. Part of the plan is to encourage cleaner-running vehicles in Arvin and in Fresno. City buses and public vehicle fleets can reduce emissions, but not soon enough for everyone. They have also considered various legal actions, primarily based on issues of environmental justice and unequal enforcement of environmental laws. Whole communities feel as though their public health is threatened. Once this mass of polluted air does move out of the valley, it will continue to have environmental impacts. One of the prominent national parks, Sequoia National Park, feels the impact of this waste stream of air pollution. It has among the highest number of bad air days in the country.

INDOOR AIR POLLUTION

The air inside of buildings can be as polluted as outside air. Indoor air can accumulate gases and other chemicals more quickly than outside air. Cooking, heating, smoking, painting, new carpeting and glue, and heavy electronic equipment usage can all affect indoor air quality. Large numbers of books without adequate ventilation can cause carbon dioxide to build up. As most people spend most of their time indoors, the exposure to this air is much greater. Vulnerable populations, such as the very young and very old, spend even more time inside. Depending on the pollutants, indoor air pollution can lead to mold and fire hazards.

POTENTIAL FOR FUTURE CONTROVERSY

The controversies around air pollution show no signs of abating. Points of concentrated air pollution are getting more attention and becoming battlegrounds.

Ports are the latest example of this. On September 5, 2007, the U.S. Environmental Protection Agency (EPA) began a research project to test equipment that measures air emissions by equipment used in ports to move goods around docks and on and off cargo ships, trucks, and trains. Most of this equipment burns diesel fuel. The EPA wants to test new equipment that can recapture the energy of hydraulic brakes and thereby use less polluting fuel. They are predicting fuel savings of 1,000 gallons per vehicle per year, with decreased maintenance costs for the fleet. The EPA is working with the Port Authority of New York and New Jersey, Kalmar Industries, Parker Hannifin Corporation, and the Port of Rotterdam. Port authorities are very powerful independent legal entities that can neither tax nor be taxed. They issue bonds. Interest on bonds is not income for federal tax purposes, nor for state tax purposes if issued in that state. Wealthy individuals can reduce their tax liability and invest in the country's infrastructure. Historically, this was done in the West with railroad bonds. *Authorities* are creatures of state law, but very little is required in the way of public participation or environmental planning. Port authorities are able to resist many environmental requirements, especially if they involve several different states. The environment and ecology of ports are often toxic and unappealing. Ports are places where many ships empty their bilges of waste, often illegally. Some states have passed legislation to prevent cruise ships from dumping their wastes in their ports, such as California. Ports have also been the site of land-based waste-dumping practices. Along tidal areas many communities did this with the idea that the tide would take it away. Wastes from fishing and fish processing can also add to the mix. Ports are also the terminus of many rivers that have collected agricultural runoff, municipal sewage, industrial water discharges, and other types of waste. Ports are among the most environmentally challenging ecosystem reconstruction projects in the United States. In early 2000 many port authorities began to incorporate principles of sustainability into their long-range strategic corporate planning. The cumulative effects of waste, the increasing liability for clean up costs and its accounting as a contingent liability, and increasing urban environmental activism all undercut achieving anything sustainable in an environmental, business, or social sense. Port authorities now partner with the EPA around air pollution, expressly motivated by a concern about sustainability. New battlegrounds will also emerge from these new policies, such as how clean is necessary.

The environmental policies and laws do have the intended effect of reducing the emissions of some chemicals emitted by most industries. However, asthma rates increase and so too does community concern. It is likely that the costs of further decreasing emissions from industry, from municipalities, and from all of us will be more expensive. The current context of global warming and rapid climate change drives many air pollution controversies to center stage.

See also Acid Rain; Climate Change; Global Warming

Web Resources

Air pollution in Chesapeake Bay. www.chesapeakebay.net/air_pollution.htm. Accessed March 2, 2008.

Air pollution control costs in Texas. www.texasep.org/html/air/air_2std_brdrair.html. Accessed March 2, 2008.

Natural Resources Defense Council: Clean Air and Clean Energy. www.nrdc.org/air/pollution/default.asp. Accessed March 2, 2008.

Further Reading: Bas, Ed. 2004. *Indoor Air Quality.* GA: The Fairmont Press, Inc.; Harrap, D. 2002. *Air Quality Assessment and Management.* London: Spon Press; Lipfert, Frederick W. 1994. *Air Pollution and Community Health.* New York: John Wiley and Sons; Moussio-poulos, Nicolas, ed. 2003. *Air Quality in Cities.* New York: Springer; Simioni, Daniela. 2004. *Air Pollution and Citizen Awareness.* New York: United Nations Publications.

ANIMALS USED FOR TESTING AND RESEARCH

The use of animals for research and testing has been part of science since its inception. The lives of research animals of all kinds were often short and painful. In contrast, animal rights activists contend that the lives of animals should be protected as if they were human. They strongly oppose the pain and suffering and killing of animals.

ANIMALS USED IN TESTING AND RESEARCH

Animals are used extensively for food and clothing. They are also used in testing and research. Many researchers are most interested in the impact on human of a given chemical in the air or water. Other researchers are working on vaccines and other public health research. Some animals are used in medical diagnosis, such as seeing if a rabbit died after injected with the blood of a pregnant woman. Human subject testing is often illegal and considered unethical. Animals are used extensively and successfully in research and testing. Seventy million animals are used in this manner in the United States each year. Organizations using animals include private research institutions, household chemical product and cosmetics companies, government agencies, colleges and universities, and medical centers. Household goods and cosmetics such as lipstick, eye shadow, soap, waxes, and oven cleaner may be tested on animals. Many of these products now advertise that they do not use animals to test their products. These tests on animals are mainly used to test the degree of harmfulness of the ingredients. Animals are generally exposed to the ingredient until about half die in a certain time period. Animals that survive testing may also have to be euthanized. The primary objections to animal testing are:

- It is cruel in that it causes unnecessary pain and suffering
- It is outdated; there are more humane modern methods
- It is not required by law

Manufacturers justify the use of animal testing to make sure none of the ingredients in their products can pose human risks. By using mammals for their tests, manufacturers are using some of the best tests available. They also claim that the law and regulation almost require them to use animal testing. This is a point of controversy. According to the law, the Food and Drug Administration

(FDA) requires only that each ingredient in a cosmetics product be "adequately substantiated for safety" prior to marketing. If it cannot be substantiated for safety then the product must have some type of warning. Furthermore,

- The FDA does not have the authority to require any particular product test
- Testing methods are determined by the cosmetics and household product manufacturers
- The test results are mainly used to defend these companies against consumer lawsuits

Part of this controversy is the issue of humane alternatives, alternatives that do not use animals for testing or research. Animal rights advocates contend that humane alternatives are more reliable and less expensive than animal tests. Computer modeling and use of animal parts instead of live animals are the main humane alternatives. One controversial test uses the eyes of rabbits. The eyes of rabbits are very sensitive to their surroundings. Some Flemish hares (rabbits) were used in the storage silos of Umatilla's biochemical weapon storage facility. If nerve gas was escaping, the eyes of the Flemish hares would dilate. One test involving rabbits' eyes could be replaced by a nonanimal test. The Draize Eye test uses live rabbits to measure how much a chemical irritates the eye. Instead of using live rabbits for this test, eye banks or computer models can be used to accurately test the irritancy level of a given chemical. However, researchers contest the reliability and cost of these alternatives.

Other alternatives to using animals for research and testing include:

- Chemical assay tests
- Tissue culture systems
- Cell and organ cultures
- Cloned human skin cells
- Human skin patches
- Computer and mathematical models

ANIMAL TESTING PROPONENTS

Today, scientists are using animal research to:

- Study factors that affect transmission of avian flu between birds as well as the genetic and molecular adaptation from wild birds to domestic poultry;
- Evaluate whether ducks in Asia are infection reservoirs sustaining the existence of the H5N1 virus;
- Develop new and evaluate existing techniques to predict which mild forms of the avian flu virus might transform into more deadly forms;
- Develop improved vaccines against avian flu for birds, and evaluate vaccines for human use.

Heightened animal research is necessary to combat avian flu and other new and emerging animal-borne diseases such as mad cow disease (bovine spongiform encephalopathy [BSE]), SARS, and West Nile virus. Scientists point out that about three-quarters of animal diseases can infect humans. Some call for more collaboration between animal health and public health organizations. Animal research and testing will be required for this collaboration.

NUMBER OF ANIMALS USED

Information from regulated research facilities does establish some type of baseline data about the kinds and numbers of animals used in testing and research. Statistics from the Animal Research Database show how animals are used for testing and research.

- There are approximately 56–100 million cats and 54 million dogs in the United States.
- It is estimated that every hour 2,000 cats and 3,500 dogs are born.
- Between 10.1 and 16. 7 million dogs and cats are put to death in pounds and shelters annually.
- Approximately 17–22 million animals are used in research each year.
- Approximately 5 billion are consumed for food annually.
- Approximately 1.1 percent of dogs and cats from pounds and shelters that would otherwise be euthanized are used in research.
- Fewer than one dog or cat is used for research for every 50 destroyed by animal pounds.
- Rats, mice, and other rodents make up 85–90 percent of all research animals.
- Only 1 to 1.5 percent of research animals are dogs and cats.
- Only 0.5 percent are nonhuman primates.
- There has been a 40 percent decrease in the numbers of animals used in biomedical research and testing in the United States since 1968.

Other federal agencies have studied standards of animal care when used in testing and research. One of them is the United States Department of Agriculture (USDA). According to the USDA, approximately.

- 61 percent of animals used suffer no pain
- 31 percent have pain relieved with anesthesia
- 6 percent experience pain because alleviation would compromise the validity of the data. Much of this work is directed at an understanding of pain.

These figures apply only to those animals covered by the Animal Welfare Act, which currently excludes rats, mice, farm animals, and cold-blooded animals. Some of these are animals that are used extensively in animal research.

There are continuing concerns that this reporting undercounts the mortality and suffering.

RESEARCH AND TESTING OF ANIMALS
TO ACHIEVE PUBLIC HEALTH VICTORIES

Major advances in U.S. public health that have increased longevity and the quality of human life were based on research using animals. The decline in U.S. death rates from cardiovascular diseases, infections, and most kinds of cancer since the 1960s is the result of new methods of treatment based on research requiring animals. Researchers claim that others do not understand the long-term results of such research and how it is conducted. Researchers also claim that others do not recognize important differences between using animals for product testing and for biomedical research. Biomedical research is more justified because of the public health benefits to society, while product testing is to increase the product safety to the consumer and profit of a manufacturer.

All researchers and research facilities are not the same. Some research sponsors are also concerned about the use of animals in testing and research. There are ways to encourage whenever possible the use of alternatives to live animal testing. The American Heart Association (AHA) sponsors important heart-related research. They have specific guidelines about how research animals are to be used and treated. First, the researcher must demonstrate that the animals are needed, and that there are no viable substitutes for the animal. Second, when animals are needed for association-funded experiments, the animals must be handled responsibly and humanely. Before being approved for Association support, the researchers must show that:

- they have looked at alternative methods to using animals;
- their research cannot be successfully conducted without the use of animals; and
- their experiments are designed to produce needed results and information.

Together with other responsible and committed research-sponsoring organizations, the AHA hopes to ensure that the use of animals for testing and research will occur more carefully. Many universities have developed ethical guidelines for the use of animals in their research programs.

Not all animal testing occurs in these types of organizations. It is still much more expensive to develop new and untested alternatives than to treat some animals as expendable. A given method of drug testing may be more humane to the animals but less effective as a predictor of the drug's impact on humans.

THE ANIMAL WELFARE ACT OF 1966 (AWA)

The main law is the Animal Welfare Act of 1966. As such it has been a flash point of controversy for animal rights activists. The Animal Welfare Act is the minimum acceptable standard in most U.S. animal rights legislation. Its original intent was to regulate the care and use of animals, mainly dogs, cats, and primates, in the laboratory to prevent abuse. Now it is the only federal law in the United States that regulates the treatment of animals in research, exhibition, transport, and by dealers. In 1992 a law was passed to protect animal breeders from ecoterrorists. Other laws may include additional species coverage or specifications for animal care and use. Some state and cities have some laws that could be argued to protect against animals' use in research and testing, primarily animal abuse laws. They usually require a cooperative district attorney to file and pursue criminal charges. Because there are so many other types of animal abuse crimes than testing and research the prosecutorial discretion to investigate and enforce abuse laws puts testing and research animal abuse as a low priority. The Animal Welfare Act is enforced through a federal agency with the usual enforcement powers of investigation, searches, and fines or penalties.

The Animal Welfare Act is enforced by the U.S. Department of Agriculture, Animal and Plant Health Inspection Service (APHIS), and Animal Care (AC). There is an extensive set of rules and regulations in place. The regulations are divided into four sections: definitions, regulations, standards, and rules of practice.

The definitions section describes exactly what is meant by terms used in the Animal Welfare Act. This section is very important as the legal definition of animal is different than its generally understood meaning. For example, the term *animal* in the act specifically excludes rats of the genus *Rattus* and mice of the genus *Mus* as well as birds used in research. There are many such exemptions in the Animal Welfare Act. These exemptions are controversial among animal rights activists because they consider the exemptions as contrary to the intent of the act. The regulations section of the Animal Welfare Act is quite specific. As noted on its Web site (http://warp.nalusda.gov/awic/legislat/regsqa.htm) the regulations methodically list subparts for licensing, registration, research facilities, attending veterinarians and adequate veterinary care, stolen animals, records, compliance with standards and holding periods, and other topics such as confiscation and destruction of animals and access to and inspection of records and property. Monitoring these records from large research facilities, both those in compliance and those out of compliance, allowed the USDA to collect large amounts of information. The actual standards for treatment of animals by species are in the next section. Most of the subchapter is the third section that provides standards for specific species or groups of species. Included are sections for cats and dogs, guinea pigs and hamsters, rabbits, nonhuman primates, marine mammals, and the general category of "other warm-blooded animals." Standards include those for facilities and operations, health and husbandry systems, and transportation. This section is the one animal rights advocates most often seek to have enforced, and therefore it is a battleground in this controversy. If the animal rights advocates seek legal redress, they must first exhaust their administrative remedies before a court will accept jurisdiction. Their first step in seeking legal redress, generally for the enforcement of the above conditions, is the focal point of the final section of the act. The final section lists the rules of practice applicable to adjudicating administrative proceedings under the Animal Welfare Act. After exhausting administrative remedies under the act, animals rights activists can then go to court. One problem with the administrative agency battleground is that it is very time consuming. The administrative agency, a potential defendant, controls the process and hearings format. Many public interest groups feel this is an unfair requirement because it drains the resources of the nonprofit organization before the issue can be resolved.

ANIMAL CARE: WHAT IS HUMANE?

While there may be agreement that inhumane treatment to animals should be regulated, there is more controversy about what specifically is humane treatment. It is generally tied to the activity around the animal. Again, according to their Web site (http://warp.nalusda.gov/awic/legislat/regsqa.htm), the Animal Welfare Act requires that minimum standards of care and treatment be provided for certain animals bred for commercial sale, used in research, transported commercially, or exhibited to the public. People who operate facilities in these categories must provide their animals with adequate care and treatment in the areas of housing, handling, sanitation, nutrition, water, veterinary care, and protection from extreme weather and temperatures. Although these federal requirements do establish a floor of acceptable standards, there is controversy about whether they go far enough. There is also controversy about how well enforced the existing law is under the present circumstances. Regulated businesses are encouraged to exceed the specified minimum standards under AWA. Some animals are bought and sold from unregulated sources for testing and research, and this remains a concern.

EXEMPTIONS

The AWA regulates the care and treatment of warm-blooded animals with some major exceptions, also known as exemptions. As older legislation from the 1960s, such as the AWA, passes through subsequent Congresses, exemptions or categorical exclusions are legislatively added to accommodate powerful interests and changes in public policy. Farm animals used for food, clothing, or other farm purposes are exempt. This is a large exemption representing powerful industrial agricultural interests. If they were included, argue these interest groups, the costs of production would increase the cost of food and clothing. Cold-blooded animals are exempt from coverage under the act, but some advocates are seeking to have them covered. The use of frogs for science courses is traditional. Many cold-blooded animals are used for training, testing, and research. Retail pet shops are another major exemption if they sell a regular pet to a private citizen. They are covered if they sell exotic or zoo animals or sell animals to regulated businesses. Animal shelters and pounds are regulated if they sell dogs or cats to dealers, but not if they sell them to anyone else. The last big exemption is pets owned by members of the public. However, no one is exempt from criminal prosecutions for animal abuse under state and local laws.

PET PROTECTION FROM ANIMAL TESTING AND RESEARCH

Selling stolen or lost pets for research and testing is another aspect of this controversy. To help prevent trade in lost or stolen animals, regulated businesses are required to keep accurate records of their commercial transactions. Animal dealers must hold the animals that they buy for 5 to 10 days. This is to verify their origin and allow pet owners an opportunity to locate a missing pet. This also helps suppress the illegal trade in stolen animals for testing and research. Many pets are lost when a natural disaster occurs. Floods, storms, hurricanes, emergency vehicles, and threatening interruptions of food and shelter cause many pets to get lost. Some pets now have a computer chip implanted in them for tracking purposes. Critics point out that many commercial transactions about animals used in testing and research do not always happen with regulated businesses. The legal definition of *regulated research facilities* specifically includes hospitals, colleges and universities, diagnostic laboratories, and cosmetic, pharmaceutical, and biotechnology industries. Animal dealers complain about the cost of holding animals that long and the other costs of verifying ownership. Many animal shelters operate on a small budget and must euthanize animals to make room for new arrivals faster than the holding period allows. Rigorous enforcement against these groups could put them out of operation, with the net result of no shelter provision or adoption site for animals.

STANDARDS OF HUMANITY IN RESEARCH FACILITIES

Much of the regulation of animal use in testing and research occurs in research facilities. The standards are slightly higher for dogs and primates. All warm-blooded animals, with some major exemptions, get some veterinary care and animal husbandry. This means that a licensed veterinarian examines the health of the animal, and the animal is kept in a clean, sanitary, and comfortable condition. Some animals require a higher standard of care by law. Regulated research facilities must provide dogs with the opportunity for exercise and promote the psychological well-being of primates used in laboratories. According to their Web site (http://warp.nalusda.gov/awic/legislat/regsqa.htm) researchers must also give regulated animals anesthesia or pain relieving medication to minimize the pain or distress caused by research if the experiment allows. Regulated entities do express some concern about the cost of these additional procedures. The AWA also prohibits the unnecessary repetition of a specific experiment using regulated animals. The regulated entity itself determines how much repetition is unnecessary. One tenet of science is the ability to repeat a given chain of events, such as corneal exposure to a chemical, and get the same result, such as death or blindness. By prohibiting the repetition of the animal-based tests, some of the results may be weaker. The public protection from a chemical may be weaker, and the industry may be exposed to a large class-action negligence suit. Therefore, research procedures or experimentation are exempt from interference when designated as such. This is a large exemption in an area where animals are thought to suffer pain, and where substitutes or alternatives to animals are not used frequently. This is a continuing battleground in this controversy.

AWA has strict monitoring and record-keeping requirements. The lack of records has been a controversy in the past. By keeping records, information about animals used in testing and research can be gathered. The problem was how to require the regulated entity to comply with the AWA and produce necessary records. The solution in this case was to require a committee at the regulated entity and mandate its membership. The law requires research facilities to form an "institutional animal care and use committee." The purpose of this committee is to establish some place of organizational accountability for the condition of animals at a given research institution. They are primarily responsible for managing aspects of the AWA, especially in regard to the use of animals in experiments. This committee is the point of contact responsible for ensuring that the facility remains in compliance with the AWA and for providing documentation of all areas of animal care. By law, the committee must be composed of at least three members, including one veterinarian and one person who is not associated with the facility in any way.

ANIMAL RIGHTS GROUP CONVICTED OF INCITING VIOLENCE AND STALKING

Those opposed to animal testing have sometimes used violence to convey their point, which further inflames this controversy. Criminal prosecution is also a part of this controversy, especially with focused legislation. This in turn makes courts one of the battlegrounds.

On March 9, 2006, members of an animal rights group were convicted of two sets of criminal acts. The first was conspiracy to violate the Animal Enterprise Protection Act and the second was interstate stalking. It was a long and controversial federal jury trial in Trenton, New Jersey. The group itself was also convicted. The jury found the group Stop Huntingdon Animal Cruelty (SHAC) and six of its members guilty of inciting violence against people and institutions who did business with Huntingdon Life Sciences (HLS), a British-based research firm that runs an animal testing laboratory in East Millstone, New Jersey.

SHAC targeted HLS, as well as other companies doing business with HLS, because it uses animals to test the safety of drugs and chemicals. SHAC has claimed responsibility for several bombings and dozens of acts of vandalism and harassment in both the United States and Europe to protest the use of animals in research and testing. Its campaign against HLS has become an international cause among animal rights activists since the late 1990s.

The six defendants—former SHAC spokesperson Kevin Kjonaas; Lauren Gazzola, whom the indictment identified as SHAC's campaign coordinator; Andrew Stepanian, a longtime activist with SHAC and the Animal Defense League; Joshua Harper a self-described anarchist and SHAC activist; and SHAC members Jacob Conroy and Darius Fullmer—were all found guilty of conspiracy to violate the Animal Enterprise Protection Act.

Kjonaas, Gazzola, and Conroy were also found guilty of multiple counts of interstate stalking and conspiracy to engage in interstate stalking. In addition, Kjonaas, Gazzola, and Harper were found guilty of conspiracy to violate a telephone harassment act.

The defendants were arrested in May 2004 by federal agents in New York, New Jersey, California, and Washington. They face three to seven years in prison and fines of up to $250,000. They may also face judgments in civil trials from victims.

The defendants were convicted of conducting a very personal, no-holds-barred campaign of terror against HLS employees and their children. During the three-week trial, prosecutors showed how SHAC's campaign against HLS involved posting personal information on the Internet about its employees and about employees of firms that do business with HLS. The information posted on the Internet included phone numbers, home addresses, and, in some cases, information on where employees' children attended school. Many of those targeted had their cars and homes physically vandalized and received threats against them or their families, according to court testimony.

According to law enforcement officials, one female employee was sent an e-mail from SHAC threatening to "cut open her seven-year-old son and fill him with poison."

The Animal Enterprise Protection Act, signed into law by the first president Bush in 1992, provided animal research facilities with federal protection against violent acts by so-called animal rights extremists. The act gave prosecutors greater powers to prosecute extremists, whose attacks create damages or research losses totaling at least $10,000. Animal enterprise terrorism is defined in the act in part as "physical disruption to the functioning of an animal enterprise by intentionally stealing, damaging, or causing the loss of any property (including animals or records)."

Some critics charged that prosecutors rarely used the Animal Enterprise Protection Act because the penalties were too mild and it was difficult to prove damages of more than $10,000. An antiterrorism bill signed into law by President George W. Bush in 2002 substantially increased the penalties for such actions. Prior to the SHAC trial, there appears to have been only a single successful prosecution under the Animal Enterprise Protection Act. In 1998, a federal grand jury in Wisconsin indicted Peter Daniel Young and Justin Clayton Samuel under its provisions for breaking into several Wisconsin fur farms in 1997 and releasing thousands of animals. Samuel was apprehended in Belgium in 1999 and quickly extradited to the United States. In 2000, Samuel pleaded guilty and was sentenced to two years in prison and ordered to pay over $360,000 in restitution. Young was a fugitive until arrested in March 2005 in San Jose for shoplifting. He was later sentenced to two years in prison.

While there are controversial questions about the enforcement of the Animal Welfare Act, there is little question about the rigorousness of enforcement against those who commit terrorist acts of protest to free animals used for fur, testing, and research.

POTENTIAL FOR FUTURE CONTROVERSY

The controversy around the use of animals for testing will continue because animals will continue to be used for testing. Animal rights groups assert that enforcement of a weak law riddled with exemptions is inadequate and that animals are being abused. One environmental impact of a poorly regulated animal trade is the importation of endangered species under conditions of high mortality. Testing on animals, unlike dogfighting, is done at large institutions often receiving federal research grants. Many of these are universities. When human health considerations are thrown into the equation, animal testing is often justified by researchers despite animal mortality.

See also Endangered Species

Web Resources

Animal Concerns. Animal rights against testing. Available at www.animalconcerns.org. Accessed January 20, 2008.
Animal Welfare Act. Available at www.nal.usda.gov/awic/legislat/awicregs.htm. Accessed January 20, 2008.

National Anti-Vivisection Society. Available at www.navs.org/site/PageServer?pagename=
index. Accessed January 20, 2008.

Further Reading: Botzler, Richard George. 2003. *The Animal Ethics Reader.* London: Rout-
ledge; Carbone, Larry. 2004. *What Animals Want: Expertise and Advocacy in Laboratory
Animal Welfare Policy.* New York: Oxford University Press; Garner, Robert. 2004. *Animals,
Politics, and Morality.* Manchester, UK: Manchester University Press; Gluck, John P., Tony
DiPasquale, and Barbara F. Orlans. 2002. *Applied Ethics in Animal Research: Philoso-
phy, Regulation and Laboratory Applications.* Ames, IA: Purdue University Press; Regan,
Tom. 2001. *Defending Animal Rights.* Chicago: University of Illinois Press; Rudacille,
Deborah. 2001. *The Scalpel and the Butterfly: The Conflict between Animal Research and
Animal Protection.* Berkeley: University of California Press.

ARCTIC WILDLIFE REFUGE AND OIL DRILLING

The Arctic National Wildlife Refuge is a vast, protected wildlife habitat in
Alaska where oil and natural gas have been found. Because of this, environmentalists representing the interests of pristine wilderness and endangered species
are pitted against oil and gas corporations. Federal, state, tribal, and community
interests are heavily involved. There have been many contentious legislative sessions in Alaska and in Washington, D.C., for many years over drilling for oil in
the Arctic Wildlife Refuge and other parts of Alaska. The battleground for this
controversy will spill over to courtrooms.

BACKGROUND

Federal lands in Alaska are vast. Roads and people are scarce. Wildlife abounds
unseen by human eyes. The weather can stop most human activities for days
at a time, as well as make any travel of people or goods uncomfortable, risky,
and expensive. Economic development around most types of activities such as
agribusiness, oil or mineral drilling, logging, tourism, and shipping is equally
constrained by the weather and the expense of dealing with cold, inclement
weather. Without good roads and transportation infrastructure most economic
development suffers, making them attractive benefits to many Alaskan communities. However, the federal government was, and is, the largest landowner and
has exerted its power to create and protect its interests.

Ever since Alaska was recognized and accepted by the United States as a
state, environmental protection and natural resource use have been at odds.
Although seemingly limitless in abundant natural resources, these existed in
fragile tundra and coastal environments. Without roads, logging, mining, or
any substantial human development were very difficult. Indigenous peoples
of Alaska were self-sufficient and subsisted on the land and water. Subsistence
rights to fish, game, and plants as well as ceremonial rights to this food are very
important to many indigenous people, including bands and tribes in the United
States. Congress passed the Alaska National Interest Lands Conservation Act
(1980) and established the Arctic National Wildlife Refuge (ANWR). Then
Congress specifically avoided a decision regarding future management of the
1.5-million-acre coastal plain. The controversy pitting the area's potential oil

and gas resources against its importance as wildlife habitat, represented by well-organized environmental interests, was looming large then. This was the not the first nor the last experience pitting oil companies against environmentalists, state interests in economic development against federal interests in preserving wilderness areas, and other confrontational battlegrounds. Numerous wells have since been drilled and oil fields discovered near the ANWR. Also, the characteristics of the ecosystem and measures of environmental impacts to date have been documented in very similar places nearby.

Global warming and climate change greatly affect this particular controversy. Most scientists agree that for every 1°F of global warming, the Arctic and Antarctica will warm up by 3°F. The planet has been warming and the Arctic ice is melting. In September 2004 the polar ice cap receded 160 miles away from Alaska's north coast, opening up more and more open water. This has dramatic environmental impacts in the Arctic because many species from plankton to polar bears follow the ice for survival. Its implications for the ANWR oil-drilling controversy are developing. Environmentalists think it may make an already sensitive ecosystem even more sensitive. Mosquitoes are now seen further north than ever before. They attack nesting birds that do not leave the nest for long periods and never had exposure to mosquitoes before. There are many anecdotal reports of species impacts in the Arctic.

The focus of the controversy about oil drilling in the Arctic National Wildlife Refuge is its coastal plain. It is a 25-mile band of tundra wetlands that is of key interest to both oil interests and environmental concerns. This area provides important nursing areas for Arctic wildlife. Damaged tundra takes a long time to recover and is generally considered a sensitive ecosystem. Tundra is very sensitive to many of the activities around oil drilling. The wetlands, which can move food around the coast efficiently, can also move hazardous materials around the same way. But the refuge's coastal plain also contains oil, exactly how much and where being a subject of dispute. Controversies about the sensitive ecological character of the refuge, the amount, kind, and accessibility of oil that lies beneath it, and the environmental impact that oil development would have on it all abound. The primary concern about its impact is the threat to species and other parts of the ecosystem.

THE ENVIRONMENT OF THE ARCTIC NATIONAL WILDLIFE REFUGE

Far north of most places, the Arctic National Wildlife Refuge is part of the vast federal landholdings in the vast state of Alaska. Brutal winters and glorious summers characterize its seasonal extremes. Given its human inaccessibility, many species of wildlife thrive here. The Arctic National Wildlife Refuge is located between the rugged Brooks Mountain Range and extends to the Beaufort Sea in northeast Alaska. By covering such a large area with ecotones ranging from mountains to the coastal plain many species can adapt to seasonal extremes. The seasonal migration of the caribou plays a large part in the food chain here. The coastal plain is a rich ecosystem easily affected by both local and global environmental forces. The Arctic National Wildlife Refuge's coastal plain alone supports almost 200 wildlife species, including polar bears, musk ox, fish, and caribou. Every year, millions of tundra swans, snowy owls, eider ducks, and other birds migrate to the coastal plain to nest, raise their offspring, molt, and feed. Other species give birth there, and many others migrate through it. Some environmental scientists consider the coastal plain to be the biological heart of the entire refuge. They maintain that any oil drilling in the refuge would irreparably harm the wildlife by destroying the unique habitat. Oil development, with its drilling, road building, water and air emissions, noise, and waste, could irreparably environmentally degrade this pristine, fragile wilderness. The fact that it is exploration and drilling for oil and gas, the very substances that cause much pollution, heightens the controversy. The environmental groups are making a stand because this is one of last remaining untouched wilderness areas of this type. This battleground is characterized by aggressive environmentalist and conservationist pressures that have been fully engaged ever since oil was found.

THE CURRENT INDUSTRIAL IMPACT: THE NORTH SLOPE EXPERIENCE

The controversy about environmental impacts of drilling is not a new one in Alaska. The oil spill of the Valdez, its environmental impacts, and subsequent, protracted litigation are well known. Environmentalists point to other nearby similar areas that have allowed oil development to argue that the environmental impacts in the Arctic National Wildlife Refuge are not worth the oil. They point to the controversial Alaskan North Slope.

The Alaskan North Slope was once part of the largest intact wilderness area in the United States. In some environmental aspects it is similar to the Arctic National Wildlife Refuge. With controversy it was opened up to oil drilling and a pipeline and accompanying environmental impacts. Alaska's North Slope oil operations expanded on a large scale. It now has one of the world's largest industrial complexes. The oil operations and transportation infrastructure are vast. They cover about 1,000 square miles of land. Most of this land and environment was as wild as the Arctic National Wildlife Refuge. There are oilfields, oil tankers, a few basic oil refineries, oil storage, and oil spills. Roads or airstrips are often built. This drastically increases environmental impacts in wild areas. Sometimes local communities are in favor of infrastructure development as a means of economic development, despite environmental consequences. On the Alaskan North Slope, Native Americans also have important interests. Sites can be sacred. Rights in land may extend to a limited set of natural resources, such as timber or seasonal harvesting. They may also be in favor of infrastructure development as a means of economic development. The scale of industrial operations here affects all these interests. Prudhoe Bay and 26 other oilfields on the Alaskan North Slope include the following:

- 28 oil production plants, gas processing facilities, and seawater treatment and power plants
- 38 gravel mines

- 223 production and exploratory gravel drill pads
- 500 miles of roads
- 1,800 miles of pipelines
- 4,800 exploration and production wells

The scale of such an industrial operation is taking place in a comparably fragile region much like the Arctic National Wildlife Preserve. In the modern context of global warming and climate change, the caribou that play such an important role are affected by the retreating ice. They cannot feed on the lichen on rocks in places formerly covered by ice. Their migratory route is being altered. The same is true for many species. Its environmental impact alone is a controversy, now fueling the Arctic National Wildlife Refuge controversy. Ecosystem resiliency is defined as the length of time it takes the ecosystem to recover from stress. Because of the same factors that affect the Arctic National Wildlife Refuge the North Slope is considered fragile. These factors are a short, intense summer growing season, bitter cold in the long winter, poor soils, and permafrost. These were both wildlife areas with little human intrusion. Environmentalists contend that any physical disturbance to the land, such as roads, oil spills, and drilling, has long-term, perhaps irreparable environmental impacts. The National Academy of Sciences concluded, "it is unlikely that the most disturbed habitat will ever be restored and the damage to more than 9,000 acres by oilfield roads and gravel pads is likely to remain for centuries." Many environmentalists contend that the cumulative impacts of oil development have affected Prudhoe Bay negatively. Environmentalists use the North Slope experience to protect the Arctic National Wildlife Refuge. They use it to show it is impossible to drill for oil without irreversible environmental consequences.

Of particular concern is spilled oil and other petrochemical waste products from engine maintenance. According to environmentalists, oil operations spill tens of thousands of gallons of crude oil and other hazardous materials on the North Slope every year. Environmentalists worry that not all spills are reported, as most industry environmental impact information is self-reported. Spills can occur when drilling for new oil, storing, and transporting it. Conditions for all these activities can be physically rough in Alaska. Weather conditions can become severe for days at a time and can be conducive to spills. According to industry and government reports, from 1996 to 2004, there were at least 4,530 spills on the North Slope of more than 1.9 million gallons of diesel fuel, oil, acid, biocide, ethylene glycol, drilling fluid, and other materials. Some of these chemicals can rapidly percolate, or move through, soil to reach water tables. Conditions in Alaska can make it difficult to contain and clean up a spill of any size.

OIL OPERATIONS AND THE AIR

Coal-burning power generation and petrochemical refineries emit large quantities of regulated pollutants into the air. Diesel generators and vehicles, trucks, and airplanes all also emit pollutants. According to the Toxics Release Inventory, oil operations on Alaska's North Slope emit more than 70,000 tons of nitrogen oxides a year. Sulfur and nitrogen are air pollutants, which contribute to smog and acid rain. North Slope oil operations also release other pollutants, which are major contributors to air pollution. Each year, they admit to emitting 7 million to 40 million metric tons of carbon dioxide and 24,000 to 114,000 metric tons of methane in the North Slope. This is probably within the terms of their air permit and may exclude de minimus or fugitive emissions. Emissions caused by natural disasters such as hurricanes and tsunamis are also exempt. Sustainability advocates point out that the methane emissions do not include the methane released because of the melted permafrost. All these impacts are in the context of larger controversies such as global warming and climate change that also affect sensitive Alaskan ecosystems. Emissions will be higher in the Arctic National Wildlife Refuge as North Slope oil is transported by tanker from the site to a refinery. It is refined and distributed off the drilling site. ANWR oil and gas could still be refined on site if oil exploration and drilling is approved.

POLAR BEARS AND GLOBAL WARMING

All over the southernmost part of their Arctic range, polar bears are thinner, lower in number, and producing fewer offspring than when the ice is thick. As the ice retreats the polar bears get food from human garbage sites. Polar bears are a particular problem because they must learn fear of humans. Some say that the lack of ice due to global warming has disrupted the ability of polar bears to hunt for seals, forcing them to swim further and further out in search of food. In 2004, researchers found four dead polar bears floating about 60 miles off Alaska's north shore. Although polar bears are capable of swimming long distances, 60 miles is considered unusual. Most observers think that the lack of ice forces polar bears to swim further for food. Others think they just follow the ice northward until a split separates them from land.

The polar bear is at the top of the food chain in the Arctic. At least 200 species of microorganisms grow in Arctic ice flows. They form curtains of slimy algae and zooplankton, and when they die they feed clams at the bottom, which feed the walruses and seals, which feed polar bears.

Also, polar bears bioaccumulate chemicals and compounds in great concentrations. This is especially true for metals such as mercury and other chemicals used in industry. Some scientists closely examine apex animals, such as polar bears and humans, that are living in the places where the effects of climate change are first observable. The North Slope polar bears are now monitored by radio collaring and mapping dens.

Airborne pollution from Prudhoe Bay has been detected as far as Barrow, Alaska, about 200 miles away. The environmental impact of industrial oil operations on the North Slope is widespread. The environmental impact of these air pollutants on Arctic ecosystems remains controversial. The Canadian government has experimented with oil companies and cumulative impacts research in neighboring northern Alberta. In the sensitive Arctic environment of northern Alberta the government has allowed oil company drilling and refining operations on the condition that they account for all impacts, including cumulative impacts. Because vast areas of Alaska are undeveloped the potential exists for a large environmental impact before it can be discovered. The effects of acid rain and air pollution on migratory animals such as birds, caribou, and whales are unknown. Robins have appeared in northern Alaska for the first time, as have other warm-weather species.

HAZARDOUS WASTE AND ITS IMPACTS ON WATER AND WETLANDS

Drilling for oil includes digging large pits in the ground. As these pits become obsolete they were and are used as waste dumps. Pits holding millions of gallons of wastes from oil and gas drilling and exploration were all over the North Slope. The pits were a stew of toxic chemicals, many with long-lasting environmental impacts in any ecosystem. Deep well injection, as this waste disposal method is called, was stopped because of its impact on underground aquifers. As aquifers dry up, as around San Antonio, Texas, they pull in the waste injected in deep wells. Of the known and undisputed pit sites more than 100 remain to be cleaned. *Clean* is a relative term. In this case, it generally means pumping out the toxic materials and removing them for treatment as hazardous waste. Clean does not mean restoring the ecological integrity of the place. This is why some environmentalists claim the impacts of oil exploration and drilling cannot be mitigated and that therefore it should not be allowed. There could be many more. Many of the sites that have already been cleaned had pervasive environmental impacts because the wastes had migrated into the tundra and killed it. The oil company pit sites contain a variety of toxic materials and hazardous chemicals. Typically, they include acids, lead, pesticides, solvents, diesel fuel, caustics, corrosives, and petroleum hydrocarbons. If the pit sites are not adequately closed, they can become illegal sites for more trash. This second wave of trash can include vehicles, appliances, batteries, tires, and pesticides. Oil industry trade groups point out that deep well injection was an accepted method of waste disposal for oil operations. It was the prevailing practice in Texas and Alaska for many years. Environmentalists respond by noting that the industry may have been acting within the bounds of its permits, but the environmental impacts are still too large. Politically, the oil operations expanded revenue for the state and built some infrastructure in a large state with a low population. Communities differ greatly on aspects of this controversy. State environmental agencies do not strictly or overzealously enforce environmental laws against large corporations. In fact, Alaska voluntarily relinquished its control of the Hazardous Waste Cleanup Program, and the U.S. Environmental Protection Agency took it over. This aspect of the ANWR controversy, the hazardous waste cleanup, is a battleground for state environmental agencies and federal environmental agencies. Most state environmental agencies get most of their revenue from the federal environmental agencies such as the U.S. Environmental Protection Agency. However, in federally mandated environmental programs, such as the Clean Air Act, the state must either do it to some minimal standards, or the EPA will do for them. In most instances states are free to choose the best method to meet the federally mandated result. However, in Alaska results were not meeting federal standards. This confrontation heightens the intensity of the ANWR controversy for industry, community, and environmental interests. It is seen by some as a test of federal sovereignty over states rights, which removes some of the environmental issues from the discussion.

If oil drilling is allowed in the Arctic National Wildlife Reserve then more impacts to the environment from hazardous and toxic waste can be expected. Environmentalists point out that most past mitigation efforts were not successful or mandatory. There is no legal requirement to mitigate the impacts of mitigation, which could themselves be considerable in large-scale projects.

CURRENT AND CONTROVERSIAL POLICIES

Generally, the George W. Bush administration is facilitating processes for the energy industry to drill for oil and gas in many sensitive public lands. Across the western United States, federal agencies such as Department of the Interior are leasing these areas for oil and gas development. And the tenants are oil and gas companies setting up operations on millions of acres of previously wild and open federally owned land. Proponents of this change in public policy note that there is an energy crunch and with rising gas prices they need access to all possible U.S. sources.

According to the Natural Resources Defense Council:

> the 2006 Bush administration is granting faster, almost pro forma, drilling approvals for requests to drill for oil on public lands. They have also loosened access to oil and gas deposits on public lands, reduced royalty payments and fewer environmental restrictions. Officials from the Bureau of Land Management, the Interior Department agency that manages the vast majority of federal lands and onshore energy resources, have directed field staff to expand access to public lands for energy development and speed up related environmental reviews. BLM data show that the number of leases for oil, gas and coal mining on public lands increased by 51 percent between 2000 and 2003—from 2.6 million acres to more than 5 million acres. The BLM has also repeatedly suspended seasonal closures designed to protect wildlife and is rushing to update numerous western land use plans to permit even more leasing and drilling. In the interior West, where most of the nation's oil and gas resources lie, more than 90 percent of BLM-managed land is already open for energy leasing and development.

There is much controversy about how much oil exists in ANWR. Critics say that if it were the only source it would yield less than a six months' supply of oil. Supporters of drilling based on national security say that all resources need to be marshaled. Overreliance on foreign oil sources leaves us dependent on other countries and vulnerable while at war. The United States is a large consumer of oil. The United States has 5 percent of the world's population but consumes almost a quarter of all the oil produced every year. The United States has only 3 percent of the world's proven oil reserves, making drilling in the Arctic National Wildlife Refuge a higher-stakes battlefield. Federal agencies have assessed the issue. The U.S. Geological Survey (USGS) estimates

> the amount of oil that might be recovered and profitably brought to market from the refuge's coastal plain is only 5.4 billion barrels, based on the U.S. Energy Information Administration's (EIA) average forecast price of $28 a barrel over the next 20 years.

At $40 per barrel the USGS estimates there would be only 6.7 billion barrels that could be profitably brought to market from the coastline reserves. The United States uses about 7.3 billion barrels of oil per year. Drilling proponents claim that at least 16 billion barrels of oil could be recovered from the refuge coastal plain. They point out that there could be recoverable oil and gas in other parts near the coastal plain. But the USGS says there is less than a 5 percent possibility that the coastal plain and adjacent areas contain that much recoverable oil. They maintain that only a small part of that oil could be economically produced and transported to markets. Drilling proponents are accused of ignoring the fact that the costs of exploration, production, and transportation in the Arctic are substantially higher than in many other regions of the world. Shipping, pipelines, and rail are all challenged by rough weather, earthquake-prone landscapes, and wilderness conditions. Extreme weather conditions and long distances to market would make much of that oil too expensive to produce at current market conditions. Drilling supporters claim that once the roads are built and the infrastructure is set up, costs will decrease, and oil demand is almost always increasing. They point out that the North American continental natural gas pipeline is expanding and that technology may make oil transport cheaper and safer for people and the environment. They also consider global warming to have one positive impact in that shipping lanes will be more reliably open because of the receding ice. The ice has drastically receded at the coastal plain in ANWR. To many rural Alaskan communities getting infrastructure and the promise of an oil-company job are benefits. With new roads and airstrips and ports, other forms of economic development would be able to occur. Tourism is a growing industry without the environmental impacts of oil drilling but requires a safe transportation network. The area's Inupiat Eskimo and Gwich'in Athabaskan-speaking Native inhabitants are actively involved in the controversy. Their respective views are significantly shaped by the nature of their relationship to the economy, the land, and its natural resources. Some of the oil reserves are on tribal lands. Some tribes are in favor, some are divided, and others are against oil exploration.

POTENTIAL FOR FUTURE CONTROVERSY

The Arctic National Wildlife drilling controversy swirls around questions of how much oil is there and whether any drilling at all is acceptable in a pristine wilderness area. It may be that there is more oil and much more gas there than currently known. It may also be the case that oil cannot be reached without irreversible environmental impacts. As other global petrochemical resources dry up, the pressure to drill for oil and gas in the Arctic Wildlife Refuge will increase.

Petrochemical controversies around protected parts of nature also affect other controversies in this battleground. Declining air quality from burning petrochemicals touches all aspects of this battleground, from local neighborhoods, tribes, and communities to global warming concerns. Political concerns about oil company profits right after Hurricane Katrina and all during the Mideast conflicts also inflame oil drilling issues in the Arctic. The earlier controversy concerning North Slope oil exploration and drilling provided evidence of severe environmental impacts. The potential for future controversy in Arctic National Wildlife Refuge drilling is very great.

See also Climate Change; Endangered Species; Indigenous People and the Environment; Permitting Industrial Emissions: Air; Transportation and the Environment

Web Resources

Arctic Wildlife Refuge. Available at arctic.fws.gov/issues1.html. Accessed March 2, 2008.

Federal Agency North Slope Science Initiative. Available at www.mms.gov/alaska/regs/mou_iag_loa/2005_BLM.pdf. Accessed March 2, 2008.

Sacred Lands: Arctic Wildlife Refuge. Available at www.sacredland.org/endangered_sites_pages/arctic.html. Accessed March 2, 2008.

Further Reading: Fischman, Robert L. 2003. *The National Wildlife Refuges.* Washington, DC: Island Press; Standlea, David M. 2006. *Oil, Globalization, and the War for the Arctic Refuge.* New York: SUNY Press; Truett, Joe C., and Stephen R. Johnson. 2000. *The Natural History of an Arctic Oil Field.* London: Elsevier; U.S. PIRG Education Fund. 2001. *The Dirty Four: The Case against Letting BP Amoco, ExxonMobil, Chevron, and Phillips Petroleum Drill in the Arctic Refuge.* Washington, DC: U.S. PIRG Education Fund.

AUTOMOBILE ENERGY EFFICIENCIES

Emissions from cars and trucks have been part of the air pollution controversy since the first federal clean air laws were passed in the late 1960s and early 1970s. Pollution-control devices and lead-free gas have decreased some emissions. The onus is on the automobile industry to produce more efficient cars that use less gas and to decrease their environmental impact. Emissions from cars and trucks continue to accumulate in land, air, and water. Increased retail sales of inefficient suburban utility vehicles (SUVs) and light trucks, combined with overall increases in number of vehicles, still produce emissions that can degrade the air quality. Environmentalists want cleaner cars. Consumers want inexpensive gas and more cars and trucks to drive. The petrochemical industry claims it is moving with deliberate speed to comply with environmental standards, garnering tax breaks and profits along the way. Communities want clean air and have valid public health concerns.

FUNDAMENTAL CONFLICT: INDUSTRY AND GOVERNMENT

The battle between the government and industry over legislating the production of more efficient vehicles is a long-standing one. Most of this legislation requires minimal compliance by industry at some date years in the future. Manufacturers claim that it takes resources from research and development right now to try to change production technologies to meet those standards. Sometimes they get tax breaks and other public policy–based encouragement to do so. One battleground is the free market. Market demand is for more cars, trucks, airports, and other petrochemical-based activities. Does legislation from democratically elected representatives constitute market demand? Most economists would say it does not. Environmentalists claim the minimal requirements are not fast or stringent enough. Currently, to meet federal fuel economy standards of the Clean Air Act, fuel efficiency must be increased to over 40 miles per gallon by 2015 and 55 miles per gallon by 2025. Some states, such as California, are adopting even more stringent standards. Rising gas prices, the slowly deepening effects of rising gas prices on food and other consumer goods, and concern about air pollution all increase public involvement. Other states with large pollution problems are exploring options and developing a state legislative and administrative department for controversies automobile energy efficiency over.

Adopting fuel-efficient or alternative fuel technologies to meet the Clean Air Act standards would save enormous amounts of gas and oil, in theory. A major controversy is whether it would prevent further environmental degradation. Global warming controversies are also pushing this issue into the public view. The United States needs to do more in terms of addressing mobile emissions sources and their environmental impacts. The exploration of alternative fuels for vehicles can be controversial in terms of environmental impacts. The removal of lead from U.S. gasoline was a major step forward, not yet replicated around the world. It greatly reduced airborne lead emissions. However, with current standards and volume of driving, assuming complete environmental compliance, U.S. vehicles would still emit 500,000 tons of smog-forming pollution every year. The United States is among the leading nations for both pollution and pollution-control technology. Diesel-powered vehicles are major polluters. They emit almost 50 percent of all nitrogen oxides and more than two-thirds of all particulate matter (soot) produced by U.S. transportation. Because the United States is more reliant on trucks, which tend to be diesel fueled, for the shipment of goods and raw materials than other nations, diesel emissions can be large contributors to an air stream with many other pollutants. Some of these regulated pollutants are from industry and some from the environment. The scale of diesel usage and its known emissions make it an environmental issue. Nitrogen oxides are powerful ingredients of acid rain. Acid rain can cause nitrogen saturation in crops and wilderness areas. Soot, regulated as particulate matter, irritates the eyes and nose and aggravates respiratory problems including asthma. Urban areas are often heavily exposed to diesel fumes. While diesel is a polluting fuel, regular unleaded gasoline can also pollute. Overall, the environmental impacts of the combustion engine remain largely undisputed. What is disputed is whether the environmental regulations go far enough to mitigate environmental impacts from these sources. The controversy about automobile energy efficiencies opens this aspect of the debate.

Commercial hybrid electric vehicle (HEV) models use both batteries and fuel. In the past few years they have been produced and marketed to the public. More recently, HEV drive trains have been used successfully in heavy-duty trucks, buses, and military vehicles.

There are many scientific controversies about specific chemicals and whether they cause a given adverse effect. This is an important controversy because these concerns form the basis for environmental regulations designed to protect the public health. Fundamental questions exist about whether chronic exposure to burning oil products causes specific symptoms. Even more questions emerge about alternative fuels. If petrochemical fuel is to be replaced by alternative fuels, many people want to make sure it is safer. The following quote reports on the first-ever conference on "Air Pollution: Impacts on Body Organs and Systems," which was held in Washington, D.C., on November 18, 1994, by the National Association of Physicians for the Environment (http://www.nutramed.com/environ ment/carschemicals.htm).

Adverse Health Effects of Chronic Exposure to Petroleum Combustion Products

HEALTH EFFECTS OF CAR EXHAUST: A short list of the likely pathogens in car exhaust:

- Carbon monoxide
- Nitrogen dioxide
- Sulphur dioxide
- Suspended particles including PM-10, particles less than 10 microns in size
- Benzene
- Formaldehyde
- Polycyclic hydrocarbons

Air pollution is the source of many materials that may enter the human bloodstream through the nose, mouth, skin, and the digestive tract.... For example, lead interferes with normal red blood cell formation by inhibiting important enzymes. In addition, lead damages red blood cell membranes and interferes with cell metabolism in a way that shortens the survival of each individual cell. Each of these harmful effects can result in clinical anemia.

Benzene and other less known hydrocarbons are produced in petroleum refining, and are widely used as solvents and as materials in the production of various industrial products and pesticides. Benzene also is found in gasoline and in cigarette smoke. It has been shown that exposure to benzene is related to the development of leukemia and lymphoma. Benzene has a suppressive effect on bone marrow and it impairs blood cell maturation and amplification. Benzene exposure may result in a diminished number of blood cells (cytopenia) or total bone marrow loss.

Common air pollutants also have an affect on blood and thus on organs of the body. For example, carbon monoxide, arising from incomplete combustion of carbonaceous materials, binds to the hemoglobin over two hundred times more avidly than oxygen and distorts the release to the tissues of any remaining oxygen. Thus, CO poisoning is akin to suffocation. In addition, it has been observed that carbon monoxide can exacerbate cardiovascular disease in humans.

The toxic chemicals in environmental air pollution stimulate the immune system to activate leukocytes and macrophages that can produce tissue damage, especially to the cells that line human blood vessels. Although the damage is initially slight and may not produce significant limitation to blood flow, repetitive exposure to toxic substances interferes with the ability of these lining cells to release a substance called endothelial-derived relaxing factor (EDRF). EDRF relaxes the smooth muscle in blood vessel walls, and blocking the release of EDRF leads to systemic hypertension. At the same time, leukocytes on the endothelium's surface appear to play a part in promoting the arteriosclerotic disease process. The combined effect of these events is to accelerate the changes that eventually lead to hypertension and ischemic heart disease.

The central nervous system (CNS) is the primary target for many serious air pollutants, such as lead, which is a major environmental hazard. Research over the past 10 years has provided evidence that levels of lead exposure associated with central nervous system effects, particularly as manifest in behavioral changes, is far lower than previously realized. Fifteen years ago, blood lead concentrations in children were not considered problematic until they exceeded levels greater than 30 to 40 micrograms per deciliter ($\mu g/dL$). Since that time, more sophisticated epidemiological studies have demonstrated changes in cognitive function at blood concentrations as low as 10 to 15 $\mu g/dL$. While children are more susceptible to lead's CNS effects, adults exhibit similar deficits in learning and memory as well. Advanced aging is also a period when enhanced vulnerability to the toxic effects of lead are predicted. In Germany, a large study documented an age-related decline in bone lead concentrations with advancing age. This effect was more pronounced in women than in men, reflecting post-menopausal processes in women which contribute to bone resorption and the release of lead back into the bloodstream. These results mean that lead exposure is actually increased during a period of already heightened susceptibility due to concurrent degeneration of other physiological functions, including both CNS and renal functions.

The effects of airborne pollutants on the immune system have been most widely studied in the respiratory tract. An airborne pollutant may enter the respiratory tract as a volatile gas (e.g., ozone, benzene), as liquid droplets (e.g., sulfuric acid, nitrogen dioxide), or as particulate matter (e.g., components of diesel exhaust, aromatic hydrocarbons). These pollutants interact with the immune system and may cause local and systemic responses ranging from overactive immune responses to immunosuppression. Most airborne pollutants are small molecular weight chemicals that must be coupled with other substances (e.g., proteins or conjugates) before they can be recognized by the immune system and cause an effect.

There is clearly an underlying genetic basis for susceptibility to immunologic disease resulting from exposure to pollutants, but knowledge in this area is rudimentary at this time. For example, there is little understanding of genetically-determined susceptibility or resistance to pollutant-induced immune disorders. There is a lack of appropriate in vitro models, and it is difficult to identify specific, biologically-active substances that may be linked to immune disorders. More research is needed to learn about the effects of airborne pollutants on mucosal immunity, the local immunity that is most effective against pathogenic microbes that invade the body through the respiratory and digestive systems.

Researchers also want to move HEV technology into a more sustainable lifestyle. They would like to produce and market plug-in hybrids that can plug in to household outlets. They want them to be able to store electricity and operate as clean, low-cost, low-environmental-impact vehicles for most of their normal daily mileage. Right now electric cars are limited by the batteries. Combining engines with them and using braking power to recharge the batteries does extend their range and power but also increases their emissions. Transportation is conceptualized as part of a environmental low impact and sustainable lifestyle. These communities unite plug-in hybrids, other low-impact transportation alternatives (bicycles, mass transit stops), zero-energy homes, a range of renewable energy technologies, and sustainable environmental practices. One example of such a community is the Pringle Creek Community in Salem, Oregon.

HYBRID ELECTRIC VEHICLES

Hybrid electric vehicles, also known as HEVs, represent a new kind of fuel efficiency. Present-day hybrids come with internal combustion engines and electric motors. The source of the fuel and the electricity may differ from model to model.

The biggest current challenge plug-in hybrids face is the cost and weight of batteries. At this point, even the most rechargeable batteries lose the ability to hold a charge. They then become hazardous waste and part of the environmental impact. There is also a financial and environmental cost to the use of electrical power. Much electrical power in the United States comes from coal-fired power plants. These plants produce large emissions of regulated air pollutants. Some would argue that the environmental cost of this use should be calculated when evaluating alternative energy sources. It is also possible to recharge plug-in hybrid vehicles from renewable energy sources. Scientists are extensively researching thermal management, modeling, and systems solutions for energy storage. Scientists and engineers also research ways to increase the efficiency of the electrical power.

Researchers are also seeking to make the plug-in electric car reversible. In many areas homes and businesses can sell back energy they do not use or that they create. This is one way to protect the electric grid from brownouts, as well as conserve energy from nonrenewable resources. In hybrid vehicles it is called a *vehicle-to-grid* or V2G. These cars would have a two-way plug that allows the home and vehicle owner and local utility to exchange power back and forth. This could make the batteries accessible backup power in the event of a natural disaster or other power outage. It could also encourage citizens to buy new hybrid cars. Utilities pay for peak, backup, and unused power. Transportation analysts can quantify the potential value of such systems in terms of gas saved, air quality, and other measures. There could be substantial automobile energy efficiencies in these approaches, but many remain untried at a large level. Reversible electrical energy may not be much less environmentally harmful if the source of the electric power and the waste generated have harmful environmental impacts.

FUEL CELL VEHICLES

As research and conceptualization has moved HEVs into production, fuel cell technology is taking shape. Hydrogen fuel cells have long been used to generate electricity in spacecraft and in stationary applications such as emergency power generators. Fuel cells produce electricity through a chemical reaction between hydrogen and oxygen and produce no harmful emissions. In fuel cell vehicles (FCVs), hydrogen may be stored as a pressurized gas in onboard fuel tanks. The electricity feeds a storage battery (as in today's hybrids) that energizes a vehicle's electric motor.

An FCV may be thought of as a type of hybrid because its electric battery is charged by a separate onboard system. This underscores the importance of advancing present-day HEV technologies. HEVs help reduce petroleum consumption immediately and provide lessons about batteries, energy storage, fuel advancements, and complex electronic controls that may apply directly to future transportation technologies.

WHAT IS BIODIESEL?

Biodiesel is a catchall term used to describe fuel made from vegetable oil or animal fats. These fats are generally converted to usable fuel by a process called *transesterification*. Biodiesel fuels are usually mixed with conventional diesel. It is estimated that 140 billion gallons of biodiesel could replace all oil used for transportation in the United States. This is an enormous amount of biodiesel, which is creating controversy and innovation in the sources of biodiesel. Large-volume biodiesel use could raise concerns about land-use impacts common to all plant-based fuels. Are there enough plants, such as corn, to meet the fuel needs? Land-use impacts could be much larger if the market demand is driven by fuel needs. Alternative energy sources almost always include renewable energy sources such as solar power. Because most biodiesel is made from plant-based oils or waste stream sources, it is a renewable fuel. Is there enough of it?

Waste vegetable and animal fat resources are estimated to be able to produce one billion gallons of biodiesel per year. That prediction is considered speculative by some because it assumes adequate plant and waste production. Collecting the wastes, distilling and cleaning the fat from it, and using it as fuel may all have environmental impacts. Farmers and proponents of biodiesel claim that distribution costs should go down as the first biodiesel stations begin operations, and the price of petrochemicals increases. Use of more than in billion gallons a year of biodiesel would require more virgin plant oils and crops for biodiesel production. It would also require discovery and organization of other waste stream sources to meet larger demands. More land would be needed to plant necessary crops, such as corn. Crops grown for biodiesel can be grown in a manner that has negative environmental consequences for entire ecosystems. Just as other crops, they can require pesticides and be genetically manipulated.

ALTERNATIVE FUELS, ETHANOL, AND CORN: POTENTIAL ENVIRONMENTAL IMPACTS

With the rapid increase in demand for ethanol, more corn is being planted in the United States. Ethanol is made from corn. The U.S. Agriculture Department estimates that 90.5 million acres of corn were planted in 2007, out of about 434 million acres of cropland, more than anytime in the last 50 years. Farmers plant the crops with the most profit. The problem with growing large amounts of corn is the environmental impact. One large concern is the amount of water required for its production. A gallon of ethanol requires about three gallons of water to produce. In locations without reliable water sources it may not be a cost-efficient alternative fuel. Corn requires about 156 pounds of nitrogen, 80 pounds of phosphorus, and differing amounts of pesticides per acre to grow from seed to harvest. Corn requires large amounts of nitrogen because it cannot absorb it from the air. What the corn does not absorb runs off the land into the water table. This causes algae blooms that warm up the water and use up the oxygen, sometimes resulting in large fish kills. Nitrates in the water, largely from agricultural runoff, are also blamed for deaths of livestock, and some suspect them in some human fatalities. Algae growth can cause other bacteria to grow that are harmful to humans.

Farmers do not like to waste fertilizer. Many corn farmers use sophisticated satellite tracking measurements to make sure fertilizer levels are not exceeded. Many keep records of inventory as a condition of bank loans and can keep track of fertilizer expenses. More and more farmers are planting buffer strips between their fields and waterways. These buffer strips are areas of vegetation, usually indigenous, that filter water or runoff from fertilized fields. It is difficult for farmers to leave a buffer strip. Usually the soil near the water is better. Buffer strips can attract animals that dig holes in the fields and destroy crops. Not all farmers are owner-operators. Many commercial agribusinesses lease land to farm. While their leases often make allowances for buffers and land held in soil and water conservation programs, they are seldom enforced.

BIODIESEL AND GLOBAL CLIMATE CHANGE

One of the environmental advantages touted with biodiesel is that it has fewer environmentally degrading emissions. Critics have pointed out that biodiesel vehicles require more fuel depending on the mix and may have overall more combustion. Some biodiesel requires chemicals to start up when it is cold. Sometimes the vehicle must warm up to get the grease warm enough to flow. Pure 100 percent biodiesel results in large reductions in sulfur dioxide. However, they can have 10 percent increases in nitrogen oxide emissions. A popular mix of biodiesel is about 80 percent biodiesel and 20 percent regular diesel, which increases the pollutants emitted proportionality. These pollutants are responsible for acid rain and urban smog. Biodiesel companies are beginning to operate service stations to distribute the fuel. There is controversy about its environmental impacts. Some tailpipe emissions are reduced. However, when the entire life of the vehicle is considered, running on 100 percent biodiesel, some smog-forming emissions can be 35 percent higher than conventional diesel.

ARE THERE OTHER ENVIRONMENTAL ATTRIBUTES OF BIODIESEL?

One big concern and source of controversy are oil spills and their environmental impacts. Regular petrochemical spills can travel quickly in water and permeate land, depending on soil structure. Biodiesel is considered less harmful to the environment because it biodegrades four times faster than conventional diesel, so its environmental impacts are not as long term or ecologically pervasive. It quickly degrades into organic components. Production of petroleum diesel creates much more hazardous waste than production of biodiesel. Biodiesel produces more overall waste depending on the source, but generally twice as much as nonhazardous waste. Some of the nonhazardous wastes may be recyclable.

One developing source for biodiesel is algae, grown specifically for the purpose. The specialized algae are grown in a variety of ways. They are vastly easier to grow than most other crops and grow very quickly. While growing they absorb large amounts of carbon dioxide, a greenhouse gas. Technological entrepreneurship is very much engaged in this, and algae strains are sometimes protected trade secrets. The species used now in the United States is *Botryococcus braunii* because it stores fat that is later used as fuel. The algae must then be broken down to separate fats from sugars. Solvents are used for this, which could be a source of environmental impacts depending on by-products and manufacturing waste streams. Fats cannot be cold pressed out of algae because they are too fragile and disintegrate. The fats are made into biodiesel. One issue is whether they could produce enough to meet demand for biodiesel from vehicles. The New Zealand algae fuel company, Aquaflow, says it has achieved this.

OIL AND POWER: IMPACTS ON AUTOMOBILE ENERGY EFFICIENCY

The primary resistance to increasing the efficiency of automobile and truck engines is the petrochemical industrial complex. Large oil companies are the backbone of U.S. industry and part of a thriving economy. They are multinational corporations that exert political power here and abroad. Some have revenues larger than most nations. Their only legal motivation is to make profit from dispensing a limited natural resource. Environmental and ecological integrity and consumer quality-of-life issues are not their concern. The oil industry has been a strong industrial stakeholder and exerted power at the local, state, and federal levels of government for almost a century. Many state legislatures have passed laws exempting oil companies from releasing their environmental audits or helping oil companies avoid compliance with environmental regulation or enforcement action. Oil companies are not responsive to community concerns and can litigate any issue with vast financial resources. The petrochemical industrial complex has also become part of social institutions such as foundations, churches, schools and universities, and athletic contests. Some employment opportunities, some infrastructure, and the hope of more economic development are offered to communities by oil companies.

Oil politics and the U.S. presidency are closely intertwined as both Bush presidents were major players in the oil industry in Texas and internationally with Saudi Arabia. Since George W. Bush was elected president in 2001, the top five oil companies in the United States have recorded record profits of $342.4 billion through the first quarter of 2006, while at the same time getting substantial tax breaks from a Republican Congress. This is extremely controversial as gas prices have risen dramatically for most average citizens, and the national debt has gone from a surplus to a large deficit. With the controversial war in Iraq, an oil-producing nation, some people in the United States thought gas prices would decrease domestically. Oil company profit taking during times of natural disaster, such as hurricane Katrina, and war, has attracted much congressional attention. In one congressional hearing in 2006 major oil executives were subpoenaed to testify before Congress and refused to swear to tell the truth. Here are some of the profits in the first quarter of 2006. They have remained consistently high since then.

ExxonMobil: $118.2 billion
Shell: $82.3 billion
BP: $67.8 billion
ChevronTexaco: $43.1 billion
ConocoPhillips: $31.1 billion

No oil company seems to be turning their profits into consumer savings. Some are just starting to research more alternative energy sources, but this is controversial. Some environmental groups have recently challenged this assertion. To many U.S. consumers it seems there is a direct correlation between record prices paid by consumers and record profits enjoyed by oil companies. From 1999 to 2004, the profit margin by U.S. oil refiners has increased 79 percent.

POLITICAL CONTROVERSY: TAX BREAKS FOR OIL CAMPAIGN CONTRIBUTIONS?

The search for automobile energy efficiency lies in the political maelstrom of the oil industry in Congress. President George W. Bush and the Republican Congress gave $6 billion in tax breaks and subsidies to oil companies in 2006 alone. This is in the face of large oil company profits. From 2001 to 2006 the oil industry gave $58 million in campaign contributions to federal politicians. Eighty-one percent of that went to Republicans. Given the awkward state of campaign financing, it is likely that these numbers would be much higher if travel and other expenses are included. Environmental groups, concerned communities, and taxpaying consumers have all protested this as corruption, misfeasance, and malfeasance in office.

POTENTIAL FOR FUTURE CONTROVERSY

Continued dependence on fossil fuels guarantees increased controversy. As oil becomes depleted, multinational oil corporations exert all their huge influence on the United States to protect their sources, even if it means going to war. The dissatisfaction of environmentalists and communities with the petrochemical industrial complex, the dependence and demand of the United States for oil, the lack of governmental support for alternative energy development, and the inability to keep large environmental impacts secret all fuel this raging controversy.

See also Air Pollution; Climate Change; Cumulative Emissions, Impacts, and Risks; Global Warming; Good Neighbor Agreements; Permitting Industrial Emissions: Air

Web Resources

National Renewable Energy Laboratory. Advanced Vehicles and Fuels Research. Available at www.nrel.gov/vehiclesandfuels/. Accessed March 2, 2008.

Union of Concerned Scientists. Cleaner Cars. Available at www.ucsusa.org/clean_vehicles/big_rig_cleanup/biodiesel.html#1#1. Accessed March 2, 2008.

U.S. Department of Energy. Retail Stations Offering Biodiesel. Available at www.eere.energy.gov/afdc/infrastructure/refueling.html. Accessed March 2, 2008.

Further Reading: Clifford, Mary. 1998. *Environmental Crime: Enforcement, Policy and Social Responsibility.* MA: Jones and Bartlett Publishers; Dobson, Andrew P. 2004. *Citizenship and the Environment.* New York: Oxford University Press; Galambos, Louis, Takashi Hikino, and Vera Zamagni. 2006. *The Global Chemical Industry in the Age of the Petrochemical Revolution.* Cambridge: Cambridge University Press; Marzotto, Toni, Vicky Moshier, and Gordon Scott Bonham. 2000. *The Evolution of Public Policy: Cars and the Environment.* Boulder, CO: Lynne Rienner Publishers; Wells, Peter E., and Paul

Nieuwenhuis. 2003. *The Automotive Industry and the Environment: A Technical, Business and Social Future.* Washington, DC: CRC Press.

AVALANCHES

Fatalities due to avalanche have been increasing since the 1950s. Avalanches claim more than 150 lives each year worldwide, and hundreds more are injured or trapped as the result of an avalanche. As more roads, buildings, and towns are forced into avalanche-prone areas, controversies arise around overdevelopment in sensitive mountain terrain as well as responsibility for monitoring and rescuing avalanche victims.

Avalanches of snow occur in mountainous regions all over the world. They are very powerful, leveling everything in their path. They can move at speeds greater than 100 miles per hour, creating a wind gust ahead of them that travels at about the same speed. They are difficult to predict, but knowledge of snow formation in a particular region and the resulting avalanche risk can help recreational users make better judgments.

WHAT CONDITIONS ARE NECESSARY FOR AVALANCHES?

Ninety percent of all avalanches occur on moderate slopes with an angle of 30° to 45° except in the unusual situation of snow accumulating on steeper slopes. Avalanches happen most frequently when the gravity pushing snow at the top of the slope is greater than the strength of the snow to hold it up. Over the snow season, snow falls in different layers. Some of these layers bind well with each other, whereas other layers do not. When the weight of the snow on top is greater than the ability of the snow to hold itself together, the layers will sheer off. A change in temperature, a loud noise, a weather front, or engine vibrations are all possible triggers to start a snowfall that begins at a *starting zone*. These are generally the higher, steeper slopes but can occur anywhere on the side of a snowy mountain. The avalanche continues downslope along the *track*. To the experienced eye these avalanche tracks can be discerned by the lack of trees, or shorn trees all in the down direction. Avalanches can start other avalanches that occur minutes or days later. The avalanche fans out from the track and slows down quickly. The snow then settles in the *runout zone*. When they stop, they harden into a solid snowpack in a matter of minutes. Generally, search and rescue operations consider anybody under seven feet or more of snowpack for more than 1/2 hour as most likely dead.

The United States ranks fifth worldwide in avalanche risk. Colorado, Alaska, and Utah have the most avalanche fatalities. Of those, Colorado has the most fatalities and the most 14,000-foot or higher mountains in the United States. There are probably avalanches in many parts of unmonitored and unsettled mountain ranges of the world.

BATTLEGROUNDS ABOUT CONTROLLING AVALANCHES

Controlling avalanches can become a battleground because of the drastic impacts they can have on sensitive mountain ecotones. Local communities may consider avalanche control necessary to protect their lives and livelihood. Avalanche prevention and mitigation involves a variety of methods, all of them requiring knowledge of local climatic conditions and planning. A primary method is to prevent snow buildup on the higher slopes. Snow fences are built to prevent the buildup of snow in starting zones, and sometimes snow precipices and other avalanche-prone snow formations are exploded, setting off a controlled avalanche. Avalanches can be diverted with careful planning. Deflecting walls are built to divert avalanche flows away from buildings and even entire towns. The reforestation of slopes with trees helps to prevent avalanches. Some environmentalists claim that allowing logging on steep slopes in the mountains can increase avalanche risk, and that timber corporations should assist with the mitigation of avalanche risk. Another aspect of avalanche injury prevention is to adequately warn recreational users and all road traffic of any avalanche danger.

Snowmobilers are those most often killed by avalanches in the United States. Snowmobiles can travel much farther than a person could walk in one day. A controversial practice of snowmobilers is to race their snowmobiles as far up a wide-open track as their machines will take them. Unfortunately, some of the wide-open tracks are avalanche tracks. The loud noise and on slope vibration caused by the machines, combined with reckless behavior, can greatly increase risk of avalanche. Most avalanches in the United States occur during January, February, and March. On average, 17 people are killed per year nationwide.

WHAT IS THE PROFILE OF A TYPICAL U.S. AVALANCHE VICTIM?

According to the Colorado Avalanche Information Center (http://avalanche. state.co.us/), 89 percent of victims are men and most are between the ages of 20 and 29. Three-quarters of victims are experienced backcountry recreationists. Mountain climbers, backcountry skiers, extreme skiers and snowboarders, and snowmobilers are the most likely to be involved in avalanches.

An underlying controversy with avalanches is the use of motorized vehicles such as snowmobiles, snocats, and helicopters in backcountry areas. Environmentalists have long been opposed to opening up wilderness areas to engines because of the disruption of the environment. Modern snowmobiles are very powerful machines with noise and vibration. One activity snowmobilers do in backcountry mountainous regions is to see how high up a steep slope they can go. After an afternoon competing on a steep slope in this fashion, it is more likely that an avalanche could occur. If it does occur and a search and rescue operation is mounted, who pays? If the search and rescue operation is faulty and one of victims dies, who is responsible? The battleground for this controversy moves then to monitoring and rescue responsibilities and costs.

WHEN AND WHERE AVALANCHES HAPPEN

Avalanches happen more frequently and predictably at different times in different locations. Wintertime is when most avalanches will slide down a slope. The highest number of fatalities occurs in January, February, and March. This is when the snowfall amounts are highest in most mountain areas in the United States. A significant number of deaths occur in May and June because of spring snows, runoff, and the melting season. Risk from avalanches is dependent on the type of activity. Mountain climbers can experience avalanches all through the summer. Each mountain is different with regard to avalanche formation. Unexpected seasonal variations in weather can create avalanche zones in predictable areas. Many mountain ranges can create their own weather systems with little or no warning. If climate change does occur this type of environmental information can mitigate the impact of natural disasters such as avalanches.

POTENTIAL FOR FUTURE CONTROVERSY

As development moves more into the mountains, the natural paths of avalanches can be crossed and avalanche zones can be created. There are ways to mitigate some avalanches when a town is directly in its path. The cost of and responsibility for these mitigation measures are part of this controversy. The underlying controversy about allowing engines and mechanized vehicles in wilderness and national park areas is still simmering. It gets added fuel from outdoor recreationists who create avalanches. The controversy surrounding the unanswered question of who bears the cost of monitoring for avalanche safety and for search and rescue operations is likely to continue.

See also Mountain Rescues; Ski Resort Development and Expansion

Web Resources

Colorado Avalanche Information Center. Annual avalanche statistics. Available at avalanche. state.co.us/. Accessed January 20, 2008.

Further Reading: National Research Council (U.S.) Panel on Snow Avalanches and National Research Council (U.S.) Committee on Ground Failure Hazards Mitigation Research. 1990. *Snow Avalanche Hazards and Mitigation in the United States.* Washington, DC: National Academies Press; Tremper, Bruce. 2001. *Staying Alive in Avalanche Terrain.* Seattle: The Mountaineers Books.

BIG-BOX RETAIL DEVELOPMENT

Big-box retail development refers to large-scale retail stores surrounded by acres of parking lots. It is an environmental controversy because of the noise and pollution they generate. Large impervious surfaces such as parking lots also contribute to runoff problems.

BACKGROUND AND CONTEXT

Big-box retail development is a very controversial issue for most communities. For many citizens it is one of their few interactions with the actual land-use planning processes that surround large commercial transactions. These developments involve large tracts of land, generate large amounts of traffic, and are often touted as economic development by local government. Lack of meaningful public notice and public participation before decisions are made is often a controversy in big-box retail development. Some have accused big-box retailers of targeting less-savvy communities with informal land-use processes. Big-box retailers try to go with market demand for their goods and services. Some communities welcome any economic development, whereas others do not. Some communities have little experience dealing with sophisticated international retail corporations and do not know what types of environmental mitigation to request in negotiations. Many communities have little interest in local land-use issues until after all the notices and processes are done. Most notices are very poor at informing citizens, especially tenants, of these events. They are generally listed in the classified section of the local newspaper (if there is one) under legal notices. If there is a legal lack of notice and meaningful participation then the legal remedy is to do the whole process again with the proper procedures. This seldom changes the result but does increase negotiation, if there is any room left for that by this time. Big-box retailers can wait for local land-use processes and can mitigate environmental impacts in new construction and in current operations. This commercial sector does not have the environmental air and water emissions of heavy industry, although they can have other environmental impacts. For example, Wal-Mart, the largest big-box retailer in the world, recently began a sustainability campaign. In the battleground for this controversy, the local land-use planning process and environmental concerns merge with health care cost concerns. Labor unions play a strong role in some of the battleground states, such as California. California also has an initiative process powered by voters, which is used in this battle. For that reason it is a good example of the parameters of this complicated and intense controversy, with a representative sample of cases discussed further on in this entry.

Big-box retailing consists of oversized stores, especially in suburban and exurban areas. Big-box retailers vary in market niches. Some big-box retail stores specialize in one kind of retail good. For example, Best Buy and Circuit City sell electronics while Home Depot and Lowe's sell home improvement products. These big-box retailers are sometimes also referred to as category killers. This means that they will eliminate any local competition that sells goods or services in their category. They can spread losses among stores until the local competition can no longer compete. At the same time, some big-box retail stores sell a variety of products with no particular niche, for example, discount department stores such as Wal-Mart and Target and warehouse clubs such as Sam's Club.

SPRAWL: THE CONTRIBUTION OF BIG-BOX DEVELOPMENT

Big-box retail development is often connected with controversies around sprawl, that is, unrestrained and environmentally consumptive growth of human habitat. Big-box retail exacerbates sprawl. One of the main environmental impacts of sprawl is the consumption of large tracts of land for development. Because big-box retailers need a large space to build their stores, they rarely choose urban infill locations for new development and choose to locate in suburbs and exurbs. This pulls economic activity away from dense urban cores, spreading out the population and vehicle traffic over more and more land. Also, urban residents without cars cannot easily access these stores. Mass transit systems in the United States are poor at best, but big-box stores support automobile dependence. They usually have very large parking lots and a large ecological footprint. However, they may try to mitigate traffic and noise concerns of nearby or contiguous residents. When big-box retail stores locate in farmland, wetlands, or green space, they eliminate natural resources and open space. According to the American Farmland Trust, the United States loses 3,000 acres of productive farmland to sprawl every day. Some of this loss is fueled and caused by big-box retail trade. The loss of acres of farmland per year equals about the size of Delaware.

Urban residents with cars may choose to shop at big-box stores because of their low prices and because of the convenience of a large variety of products housed in one building. Research has shown that a shopper will park further away from a suburban mall retail shop and thus walk further than if they had parked closer to urban retail shops. In this way, big-box retail takes business from local so-called mom-and-pop shops, not just downtown shops. In small towns with little downtown area, small family-owned businesses face stiff competition. Big-box stores have the ability to undercut local, homegrown retailers with lower prices because of their sheer size and economies of scale. If a particular big-box retail store chooses to, it can lower the price on the goods sold by the smaller stores, one or more at a time. Because they have larger inventories, they can raise the price on other goods to make up for their loss leader item. Industry representatives say that this is the free market model; if small stores want to compete they can lower prices, increase personal service, or offer specialty items. In many towns in the United States, big-box retailers moving into a community have eliminated the competition. The concern then is loss of jobs, lack of consumer price control, and sometimes an unanticipated change in the character of the community.

FAMILY-WAGE JOBS

One of the biggest limitations of retailing as a form of economic development is the fact that retail jobs can seldom sustain a family. Family-wage jobs are sought because they represent stable communities and property ownership. Retail jobs most often pay lower than a living wage. In fact, many retail jobs have pay scales that hover near minimum wage. Retail jobs also are most commonly part-time jobs with fewer hours and no medical benefits. Also, retail jobs typically lack career tracks. The chances of significant advancement from a retail job are slim. There have also been concerns about race and gender employment discrimination at the big-box retail outlets. Unionized grocery stores are the only exception to the poverty-wage problem of retail economic development.

GHOSTBOXES AND GRAYFIELDS

Dead malls, grayfields, and ghostboxes are nicknames for vacated retail space, which generally consists of abandoned structures and parking lots. While it is not immediately environmentally harmful, it is not particularly environmentally beneficial. It is an indication of the level of land-use and business planning in a community, as well as changing circumstances and site obsolescence. The United States has excess retail space. The National Trust for Historic Preservation estimates that there is 38 square feet of store space for every man, woman, and child. The program director of the National Trust for Historic Preservation's Main Street Center has testified that cities with too much retail space suffer all kinds of hidden costs—in addition to whatever subsidies they grant. When just one Main Street store, with two floors of 2,000 square feet, goes from being occupied and busy to being vacant, the total cost to the local economy is almost $250,000 a year. That includes losses in property taxes, wages, bank deposits and loans, rent, sales, and profits. A 2001 study by the Congress for the New Urbanism and PriceWaterhouseCoopers about grayfields found that 7 percent of regional malls were already grayfields and another 12 percent are potentially moving toward grayfields status in the next five years. That would be 389 dead malls.

Because vacant or underutilized properties usually get reassessed and pay much lower property taxes, dead malls mean big tax revenue drops. When tax revenue decreases, the quality and quantity of municipal services also decrease. As new big-box retail stores are developing, old retail stores and malls are vacating. When this happens they are called grayfields.

BIG-BOX DEVELOPMENT AND COSTS TO LOCAL ECONOMY

When Wal-Mart and other low-wage big-box retailers fail to provide their workers with a decent wage and full-time hours, many employees and their families qualify for safety-net help such as Medicaid, State Children's Health Insurance Program, Earned Income Tax Credits, Section 8 housing assistance, low-income energy assistance, and free or discounted school lunches. These programs cost taxpayers money. A 2004 report by congressional staffers tallied all of these hidden costs; they estimate that each Wal-Mart store with 200 employees costs federal taxpayers $420,750 a year in safety-net costs. Multiply that by the 3,500 stores Wal-Mart already has in the United States and by the 300 more stores it plans to open every year and the safety-net costs to the community become staggering. To date, 19 states have disclosed the names of employers who hire the greatest number of workers that depend on taxpayer-funded health care programs.

Many big-box retailers receive massive economic development subsidies to locate in new areas. Sometimes property and utility taxes are reduced to tempt a big-box retailer into a community. An environmental controversy can ensue if the site selected needs to be cleaned of waste and the city or state agree to wave environmental cleanup responsibilities. A prospective land buyer is supposed exercise due diligence to find out about environmental liabilities, and most buyers are required to disclose most of them.

Sometimes the site is a wetland or a soil and water conservation district. These are not environmental problems unique to big-box retail development.

Subsidizing retail trade is not a very effective form of economic development. Big-box retail economic development ranks among the least effective. Economic development success is generally measured by the increased wealth of a community, indicated by income and increased property values. Stable, high-paying or at least family-wage jobs are the most sought-after types of economic development. Manufacturing-based industrial development with local suppliers, warehouses, and distributors is considered very effective economic development by these crude measures. Environmental concerns now enter into many more local land-use decisions, and communities that can afford to seek only clean industrial bases for their economic development. Big-box retail economic development does not create many jobs for the local economy from the supplier side as most

PROPERTY ASSESSMENT: APPRAISALS VS. ENVIRONMENTAL ASSESSMENTS

In the private real estate market where land is bought and sold, appraisers assess value. Traditionally, to determine value residential and commercial land appraisers looked at the sale of comparable property, if there was any.

As environmental law enlarges the liability for cleanup of contaminated land sites, sellers tend to keep this information quiet. Buyers want to know more about any environmental issues. Banks and mortgage lenders especially want to know whether they could be liable for environmental contamination. Buyers and sellers are supposed to use due diligence to disclose and find out about environmental contamination. This can mean a number of activities such as interviewing contiguous neighbors, sampling soil and water, and searching public records.

The traditional real estate appraisal does not provide enough security for banks and mortgage lending institutions to know about potential environmental liability. They therefore require an environmental evaluation before approving loans to buy questionable land. Unlike an appraisal, the actual environmental condition of the property is analyzed. Both appraisals examine comparable market data, but environmental assessments are required to check all agency information and to verify it with other market participants. Extensive environmental agency research at the federal, state, and local levels is also required. If the land is found to be contaminated, an environmental assessment must include the costs of remediation in the assessment. Environmental risks are analyzed, but generally no human health risk assessment is required. In the end, whereas a traditional residential or commercial real estate appraisal may not reflect actual value, an environmental assessment does by including actual environmental conditions.

Environmental assessments are expensive and time consuming. They may also disclose environmental liability of the seller that remains if the land remains unsold. Industry resists the costs in money, time, and potential liability. Communities applaud the disclosure and cleanup of environmental contamination. Government agencies work with national lending institutions to develop enforceable public policies that clean up contaminated land for the public health safety and welfare and to develop profits for private financial institutions.

---

goods come from overseas or elsewhere. The jobs created are few, do not provide a family wage, and lack health care. That means most retail workers have very small disposable incomes and therefore little buying power to stimulate the local economy. Some contend that there is only one justifiable time for government to subsidize retail economic development. That is to help neighborhoods that lack access to basic retail goods such as food, drugs, and clothing.

BATTLES OF THIS CONTROVERSY

Big-box retail development is engaged in land-use battles all across the contiguous United States and Hawaii. The state of California has a growing, dynamic population and expanding market for retail trade big-box development. Cities in California are considering passing laws to stop or restrict additional big-box retail development. Several other cities in California have approved plans for the construction of Wal-Mart supercenters after some negotiation. Following are examples of the battleground of this fierce controversy from California. Examples include current laws that restrict big-box or supercenter development as well as information on recent approvals in several California cities. This research was published by the Public Law Research Center at Hastings Law School in the spring of 2006, edited by Jodene Isaacs, with the title *California Responses to Supercenter Development: A Survey of Ordinances, Cases, and Elections.*

Alameda County

Statute/Proposal: On January 6, 2004, the county board unanimously passed an ordinance that bans retailers of more than 100,000 square feet that devote more than 10 percent of their floor space to groceries and other nontaxable goods in unincorporated county areas.

Players/Purpose: County supervisors say that the ban protects small business and will help minimize traffic concerns in unincorporated areas.

Legal/Political Issues: January lawsuit by Wal-Mart. Wal-Mart argues that the ban unfairly targets Wal-Mart supercenters. The petition claims county supervisors violated the California Environmental Quality Act when they said review was not needed before approving the ordinance, and also asserts that the public was not properly notified the board was considering the action. Wal-Mart also claims the county overstepped its authority "by enacting a law that imposes unusual and unnecessary restrictions on lawful business enterprises," and didn't follow state-mandated procedures in enacting the ordinance. Wal-Mart claims the ordinance should have been reviewed by the county Planning Commission prior to adoption. The petition asks the court to block the ordinance from taking effect in February and to deem it invalid.

Comments: The Alameda County suit is the first Wal-Mart has filed in California. Wal-Mart spokeswoman Amy Hill commented that a lawsuit was more appropriate than a referendum. "They were so determined to get this passed immediately," Hill said, "we felt a lawsuit was a more appropriate course of action."

City of Beaumont

Statute/Proposal: City Wal-Mart approval. Wal-Mart plans to build a 149,500 square-foot building that could be expanded by 71,500 square feet.

Legal/Political Issues: Two hundred residents turned out for a January public hearing about the Wal-Mart and raised questions about traffic, air pollution, and urbanizing the San Gorgonio Pass.

Comments: After listening to residents, planning commissioners approved the store for a site south of Interstate 10. They sent the project to the city council, which has the final say.

City of Calexico

Statute/Proposal: June 2001 ban on any store in excess of 150,000 square feet that dedicates 7.5 percent of its floor space to nontaxable items. Ban overturned.

Players/Purpose: Wal-Mart and labor unions.

Legal/Political Issues: Wal-Mart spokesman Peter Kanelos noted that in March 2002 voters in the border town easily overturned a measure the city had passed to block Wal-Mart and similar retailers from doing business.

Comments: Measure B lost 1,381 to 2,651 with Wal-Mart in campaign spending roughly $56 per vote.

Contra Costa County

Statute/Proposal: June ban in unincorporated areas on stores larger than 90,000 square feet that devote more than 5 percent of floor space to selling nontaxable groceries. Overturned by Ballot Measure L—53.8 percent to 46.2 percent, March 2, 2004.

Players/Purpose: Board of Supervisors stated that traffic and controlling urban sprawl in unincorporated county areas was the major purpose for the ordinance. John Gioia, a Contra Costa county supervisor, argues that there are parts of Contra Costa County outside of city limits that already have traffic congestion and are at risk of losing open space.

U.S. Representative George Miller contends that Wal-Mart workers could also be a drain on county health resources because the company does not offer adequate health insurance, an allegation Wal-Mart disputes.

Legal/Political Issues: Ballot Measure L: Supporters of Measure L, who included elected county officials, community members, environmentalists, the United Food and Commercial Workers Union, and Safeway say that without sales tax revenue, the county cannot make the road improvements needed to handle increased traffic and other related impacts that the huge stores would have. Opponents of Measure L pointed to the negative impact on consumer choice. Opponents argued that Measure L would unfairly restrict Wal-Mart from selling goods at lower prices to working families.

Comments: Wal-Mart spokesperson Amy Hill has acknowledged that Wal-Mart has contributed about $500,000 to the "No on Measure L" campaign before the end of January and could easily spend more than $1 million. Hill said Wal-Mart does not have any plans to open a supercenter in Contra Costa County.

City of Gilroy

Statute/Proposal: March 2004 city approval of Wal-Mart supercenter.

Players/Purpose: The United Food and Commercial Workers union and Councilman Paul Correa worked to overturn the council decision. Correa notes that "if it's going to happen for sure, maybe we can sit down with Wal-Mart and talk to them about having a positive impact on this community other than delivering low-cost goods." Correa said he would like to discuss with Wal-Mart issues such as hiring workers from Gilroy first, using local companies to do construction work at the new site, paying livable wages to its employees, and donating more to local charities.

Legal/Political Issues: Wal-Mart campaign funding of supportive council members. Wal-Mart sent out last-minute mailers urging residents to vote against union-friendly candidates.

Comments: A nearly 220,000-square-foot Wal-Mart may become one of California's first.

Supercenter Sites

City of La Quinta

Comments: Home of California's first Wal-Mart supercenter, opened March 2, 2004.

City of Lodi

Statute/Proposal: City has considered a supercenter ban.

Players/Purpose: Wal-Mart officials would like to replace the store on Kettleman Lane and Lower Sacramento Road in Lodi with a 219,000-square-foot supercenter. The supercenter project has not yet reached the planning commission.

Legal/Political Issues: When the Lodi Planning Commission considered a size limit as part of its design standards in February 2004, Wal-Mart submitted a petition against the action containing more than 1,000 signatures collected at its current Kettleman Lane location. The city also received a letter from a law firm representing Wal-Mart, claiming a size limit without proper research would be a violation of state law.

Comments: More than 100 Lodi residents turned out for the commission's January 28, 2004, meeting, asking that a size limit—100,000 square feet was the most common number—be adopted by the city. At its February 11, 2004, meeting, the commission considered such an option but decided not to include it among the standards.

City of Los Angeles

Statute/Proposal: Proposal to ban any store whose stock includes grocery items from exceeding 100,000 square feet.

Players/Purpose: Los Angeles councilman Eric Garcetti has spearheaded the campaign against Wal-Mart. Garcetti argues that supercenters would drive down local wages, as rival businesses struggle to survive; wipe out more jobs than they create; and leave more residents without health insurance—and with no choice but to use public hospitals and clinics that are already overrun by demand.

Comments: Garcetti states that: "We don't believe their business model is good for the kind of economic development that we want in the places where we need it most. And we want people to realize that the 10 cents they may save on a jar of pickles could mean paying another $5 in taxes for all the extra visits to local emergency rooms."

City of Manteca

Statute/Proposal: Wal-Mart has proposed a supercenter project, and city officials contend that they have no plans to adopt a ban.

City of Martinez

Statute/Proposal: A ban on supercenter-format big-box retail.
Legal/Political Issues: Wal-Mart has not challenged the ban.

City of Moreno Valley

Statute/Proposal: City approval.
Comments: The Moreno Valley City Council voted on November 25, 2003, to approve a shopping center, despite objections from a number of residents and environmental activists. LJC Enterprises plans to build the shopping center, including a 227,194-square-foot Wal-Mart. The project also will include as many as 300 multifamily housing units to be built later.

City of Oakdale

Statute/Proposal: City staff will tailor a proposed ordinance after that adopted in Turlock, says community development director Steve Hallam (limits stores selling nontaxable goods on 5 percent or more of their floor space to 100,000 square feet or less).
Players/Purpose: Oakdale city administrator Bruce Bannerman said the city's streets "are not designed to accommodate this kind of traffic.... And there is only so much retail trade in any community. And if a retailer like this comes into a small community, and Oakdale is a small community, it takes away from others."
Legal/Political Issues: Amy Hill, a spokeswoman for Wal-Mart, said Oakdale is not a planned supercenter location.
Comments: The city council voted 5–0 to refer a proposed ban on Wal-Mart supercenters to the planning commission, which will be heard in early April 2004. Any planning commission decision will serve as an advisory vote for the city council, which would ultimately decide on the ordinance.

City of Oakland

Statute/Proposal: City Code § 17.10.345: October 2003 ban on stores more than 100,000 feet with more than 10 percent of sales floor area devoted to non-taxable merchandise, but excluding wholesale clubs or other establishments selling primarily bulk merchandise and charging membership.
Legal/Political Issues: Wal-Mart has not challenged the ban.

City of Redding

Statute/Proposal: City supercenter approval.

Comments: Planning commission has approved a Wal-Mart supercenter. Proposed development would add 93,000 square feet to an existing discount store, expanding it to 220,000 square feet, including 60,000 square feet of grocery space. May be location of second California supercenter.

City of Redlands

Statute/Proposal: No final action to develop or prohibit.

Comments: Community development director Jeff Shaw has said that Wal-Mart officials approached city officials about two years ago for preliminary talks about supercenter sites. The most recent talks took place in November 2003 and Wal-Mart has not submitted a formal proposal or requested a preliminary review. Two Redlands sites discussed were vacant parcels on San Bernardino Avenue and Tennessee Street and California Street and Lugonia Avenue. A site in the unincorporated Donut Hole also was discussed. Redlands Mayor Susan Peppler said she was aware of Wal-Mart criticisms but has not heard any local outcry. "We'll always be mindful of that but because discussions are very preliminary, it's a little too early and we're not at the point where it should be a major concern," she said.

City of Sacramento

Statute/Proposal: Proposed ban on stores larger than 100,000 square feet with between 5 percent and 10 percent of nontaxable sales items.

Players/Purpose: City Councilwoman Sandy Sheedy said when smaller stores are pushed out of business by the large chains, it can result in blighted property. Sheedy has said that big box stores are "too big for an urban setting."

City and County of San Francisco

Statute/Proposal: Retail business size-caps: The city and county created "Neighborhood Commercial Individual Area Districts" (NCDs) in 1987 in San Francisco requiring neighborhood zoning size-caps, and subjecting structures greater than a given size to conditional uses as deemed appropriate by the characteristics of each district in the city (e.g., North Beach).

Players/Purpose: The city contends that the ordinance creates greater regulatory control over the size of nonresidential uses within the NCDs, and therefore, preserves and enhances the existing neighborhood-serving uses and enhances future opportunities for resident employment and the ownership of other neighborhood-serving business.

City of San Marcos

Statute/Proposal: August approval of Wal-Mart store by city council.

Legal/Political Issues: Wal-Mart tried unsuccessfully to stop a March referendum by voters wanting to overturn their city council's approval of a regular Wal-Mart store.

These are a few examples of the bitterly fought land-use and environmental battles over big-box retail trade in California. Not every state has an initiative process, but these big-box retail controversies occur in every state and in other countries.

POTENTIAL FOR FUTURE CONTROVERSY

This controversy has grown from community resistance to state and federal court litigation and state legislation. Environmental controversies are part of the concerns of many citizens when considering big-box retail development issues, but it is mixed with many others. Local retail trade may fear the inability to compete with chain giants, especially if the chain is a category killer in the same category as the local store. Unions fear loss of family-wage jobs. Environmentalists and land preservationists fear the loss of environment due to sprawl. The neighborhood in the site where the retail trade development operates will have noise and traffic concerns. These factors can lower property values and erode municipal tax bases. All citizens get annoyed and angry when they experience a land-use planning process that does not value public participation or that makes secret land-use deals with big-box retail corporations. Many of these retail trade corporations seek entitlements to use the property when they acquire the site. These entitlements are generally designed to help ensure profits for the store, and high-volume traffic is desirable from that point of view. Nearby neighborhoods almost always object to increased traffic and parking concerns. Therefore, when acquiring sites, big-box development prefers entitlement negotiation be kept confidential for fear of losing legal rights after public disclosure. The retail big-box trades contend they are providing a needed service and paying their labor fairly.

This controversy shows no sign of diminishing. Outcomes are uncertain and political saliency is high. Sprawl can affect big, small, and medium-sized cities. Small communities desperate for any economic development create strips of retail trade near interstates and large state roads. When a big-box retailer comes in and develops a new site, many of these older retail strips become vacant.

A fundamental aspect of this particular controversy is whether government should help local businesses compete. Most would answer yes. With big-box retail development the land-use question becomes what can we do to keep out business that competes better than local business. Environmental concerns and controversies can become part of the mechanizations communities endure when trying to exclude certain land uses. Many low- and moderate-income citizens like the low prices and one-stop shopping big-box retail trade offers them, so community resistance to them across the board is uneven.

See also Citizen Monitoring of Environmental Decisions; Public Participation/ Involvement in Environmental Decisions; Sprawl

Web Resources

AlterNet. Big Box Swindle: The Fight to Reclaim America from Retail Giants. Available at www.alternet.org/stories/45166/?comments=view&cID=372780&pID=372682. Accessed January 20, 2008.

Dead Malls.com. Available at www.deadmalls.com/. Accessed January 20, 2008.

Good Jobs First. Disclosures of Employers Whose Workers and Their Dependents Are Using State Health Insurance Programs. Available at www.goodjobsfirst.org/corporate_sub sidy/hidden_taxpayer_costs.cfm. Accessed January 20, 2008.

Good Jobs First. Shopping for Subsidies: How Wal-Mart Uses Taxpayer Money to Finance Its Never-Ending Growth. Available at www.goodjobsfirst.org/pdf/wmtstudy.pdf. Accessed January 20, 2008.

Further Reading: Cohen-Rosenthal, Edward, and Judy Musnikow. 2003. *Eco-Industrial Strategies: Unleashing Synergy between Economic Development and the Environment.* Sheffield, UK: Greenleaf Publishing; Nevarez, Leonard. 2002. *New Money, Nice Town: How Capital Works in the New Urban Economy.* New York: Routledge; Rubinfeld, Arthur, and Collins Hemingway. 2005. *Built for Growth: Expanding Your Business Around the Corner or Across the Globe.* PA: Wharton School Publishing; Satterthwaite, Ann. 2001. *Going Shopping: Consumer Choices and Community Consequences.* New Haven, CT: Yale University Press.

BROWNFIELDS DEVELOPMENT

Policy aimed at cleaning up contaminated land is called *brownfields*. It is relatively new, with a distinct urban focus. It is controversial because of the displacement of current residents and reliance on market forces to rebuild some sites. Whether the site is cleaned up to a level safe for residential development, or just safe enough for another industrial use, is a community controversy.

BEGINNING OF BROWNFIELDS POLICY

Since the disaster of Love Canal and the Hooker Chemical Company, U.S. environmental policy has developed a distinct clean-up aspect. It is very controversial. Extremely hazardous sites were prioritized as part of a National Priorities List under the Superfund program. Superfund can assess liability for the cost of a cleanup against the property owner if no other primary responsible parties are around. Huge amounts of unaccounted wastes were produced before the U.S. Environmental Protection Agency was formed in 1970, and huge amounts have continued to be produced. As much as 80 percent of Superfund budget allocations have gone to the litigation that can surround these sites. As knowledge and public awareness increased, many more sites were located. Most states had adopted some type of landfill or waste management environmental policy by the late 1980s. Controversies that still simmer today over what is hazardous pushed the EPA into a new phase of cleanup policy. There was a need to prevent nonhazardous sites from becoming hazardous. This can happen at illegal dumps over time. Metals from refrigerators, stoves, cars, cans, and roofing can leach into the water, depending on the site. Many of these sites were in or near areas densely populated, not a traditional area for the EPA at this time. In some states this was legal if operated as a dump, no matter where it was located. There were many sites waiting to be verified as hazardous or not for Superfund consideration. Even if the community was successful in getting the site designated as hazardous there was a complicated and political process of getting on the National Priorities List, a list of about 1,200 or so of the most important sites for the EPA. Getting a site designated as hazardous was not always considered the best thing for the community because it could suppress property values. Sometime local government fought against such a designation, against the community and the EPA. This is often the case in many communities seeking environmental justice. As environmental justice advocacy increased within the EPA in the early 1990s cleanup policy changed to include more of these sites. In 1993 the EPA first began to address sites that may be contaminated by hazardous substances but that did not pose a serious enough public health risk to require consideration for cleanup under the Superfund program.

The cleanup policy that evolved was conceptualized regionally at first, partially motivated by concerns about sprawl. The idea was to save as much green space as possible by reusing, or infilling, some of these polluted sites. Infill is often proposed as a mitigating solution to sprawl. Municipal boundaries are not related to bioregions or ecosystems, and one municipality or city may not want infill. Municipalities have different and sometimes competing priorities. Combating sprawl or environmental protection and cleanup are generally not as important as economic development at this level of government. As a result, there are many polluted sites. When they become abandoned and are foreclosed on by the municipality or city for failure to pay property taxes, the city owns it. Cities, such as Milwaukee, Wisconsin, then become liable for the cleanup of the polluted sites, as well as losing any tax revenue. It is extremely difficult to sell polluted land, and all efforts are made to escape environmental liability in the process. According to the EPA,

> Brownfields are abandoned, idled, or under-used industrial and commercial facilities where expansion or redevelopment is complicated by real or perceived environmental contamination. They range in size from a small gas station to abandoned factories and mill sites. Estimates of the number of sites range from the tens of thousands to as high as 450,000 and they are often in economically distressed areas.

Portland, Oregon, estimates that it has about 1,000 brownfield sites. Developers avoid them because of cleanup costs, potential liability, or related reasons.

RECENT DEVELOPMENTS

In 2001, new brownfields policy development authorized granting a liability exemption to prospective purchasers who do not cause or worsen the contamination at a site. It also gave this exemption to community-based nonprofit organizations that seek to redevelop these sites. Most states now have their own brownfields programs. There were substantial differences between some state approaches and the EPA brownfields policy. Some of this has to do with the level of cleanup required for a site to be considered clean. An industrial level is cheaper but still polluted. A residential level is very expensive but not polluted. It is still very controversial. Often no developers or nonprofits are willing to clean up the site. Unlike Superfund, brownfields policy does not attack primary responsible parties for liability. State policy approaches are given some leeway in the 2001 policy changes. The new policy stops the EPA from interfering in the state cleanups. There are three exceptions written into the law:

1. a state requests assistance,
2. the contamination migrates across state lines or onto federal property, or
3. there is an "imminent and substantial endangerment" to public health or the environment, and additional work needs to be done.

U.S. URBAN ENVIRONMENTALISM

The United States is still in the early stages of urban environmentalism, a complex subject with intricate and important histories. The potential for unintended consequences for people, for places, and for policy is great. Solid wastes are accumulating every day, combined with a century of relatively unchecked industrial waste that continues to pollute our land, air, and water on a bioregional basis. The wastes in our ecosystem respect no human-made boundary, and the consequences of urban environmental intervention through policy or other actions, intended or not, affect us all.

TERMS OF ART

Brownfields Site

According to the most recent law and policy Public Law 107–118 (H.R. 2869), "Small Business Liability Relief and Brownfields Revitalization Act," signed into law January 11, 2002, the definition of brownfield is:

> With certain legal exclusions and additions, the term "brownfields site" means real property, the expansion, redevelopment, or reuse of which may be complicated by the presence or potential presence of a hazardous substance, pollutant, or contaminant.

Superfund Site

A Superfund site is any land in the United States that has been contaminated by hazardous waste and identified by the EPA as a candidate for cleanup because it poses a risk to human health and/or the environment. There are tens of thousands of abandoned hazardous waste sites in our nation. The implementing edge of the Superfund program is a system of identification and prioritization that allows the most dangerous sites and releases to be addressed, called the National Priorities List.

When outcomes from cleanup and revitalization projects are assessed, the EPA may have unintentionally exacerbated historical gentrification and displacement. EPA funds may have been used to continue private development at the expense of low-income residents.

Urban Environments

Urban areas are complex. For at least a century, urban areas in the United States experienced unrestrained industrialization, with no environmental regulation and often no land-use control. U.S. environmental movements have focused on unpopulated areas, not cities. In addition, U.S. environmental movements did not consider public health as a primary focus. Rather, they emphasized conservation, preservation of nature, and biodiversity. In addition to being the dynamic melting pot for new immigrants, cities became home to three waves of African Americans migrating north after the Civil War. These groups faced substantial discrimination in housing, employment, education, and municipal services. African Americans are the only group in the United States to not have melted into equal opportunity for employment, housing, and education. In addition, people of color and low-income people faced increased exposure to the pollution that accompanied industrialization.

Citizens living in urban, poor, and people-of-color communities are currently threatened by gentrification, displacement, and equity loss on a scale unprecedented since the urban renewal movement of the 1960s. Market forces appear to be the primary drivers of this phenomenon. Spurred by local government attempts to reclaim underutilized and derelict properties for productive uses, residents and businesses who once abandoned the urban core to the poor and underemployed now seek to return from the suburbs. By taking advantage of federal policies and programs, municipalities, urban planners, and developers are accomplishing much of this largely beneficial revitalization. However, from the perspective of gentrified and otherwise displaced residents and small businesses, it appears that the revitalization of their cities is being built on the backs of the very citizens who suffered, in place, through the times of abandonment and disinvestments.

While these citizens are anxious to see their neighborhoods revitalized, they want to be able to continue living in their neighborhoods and participate in that revitalization.

In addition to facing tremendous displacement pressure, African Americans and other people of color also face difficult challenges in obtaining new housing within the same community (or elsewhere) after displacement. For example, when these populations are displaced they must often pay a disproportionately high percentage of income for housing. Moreover, they suffer the loss of important community culture. While it is not fair to suggest that federal reuse, redevelopment, and revitalization programs are the conscious or intentional cause of gentrification, displacement, and equity loss in these communities, it is apparent that the local implementation of these programs is having that net effect. These then become the unintended impacts of these well-intended and otherwise beneficial programs. Brownfields is a pioneering urban environmental policy, and unintended impacts could easily occur.

Community activists should have an educated perspective to decide if brownfields programs will provide hope and opportunity to their distressed neighborhoods, or whether they will exacerbate environmental contamination and/or provide little or no opportunity for their own families to benefit proportionately. Brownfields redevelopment is a big business. Profits are generally more important to brownfields entrepreneurs than community concerns about displacement or reduced cleanup standards. In fact, at EPA's 2004 National Brownfields Conference, developers reinforced this notion by highlighting their perspective that in order for communities to be players in the redevelopment and revitalization process, they need to be financially vested in the process. This view clearly speaks to the need for EPA intervention to ensure meaningful community involvement irrespective of financial status.

The EPA provides some funding for brownfields to state and local government and to some tribes. As of July 2007, about 2.2 million dollars was awarded to brownfields revolving loan fund recipients. The EPA claims that since 1997 they have awarded about 55 million dollars for about 114 loans and 13 subgrants. The EPA states these loan funds have leveraged more than $780 million dollars in other public and private cleanup and redevelopment investment. Some criticize the program as being underfunded and underresourced. They say the need for cleanup of the places where we live, work, and learn is paramount for any environmental cleanup policy.

HOW CLEAN IS CLEAN?

Cumulative impacts concern EPA because they erode environmental protection and threaten public health, safety, and welfare. They cross all media—land, air, and water. Independently, media-specific impacts have been the focus of the EPA's work for years. However, if the combined, accumulating impacts of industrial, commercial, and municipal development continue to be ignored, the synergistic problems will only get worse. The cleanup of past industrial practices must be thorough and safe for all vulnerable populations, say most communities. Another community concern is that long-term industrial use of a given site may decrease the overall value of property in the area, resulting in a loss of wealth over time. However, to clean up the site to a level safe enough for residential development is much more expensive. It is also fraught with uncertainty, which translates into risk for most real estate financial institutions. The state of the law of brownfields cleanup is also very uncertain and dynamic. One thing is certain though: the United States is dotted with contaminated sites generally concentrated in urban areas and multimodal transit nodules (e.g., ports, depots).

By far, the populations most impacted by brownfields decisions are those who live, work, play, or worship near a contaminated site. These people are already in areas with a high pollutant load, with generally higher rates of asthma.

Vulnerable populations such as pregnant women, the elderly, children, and individuals with preexisting health problems are at increased risk. In many environmental justice communities, a brownfields site may be the only park-like setting available, so it can attract some of the most vulnerable populations.

To the extent members of the community are forced to leave because of increased housing costs, the community loses a piece of its fabric, and sometimes knowledge of its history and culture. This adverse impact needs to be addressed as part of a cumulative assessment. The sense of identity common to many environmental justice communities is threatened when communities are displaced.

POTENTIAL FOR FUTURE CONTROVERSY

As part of the first and very necessary wave of urban environmentalism, brownfields unearths many deep-seated environmental and political controversies. U.S. environmental policy and the U.S. environmental movement have ignored cities, where most of the pollution and most of the immigrants and people of color reside. The environment, urban or not, is difficult to ignore as population expands and concepts of sustainability are developed. Citizen monitoring of the environment, environmental lawsuits, and the need to enforce environmental laws equally have driven environmental policy to urban neighborhoods. Cleanup of the environmentally devastated landscape is usually an early priority for any governmental intervention in environmental decision making.

A hard uncertainty underscores the current methods of holding private property owners liable for waste cleanup. What if they cannot afford it? What if they manipulate bankruptcy or legitimately cannot afford it? What if the contamination is so extensive that no one stakeholder alone can afford to clean it up? Ecosystem risk assessment, now mandated at Superfund sites, will unearth only more contamination. The levels of contamination themselves are highly controversial because some believe they do not protect the public enough. How much real estate corporations and banks should be supported by government in developing market-based cleanup strategies is a big policy controversy.

Yet, without any intervention these sites accumulate wastes that can spread to water and land. They do not go away but generally get worse. Over time there will be no hiding any of them. With the new and rapidly developing global consensus on sustainability, cleanup of contaminated sites is a natural and necessary first step. This step takes place in a political context of race, class, and awkward histories of human oppression. The immigrants and migrants always lived in the tenements or on the other side of the tracks. (Train toilets dumped directly on the tracks until the late 1990s.) Success was defined as leaving the city for a house in the suburbs, with a better school district. Many but not all immigrant and migrant groups came through polluted and unhealthy urban neighborhoods.

This political step is also a necessary step, and one that remains very controversial in the U.S. context. Currently, brownfields is the policy face of that step forward.

BROWNFIELDS PLAYING A ROLE IN THE BIOFUELS INDUSTRY

Many contaminated sites are old gas stations, many of which have leaking underground storage tanks. The cleanup costs and liability are much larger when the contamination has spread, especially if it has spread to water. However, without cleanup the contamination can spread. From the city's point of view it is an unproductive piece of taxable property. This common scenario has repeated over and over again over the last 30 years. Creative new solutions to this difficult and controversial policy issue require collaboration by local government, the EPA, and the property owner.

SeQuential Biofuels opened the first alternative fuel station in Oregon. The country has a renewed interest in gaining energy security and independence by moving toward producing more of its fuels, including biofuels such as ethanol and biodiesel. Biofuels are cleaner fuels that produce fewer pollutants than mainstream fuels. There also is much potential for homegrown economic development in this rising new industry.

Brownfields redevelopment can play an important role in this emerging industry. Brownfields are a good fit because redeveloping these contaminated lands protects green space; the sites often are an opportunity to reutilize unused urban and industrial space. And often these former gas stations are ideal for such development because they already sit on properties close to roadways.

SeQuential Biofuels has 33 branded pumps around the state with independent retail sites. The company, which owns 60 percent of the biodiesel market share in Oregon, has a large commercial biodiesel production facility that may serve as for a model gathering the fat necessary for biofuel production. The facility produces one million gallons of biodiesel made from used cooking oil collected from regional restaurants and food processors. It also uses virgin canola oil grown in eastern Oregon. By gathering resources and wastes locally, recycling and processing them, and distributing them locally, the overall ecological production footprint is smaller because of lower energy costs through less transportation. The retail fuel station sits along a commercial corridor adjacent to Interstate 5. The former Franko facility sold gasoline from 1976 until 1991. At that time, the property was turned over to a bankruptcy trustee. Also in 1991, petroleum contamination from the site was observed during trenching along the highway east of the site. Contamination also had migrated to a residential well west of the facility.

In 1996, a private party purchased the property and removed the five underground storage tanks and some contaminated soil. Subsequent assessment identified the former fuel pump islands as the primary source of contamination. Lane County then acquired the property through tax foreclosure and in January 2005 removed more than 400 tires and 15 drums of waste.

SeQuential purchased the property later that year after entering into a prospective purchaser's agreement with the Oregon Department of Environmental Quality (DEQ). The retail fuel station, which sells ethanol and biodiesel blends, opened last fall.

Renewable energy, energy efficiency, and sustainable design elements are all part of the business plan. Covering the fueling islands are 244 solar panels that will provide 30 to

50 percent of the electrical power the station requires annually. On the roof of the convenience store is a garden. There are 4,800 plants in five inches of soil.

SeQuential took advantage of several state incentives on this project. Oregon and Washington have played active roles in providing tools to advance biofuels in the private sector. On this project, the Oregon DEQ provided $19,600 for site assessment, the EPA awarded a $200,000 brownfields cleanup grant to Lane County, and the Oregon Economic and Community Development Department provided a $50,000 loan as matching funding for the EPA assessment grant through its Brownfields Redevelopment Fund. The project also qualified for the Oregon Department of Energy's Business Energy Tax Credit, which equals 35 percent of an eligible project's costs, and its Energy Loan Program, which provides low-interest loans.

In the first six months in business the retail station exceeded volume projections. Its biggest obstacle now is teaching consumers that these biofuels are appropriate for any vehicle.

See also Air Pollution; Cumulative Emissions, Impacts, and Risks; Ecological Risk Management Decisions at Superfund Sites; Ecosystem Risk Assessment; Environmental Justice

Web Resources

Municipal Research and Services Center of Washington. Brownfields & Brownfield Redevelopment. Available at www.mrsc.org/Subjects/Environment/brownfields.aspx. Accessed January 20, 2008.

Tarr, Joel A. "Urban History and Environmental History in the United States: Complementary and Overlapping Fields." Available at www.h-net.org/~environ/historiography/usurban.htm. Accessed January 20, 2008.

Further Reading: Cohen-Rosenthal, Edward, and Judy Musnikow. 2003. *Eco-Industrial Strategies: Unleashing Synergy between Economic Development and the Environment.* Sheffield, UK: Greenleaf Publishing; Collin, Robert W. 2006. *The U.S. Environmental Protection Agency: Cleaning Up America's Act.* Westport, CT: Greenwood; Russ, Thomas H. 1999. *Redeveloping Brownfields: Landscape Architects, Planners, Developers.* New York: McGraw-Hill Professional; Thomas, June Manning, and Marsha Ritzdorf, eds. 1997. *Urban Planning and the African American Community: In the Shadows.* London: Sage; Thompson, J. William, and Kim Sorvig. 2000. *Sustainable Landscape Construction: A Guide to Green Building Outdoors.* Washington, DC: Island Press; Witkin, James B. 2005. *Environmental Aspects of Real Estate and Commercial Transactions: From Brownfields to Green.* Chicago: American Bar Association.

CANCER FROM ELECTROMAGNETIC RADIATION

The controversy about power lines and cancer involves farmers, electrical workers, schools, and residents because animals, children, and homes might be affected. Utilities, bond markets, power line and pole manufacturers and distributors, and their trade associations all insist on the necessity of power lines. Government allows utilities to operate. Some research shows a correlation between environmental emissions of power-frequency magnetic fields and the incidence of some cancers.

Electromagnetic sources of cancer are very controversial on several levels. Scientists disagree with each other, and the U.S. public questions the credibility of the science to date. Allowing high levels of electromagnetic energy benefits industries with high needs for power and communities seeking economic development. If or when a scientifically validated cause of cancer is established it will open a floodgate for lawsuits from people with cancer, ranchers and farmers, and parents with exposed children. It may require a more direct accountability for industrial impacts on the environment to the exposed community. A possible link to cancer is raising the profile of utilities and power supply grids. The background of this controversy is scientific, and communities, schools, and farmers are rising to the challenge of understanding this science because of a fear of direct, life-threatening impacts. Local governments seek economic development and a stable tax base but wonder if the public health risks and costs are worth it. Local landowners are very upset when power lines force their way through their private property because of the potential health risk and the risk of suppressed property values.

BACKGROUND

The specific frequency of the source of electromagnetic frequency (EMF) has a direct impact on how it affects the environment. At the very high frequencies (less than 100 nanometers), electromagnetic particles (photons) have enough power to break chemical bonds in living matter. This is very destructive. This breaking of bonds is called ionization; X-rays are an example of the ionization of molecules. At lower frequencies the power of a photon is generally considered too low to be destructive. Most visible light and radio frequencies fall into this range.

IONIZING ELECTROMAGNETIC SOURCES AND BIOLOGICAL EFFECTS

Many of these impacts on plants or animals have not been studied. The electric fields around the power-frequency sources exist whenever voltage is present. These electric fields are generally considered too weak to penetrate buildings or skin. However, there is controversy about this on several battlefields. The scientific community contests this on grounds of failure to consider cumulative effects and synergistic or antagonistic chemical interactions, and failure of sample size. Communities are not reassured by this. When children or livelihoods are at risk from EMF exposures, communities can move the battleground to the local legislative and land-use areas of government. Utility rights-of-way have long been contested by environmentalists. The possibility that utility lines could degrade the environment around them only pushes the battleground even further into the litigation area.

How do you measure a magnetic field? Measurement forms the basis of scientific observation. Dynamics that are difficult to measure remain unproven under scientific principles. Without a measure of exposure it is difficult to know who is actually exposed, how much they are exposed, and how much this risk contributes to cumulative risks; risk will usually be underestimated.

In the United States magnetic fields are still measured in gauss (G) or milligauss (mG), where:

1,000 mG = 1 G.

In the rest of the world, magnetic fields are measured in tesla (T):

10,000 G = 1 T
1 G = 100 microT (µT)
1 microT = 10 mG

Electric fields are measured in volts/meter (V/m).

Another crinkle in this controversy is about the magnetic fields around power-frequency sources. Electrical currents generate magnetism, which only occurs when current is flowing. These magnetic fields are difficult to contain, and they penetrate buildings and people. The concern is whether residential exposure to power-frequency fields may be because of the magnetic field. There is much less conclusive scientific research in this area. Belief in the healing power of magnets has risen in the popular culture.

POWER LINES AND CANCER RATES

This aspect of the battleground awkwardly incorporates public health and land use. Because it involves cancer and children it is very powerfully felt among affected stakeholders. Many areas of the United States are underserved by public services competent to measure EMFs, as are their land-use decision-making bodies. They turn to published scientific research. There are hundreds of published research studies, and more that remain unpublished. Scientists from industry, research institutions, and government are researching this issue. Some studies indicate that children living near high-voltage transmission power lines have higher than average rates of leukemia, brain cancers, and/or overall cancer. They are correlations, not causal models. The correlations are weak. Other studies have shown no correlations or causality between residence near power lines and risks of childhood leukemia, childhood brain cancer, or overall childhood cancer. Many of these studies themselves are controversial. They are criticized for requiring very high levels of proof, ignoring cumulative and real-life multiple exposures, and focusing on animals rather than humans. Some, as in other environmental controversies, challenge the independent judgment of the scientists if their study is funded by utilities or others who profit from power line expansion. This would include industrial manufacturers, distributors, retail trade, and the trade associations affiliated with these groups and with big business, generally like the U.S. Chamber of Commerce.

POTENTIAL FOR FUTURE CONTROVERSY

The scientific perspective is very important in this controversy. Scientists fundamentally contend that the results of epidemiological studies cannot be used as a basis for land-use restrictions. Citizens at home and work and in hospitals and schools expect their government to protect them from serious public health risks. The mere perception of risk affects property values. Homes near power lines may be less expensive now and much less attractive to young home buyers with children.

An additional problem is the lack of knowledge of the correct dose metric. How do you measure how it affects humans, animals, or plants? This creates complex multiple-comparison problems for researchers, regulators, litigators and courts, and scientists. This basic research issue will continue until a new way of measuring EMF doses and methodology is developed. It hinders communities from preventing exposure because it is difficult to develop land-use and public health regulations without a metric of exposure. Absence of a way to measure impacts allows industrial and municipal power generators and users to continue with business as usual. It also neglects an important aspect of environmental impact assessment.

There is a broad consensus that exposure to these fields cannot be proven to be safe or dangerous. Scientific approaches vary, differing in methodology and result. The public controversy increases daily. As we increase in population, increase our electrical power needs, and develop along the electrical power grid, this controversy is going to increase. There is a marked divergence in conclusions between U.S. and British research on the danger of EMFs. When a measure and dose metric are developed it could turn out that doses were very dangerous, or that they were completely harmless, perhaps even good for you. Technological advancements in the near future will form the early contours of this battleground, and they could open it up to federal legislation, class action lawsuits, and thousands of land-use ordinances designed to prevent the siting of power lines. These advancements could also conclusively prove that EMFs do not contribute to cumulative risks, pose no threat to any life-form, and efficiently and safely move electrical energy.

See also Children and Cancer; Cumulative Emissions, Impacts, and Risks; Environmental Impact Statements: United States; Pesticides

Web Resources

Electromagnetic Radiation and Public Health. Available at http://www.math.albany.edu:8008/EMF.html. Accessed January 20, 2008.

Farley, John W. Powerlines and Cancer: Nothing to Fear. Available at www.quackwatch.org/01QuackeryRelatedTopics/emf.html. Accessed

Stony Brook News. Breast Cancer and Electromagnetic Fields Study. Available at http://commcgi.cc.stonybrook.edu/cgi-bin/artman/exec/view.cgi?archive=3&num=481. Accessed

Further Reading: Doll, Richard. 1992. *Electromagnetic Fields and the Risk of Cancer*. Documents of the National Radiological Protection Board, United Kingdom; Norden, Bengt. 1992. *Interaction Mechanisms of Low-Level Electromagnetic Fields in Living Systems*. Oxford: Oxford University Press; Wilson Bary W., ed. 1990. *Extremely Low Frequency Electromagnetic Fields: The Question of Cancer*. Columbus, OH: Battelle Press.

CARBON OFFSETS

The increase in carbon dioxide in the atmosphere and its contribution to global warming and climate change have motivated individuals and corporations to purchase ways to offset the ecological footprint of their carbon dioxide emissions. A number of companies offer to perform environmental activities to offset the carbon dioxide emitted into the atmosphere by certain activities. Controversy is evoked when these activities are scrutinized, and the actual environmental contributions of the carbon offsetting activity are analyzed.

WHAT ARE CARBON OFFSETS AND HOW DO THEY WORK?

Carbon offsets are provided by profit and nonprofit organizations to offset the carbon dioxide emitted, generally by a specific activity. The profile of carbon dioxide offsets was raised when mortgage companies in the United Kingdom used them in their mortgage advertising, and when carbon offset organizations specifically marketed them to air travelers. Air travelers could elect to pay for their share of the carbon dioxide on a particular flight. Airplane emissions are substantial. Besides fuels and lubricants, airports use wing deicer and other toxic solvents. With acres of paving the runoff of these pollutants usually affects local water supplies unless treated. The money paid is supposed to go to an activity that uses carbon dioxide, to offset the carbon dioxide emitted on the flight, such as tree plantings. Some have estimated the carbon offset market could be as high as 100 million dollars.

U.S. businesses have also been buying carbon dioxide offsets in order to engage in international business. Many other industrialized and nonindustrialized nations signed the Kyoto Treaty on global warming. The United States has refused and is one of the largest emitters of greenhouse gases. The Kyoto Protocol set global caps on emissions of greenhouse gases, like carbon dioxide. Many nations devoted substantial resources for many years to the Kyoto process, as did international bodies like the United Nations and the Union of Concerned Scientists. National and international environmental groups, along with community groups and labor unions, all also devoted considerable re sources to this process. There is an international movement of cities that sign on with the Kyoto Protocols, including many of the major U.S. cities. U.S. businesses feel strong pressure to reduce the emission of greenhouse gases in order to continue international business transactions where higher standards are required.

EMERGING BATTLEGROUNDS

There are still some questions about how carbon dioxide emissions are calculated, although the emissions estimates for most major activities are known. The big battleground is about how the money for carbon offsets is spent. The range of carbon offset projects has attracted criticism of them. There is no well-defined offset protocol or policy, so there are many gray areas. If the money goes to develop alternative renewable energy sources like wind and solar power, is the carbon dioxide from the petrochemicals that would have otherwise been used offset? Does it make a difference if the companies assisted make a profit, are nonprofit, or are state operated? Another gray area is home weatherization to save energy costs as a carbon offset. Does it make a difference to an offset program if a single homeowner is benefited? The argument for it counting as a carbon offset is that decreased energy use through conservation measures reduces carbon dioxide emissions by lowering consumption of pollution-causing energy sources. These differences can easily become battlegrounds.

NOT ENOUGH REGULATION TO BE RELIABLE?

The biggest upcoming battleground for this controversy is government regulation of it. Currently there is very little. There are big differences in costs and projects of the carbon offset programs. There is a private, nonprofit effort to create a Green-E certification requiring the carbon offsets that meet some type of standard. Consumers of carbon offsets want their offsets to truly offset carbon dioxide. There are also concerns that without regulation some offsets may be sold many times or go to projects that occur anyway.

POTENTIAL FOR FUTURE CONTROVERSY

Carbon offsets are reaching businesses, individuals, and some cities. Carbon offsets are a voluntary market. There are many ways to mitigate carbon dioxide use, and some feel that buying offsets just uses money to justify pollution. Some fear it favors big polluters that just pay for their pollution, with the degrading environmental impacts such as global warming. Most of the carbon offset programs would take many years to offset the carbon dioxide used in one airplane trip. Realistically, trees take about 100 years to mitigate the carbon dioxide emissions of one person on one long airline trip. Others feel that the criticisms of the carbon offset market are inflated.

The potential for future controversy around carbon offsets is unknown. It depends on how much they are used, the project selection, and looming governmental regulation. It also depends on whether it stays limited to just carbon dioxide emissions. Some environmentalists question the underlying premise of paying for your pollution. It allows rich nations and rich people to pollute. If the environmental impacts are not negligibly reduced, then they question the overall efficacy of it. Carbon offsets do engage the public imagination and give business an avenue to express environmental concern in a voluntary market. Some have argued that overregulation can limit the ability of industry to make pollution reduction and prevention changes. However, governmental environmental regulations are generally phrased in ways that induce compliance to minimal standards. Most times these standards are simply require industry to report their own emissions to the government. Large emissions are permitted, and industries self-report whether they are under a regulatory threshold necessary for a permit. Government regulators and environmentalists claim that industry is always free to do more for the environment. Purchasing carbon offsets is one way they are beginning to do just that.

See also Air Pollution; Climate Change; Global Warming

Web Resources

EcoBusiness Links. Carbon Emissions Offsets directory. Available at www.ecobusinesslinks. com/carbon_offset_wind_credits_carbon_reduction.htm. Accessed January 20, 2008.
Environmental Defense Fund Evaluation of Carbon Offset Programs. Available at http:// www.environmentaldefense.org/page.cfm. Accessed January 20, 2008.

Further Reading: Bass, Stephen, and D. B. Barry Dalal-Clayton. 2002. *Sustainable Development Strategies: A Resource Book.* London: James and James/Earthscan; Bayon, Ricardo, Amanda Hawn, and Katherine Hamilton. 2007. *Voluntary Carbon Markets: An International Business Guide to What They Are and How They Work.* London: James and James/Earthscan; Follett, Ronald F., and John M. Kimble. 2000. *The Potential of U.S. Grazing Lands to Sequester Carbon and Mitigate the Greenhouse Effect.* Boca Raton, FL: CRC Press; Smith, Kevin. 2007. *The Carbon Neutral Myth: Offset Indulgences for Your Climate Sins.* Amsterdam: Transnational Institute.

CARBON TAXES

As concern about greenhouse gases increases because of concerns about climate change, more policy focus is brought to bear on taxing the gases that cause it. Carbon dioxide is one such gas. Industry and others resist reducing carbon dioxide emissions because of the cost.

BACKGROUND

Carbon taxes are taxes on those who contribute to climate change by emitting carbon dioxide into the environment, including homes, businesses, and industries that use petroleum products for energy. Carbon taxes are seen as one way of making those who contribute to pollution pay the cost of doing so. Nations around the world now have carbon taxes. In September 2007 the Vatican in Rome officially became *carbon neutral.*

Carbon dioxide emissions are a global concern, with the United States leading the world in emissions. In the United States, carbon dioxide makes up 82 percent of all greenhouse gases when weighted by climate impact. The United States emits almost 22 percent of global carbon dioxide emissions but has only about 5 to 6 percent of the world's population. These emissions are large and current environmental regulatory regimes do not adequately prevent them from entering the atmosphere. Some scientists are beginning to examine ways to sequester carbon dioxide under the ground or sea, such as the state oil company of Norway. However, this research is just starting, with no battlegrounds developing as of print. Economists and sustainability proponents have traditionally embraced the idea that the polluter pays, which means that the source of the pollution should pay the costs of cleaning up the pollution. While in theory this principle is attractive, it is very difficult to implement as an environmental protection policy. Enforcement is weak, coverage of polluters incomplete, and questions about whether money can ever compensate for environmental degradation all prevent environmental regulations from reducing carbon dioxide emissions quickly enough to avoid a global tipping point. The *tipping point* concept is itself debated in the scientific community. It means that once the atmosphere reaches a certain level of carbon dioxide there will be no way to turn back to lower levels. Many carbon tax advocates feel that the only way to reduce carbon dioxide emissions quickly enough to prevent a tipping point problem is to tax carbon dioxide emissions. Sustainability advocates also like carbon taxes because they could help move society away from nonrenewable fuels like gas and oil. Critics point out that many of the results sought in terms of carbon emissions could be approached by reducing subsidies for nonrenewable energy sources like oil. Politically, a policy of reducing subsidies to nonrenewable energy sources has simply not happened, argue carbon tax proponents, and results are needed. Critics also say that the same results could be attained by use of a cap and trade program. Carbon proponents argue back that traditional cap and trade programs only apply to generators of electrical energy, which account for only about 40 percent of carbon emissions; that carbon taxes are more transparent and easily understood than cap and trade arrangements between industry and government; and that carbon taxes will get quicker results in terms of decreased carbon dioxide emissions.

CARBON TAXES AND GOVERNMENT

The first country to put a tax on atmospheric carbon dioxide emissions was Sweden in 1991, followed by Finland, Norway, and the Netherlands. At first it was applied to combustion-only point sources over a certain size. The tax was equivalent to about $55 (U.S.) per ton of carbon dioxide emitted. Norway has extended the tax to offshore oil and gas production. They found it effective in spurring industry to find cost-efficient ways to reduce carbon dioxide emissions. The European Union just began a carbon tax in 2007 and plans to increase the rate of tax in the near future. It is clear that other industrialized nations are taxing carbon to control pollution. The more the United States resists these efforts to meaningfully slow global warming and radical climate change the more the battleground for carbon taxes will become international. Radical climate change could have drastic impacts on agricultural food production. One recent climate change model has suggested that severe drought could occur in the same areas that feed and shelter 80 percent of the global population. The global community is highly motivated to reduce the risk of environmental uncertainty rapid climate change would cause.

THE UNITED STATES AND THE CARBON TAX

The United States emits about 4.5 tons of carbon dioxide per capita; Denmark, 2.8 tons; Norway, 1.2 tons; Sweden, about 1.2 tons; Finland, about 2.8 tons; West Germany, about 2.8 tons; Japan, about 2.6 tons; France, about 1.8 tons; the United Kingdom, about 2.7 tons; and Italy, about 2.1 tons per capita in 1991. Most analyses conclude that the United States has higher emissions from primary energy used, larger houses, and much more driving per capita.

Carbon taxes have been battlegrounds at the city, state, and federal level. Some environmentally conscious communities have passed their own carbon tax, such as Boulder, Colorado. Boulder's carbon tax is about $7 a ton and applies only to electricity generation. The mean electric bill for residential consumers is $16 per year, and $46 per year for businesses. Advocates there claim it will raise almost seven million dollars over five years. Direct environmental regulation of carbon by municipalities could occur rapidly in the United States, especially if state legislatures or Congress does not develop a carbon tax policy. Currently, there are about 330 U.S. mayors signed on to the U.S. Mayors Climate Protection Agreement. Carbon tax battlegrounds could occur in many different localities. Many state legislatures are being exposed to the idea of carbon taxes. In the United States, Massachusetts has formed a study commission on tax policy and carbon emissions reduction. Advocates for a carbon tax say that it will reduce U.S. oil dependence. Nonrenewable energy sources account for much of the U.S. carbon waste stream. Natural gas is responsible for 22 percent, coal burning for 36 percent, and petroleum products for 42 percent of U.S. annual carbon emissions. If the carbon tax is enough to change pollution behavior, then advocates claim it will generate a large amount of revenue, some estimating between \$55 billion and \$500 billion a year depending on the carbon tax program.

AL GORE

The First National U.S. Carbon Tax Proponent and Environmental Leader

In 1994 Vice President Al Gore presented a carbon tax plan to President Clinton. It did pass the House of Representatives, but failed in the Senate. Republicans claimed that gas prices would increase 20 percent if the bill passed. It was an extremely controversial proposal in the battleground of national politics.

Al Gore has held many leadership posts in addition to vice president of the United States. In all his national positions he has steadfastly advocated cleaner, cheaper, and smarter ways of increasing environmental protection. Many now-accepted environmental policies about hazardous waste cleanups owe their existence and continued development to Representative and then Senator Al Gore. Intelligent, sophisticated, well-informed, and well-known, Mr. Gore continues to advocate for environmental protection to many public audiences. Although he received the most votes for president in 2000, the Republican-dominated Supreme Court selected George W. Bush as president, leading to a continuing controversy on election fraud by political parties. One dissenter in the decisive *Bush v. Gore* case, Justice John Paul Stevens, wrote:

> Although we will never know with complete certainty the identity of the winners of this year's presidential election, the identity of the loser is perfectly clear. It is the nation's confidence in the judge as the impartial guardian of the law. (*Bush v. Gore*, 531 U.S. 98 [2000], available at: www.oyez.org/cases/2000–2009/2000/2000_00_949/)

With characteristic strength and perseverance, Gore has continued to lead the charge against global warming and coming climate change and now commands an international constituency committed to environmental protection and sustainability. In 2006 he was awarded the Nobel Prize, among his many other honors. He is also the owner of a new media corporation.

Al Gore was born in 1948 in Washington, D.C. His father was a senator, and many in his family were very politically active. In high school he was very athletic. At age 17 he began college at Harvard University and in 1969 received a BA in government. Although he opposed the Vietnam War, he served in Vietnam. From 1971 to 1972 he studied theology at Vanderbilt University. He switched over to the law school from 1974 to 1976. From 1977 to 1985 he served in the House of Representatives. He then served as senator and vice president, and in other leadership positions in government. Besides his strong advocacy for the environment, Gore, and his father, Senator Gore, Sr., distanced themselves from traditional separate but equal southern racial sentiments.

In these positions Gore began to reinvent government. He focused on the U.S. Environmental Protection Agency and created various initiatives. These initiatives helped focus policy to achieve substantive environmental results. He encouraged scientific research and exploration into new pollution abatement and control technologies and included stakeholders like Environmental Justice for the first time. Under his vision, the United States slowly began the process of switching some permitting over to different industrial sectors, instead of the one-size-fits-all permit. Tighter fits between industries and their environmental permits allow for better environmental regulation. Modern industrialized nations recognize between 20 and 60 industrial sectors, identified with standard industrial codes or SICs. The International Standards Organizations (ISO) uses these sectors to evaluate compliance with international treaties. Gore's initiatives helped push the United States up to international standards by creating a policy focus on sector-based permitting. By doing this he divided industry stakeholders into their respective industrial sectors, fundamentally changing the battleground. His vision also included many parts of the public not usually included in EPA processes, including cities and people of color. He is a firm believer in robust public participation practices for everyone. He was a strong advocate for the formation and expansion of the Toxics Release Inventory, a formidable tool in today's environmental decision-making processes involving citizens. Al Gore did reinvent government from the technical levels to the politically inclusive processes of public policy.

Source: Al Gore, *Earth in Balance: Ecology and the Human Spirit* (New York: Houghton Mifflin, 1992).

Carbon taxes have made their way into congressional discussions and legislative proposals to reduce pollution. In September 2007 Representative Peter Stark from California and Jim McDermott from Washington introduced a bill to reduce global warming by taxing carbon in fossil fuels. It is called the Save Our Climate Act. He wants it discussed along with controversial and complicated cap and trade programs that are under legislative consideration. Their bill would charge $10 per ton of carbon in coal, petroleum, or natural gas. It would increase by $10 every year until U.S. carbon emissions decreased by 80 percent of 1990 levels. He maintains that this is the level most scientists claim is necessary to slow rapid climate change. Some environmental groups do not endorse it, some influential newspaper columnists (self-described as liberal and conservative respectively) do endorse it, and many are leery of anything called a *tax*.

Accepting the concept of carbon taxes requires acceptance of global warming and climate change. These are environmental controversies themselves. In the United States they are also political controversies. As a political controversy, taxes for anything environmental may have difficulty getting enough support to become law.

There is now academic discussion of a global carbon tax. Economists focus on carbon taxes in some of their environmental and policy analyses. It is likely that many nations would find this appealing. Huge issues of enforcement in rich and poor nations would prevent this. No nation likes another nation to tax its citizens in their own country. Nonetheless, the small economics battleground of carbon taxes as environmental policy is receiving serious attention.

POTENTIAL FOR FUTURE CONTROVERSY

Carbon taxes shift environmental policy into the tax policy regulatory arena. Taxes are very strong policy devices and are used here to change the behavior of polluters. As such there are distributional impacts and other taxes to consider. Some have wondered, why not just prohibit carbon dioxide emissions, or regulate them out of existence? Carbon taxes do appeal to the basic polluter-pays principle but suffer some of the same problems. Who does the polluter pay? Who gets the benefit of the new carbon tax revenue? These questions fall into the same category of questions as avoiding carbon taxes by subsidizing renewable energy or withdrawing nonrenewable energy subsidies (for oil companies, for example). The political will of regulatory agencies and federal courts is not strong enough to reduce the power of industry enough to tax carbon dioxide emissions. The fundamental question is whether carbon taxes will actually reduce carbon dioxide emissions.

See also Climate Change; Global Warming; Sustainability

Web Resources

Carbon Tax Center. Available at www.carbontax.org. Accessed January 20, 2008.
EcoNeutral. Neutralize Your Footprint. Available at www.econeutral.com. Accessed January 20, 2008.
The Idea of Carbon Tax. Available at http://deq.utah.gov/Issues/climate_change. Accessed January 20, 2008.

Further Reading: Carraro, Carlo, and Domenico Siniscalco. 1993. *The European Carbon Tax: An Economic Assessment.* New York: Springer; Dellink, Rob B. 2005. *Modeling the Costs of Environmental Policy: A Dynamic Applied General Model.* Northampton, MA: Edward Elgar Publishing; Park, Patricia D. 2002. *Energy Law and the Environment.* Boca Raton, FL: CRC Press; Serret, Ysé, and Nick Johnstone. 2006. *The Distributional Effects of Environmental Policy.* Northampton, MA: Edward Elgar Publishing.

CELL PHONES AND ELECTROMAGNETIC RADIATION

Causes of cancer are controversial. When an important, new, and ubiquitous piece of communications technology such as the cell phone is suspected of cancer-causing emissions, the environmental aspect of the controversy heats up.

The rapid development of communication technology has moved large amounts of information into the palm of our hand. Although initially starting as cellular telephones, these handheld devices now include cameras, computers, voice recorders, televisions, and video-calling capabilities. Industry has invested billions of dollars in these technologies, and consumers demand them. The federal government has taken a hands-off approach to their development. But there is a problem. Tests have found a possible link between cell phone emissions and cancer. The studies have two disturbing findings: (1) biological indications of cell changes seemed to occur with U.S. cell phone exposures, and (2) there is a statistical link correlating cell phone usage with certain types of brain cancer. However, other research has found no such link. Some of the early tests were funded by the cellular telephone industry, others by the government. There are differences in both research results and cell phone technology by country. The United States uses microwave technology via cell towers. Microwave ovens can emit radio frequency waves that threaten heart pacemakers. Western Europe, in contrast, uses satellite transmissions more in the cell phone systems. The U.S. Food and Drug Administration interpreted the brain cancer test results as worthy of more research. Some consumer protection groups wanted more consumer protection, interpreting the test results in the light of cell phones being more risky for the average consumer. Consumer groups and public health advocates want stricter regulation to ensure public safety. Brain cancer victims may sue cell phone manufacturers for negligence if causality and liability can be established, shifting this aspect of the controversy to the courtroom.

A part of this controversy involves the amount of exposure to radio frequency emissions. Some occupations require exposure to these emissions and are generally regarded as dangerous. Workers generally try to avoid prolonged exposure to transmitting antennas. The safety, power, and technology of transmitting antennas have improved rapidly in the past 50 years. Cell phones in the United States operate in a radio frequency of 850 to 100 megahertz and emit about .6 watt of power, although new handheld communications devices may be more powerful. The radio frequency is nonionizing at that range. Ionizing radiations, like X-rays, produce more risk. The antenna produces the most energy, which comes in the form of heat. Given that the power is so low, most assume that it is not enough to cause damage. The ability to use satellites to transmit has greatly increased their communication potential. Handheld cell phones and communication devices place the transmitting antenna directly against the skull. Their proximity to the brain makes concern about brain impacts of their environmental emissions a flash point in this controversy. Cell phone transmitters remain on even when the user is listening rather than talking. This increases the total amount of time the consumer is exposed to radio frequency emissions. The increased time of exposure and the proximity to the brain make cell phone usage risky, claim consumer protection advocates.

EVIDENCE OF HEALTH DAMAGE

Because much of this controversy in the United States hinges on a set of research tests, advocates on all sides have scrutinized them closely. The tests were conducted at Stanford University and Integrated Laboratory Systems in Research Triangle Park, North Carolina, a respected research center. Their research found "chromosomal changes in blood cells subjected to the same type of electromagnetic radiation emitted by hand-held cell phones."

There have been other ominous research studies. One research study compared brain cancers in people who use cell phones with those who do not. It found a large increase in brain cancer in those who used cell phones. One kind of brain cancer was noticeably increased. Neurocytomas are brain cancers that grow inward from the periphery of the brain and seem to increase with cell phone usage. Industry, government, and university researchers are now examining these findings and exploring other aspects of this controversy.

Industry has challenged some of the conclusions about cell phones causing cancer, especially brain cancer. Using science, they claim it is impossible to prove cell phone use alone caused brain cancer when there are multiple intervening and overlapping variables that could equally have caused it. It is difficult to research because longitudinal studies take time, cell phone technologies are changing rapidly all the time, and consumer use patterns differ. Science and industry, and some courts, do not accept animal studies or computer models as solid proof of damage to a given individual. However, environmental and public health regulatory agencies do accept them as a basis for public health, safety, and welfare regulations.

CHILDREN CAN BE AT HIGHER RISK FOR GENETIC DAMAGE FROM RADIO FREQUENCY RADIATION

Children are among the most vulnerable segments of any society. As cell and handheld communications devices reach more young people the concern is the environmental impact on them, given that they are still developing and that their lifetime exposure may be longer. Research from the United Kingdom has created some controversy. A British report recommended that children under 16 not use cell phones unless necessary. The panel said,

> the developing nervous system is more vulnerable to functional genetic damage both because of the higher rate of cell division and the thinness of the skull, which allows more penetration of radio frequency radiation into the brain. In addition, children may be more vulnerable because they would presumably be using cell phones over a longer period of time than adults, thus potentially increasing the risks associated with exposure over time.

The British report made several recommendations, including:

- Establishment of a system to capture health complaints among cell phone phones users so that reliable data can be gathered and analyzed;

- Adult-onset leukemia should be added to the outcomes potentially related to radio frequency exposure, along with the brain cancer and salivary gland tumors now being studied;
- Specific studies of children for health concerns related to radio frequency emissions;
- Studies of the impact of radio frequency radiation on pregnant women and fetuses.

Other studies have researched effects of pulse-modulated radio frequencies on the movement of calcium ions in cells and tissues of the nervous system. Early studies were refuted by later studies. The American Cancer Society said there was no increase in brain cancer incidence or mortality in the past decade, when the use of cell phones dramatically increased. Critics point out that cell phones are a recent phenomenon and it is extremely difficult to prove that a new technology is absolutely safe. Industry claims that proving cell phones do not cause cancer is too difficult a burden.

SPERM

Because proximity to the cell phone or other handheld communication device increases risk from radio frequency emissions, where it is stored matters. Many consumers and workers keep their cell phones on their belts or in their pockets. A British study finds, "storage of mobile phones close to the testes had a significant negative impact on sperm concentration and the percentage of motile sperm," according to a report published by the British Royal Society. The study was conducted under the auspices of the University of Western Australia and based on samples collected from 52 heterosexual men. The samples were assessed using World Health Organization guidelines. The study found "that those men who carried a cell phone in their hip pocket or on their belt had lower sperm motility and a lower sperm concentration than men who carry a mobile phone elsewhere on the body."

BRAIN DAMAGE

The cautious government response has been seen as overly protective of a dangerous technology by some. The Food and Drug Administration (FDA) public policy posture is that it will review the health effects of cell phones. Consumer advocates say this is typical of the FDA, and is why they proceed to litigation.

There have been many studies on this issue in other countries. Many do not find any scientific causality. Some just observe characteristics of trends in data for any correlation. Some of the studies are longitudinal. Some involve different cell phone technologies. Two Swedish studies reported an increased risk of brain tumors linked to use of cell phones used for more than 10 years. They concluded that those who used cell phones heavily had a 240 percent increased risk of a cancerous tumor. They also found that the tumor grew on the side of the head where they used their phone. Other researchers had also noticed that the type of brain tumors associated with cell phones tended to grow on the side and near the place the phone was used. The researchers at the Swedish National Institute for Working Life had compared data from 2,200 cancer patients and 2,200 healthy patients.

The FDA's response to a set of studies replicating the Swedish studies in the United States is:

> The FDA continues to monitor studies looking at possible health effects resulting from exposure to radio frequency energy. In 1999, FDA signed a Cooperative Research and Development Agreement (CRADA) with the Cellular Telecommunication & Internet Association (CTIA). As called for by this CRADA , FDA plans to convene a meeting in the near future to evaluate all completed, ongoing and planned research looking at health effects associated with the use of wireless communication devices and identify knowledge gaps that may warrant additional research. (www.fda.gov/cdrh/wireless/braincancer040606.html)

To many consumers and self-perceived victims, this response indicates industry control of the FDA. A meeting in the near future to identify research priorities seems to underestimate risk to the public if the research studies already completed are accurate. Nonetheless, it indicates that one battleground in this controversy is whether to research it at all at the government level in the United States.

CELL PHONES AND PROFIT

In 2005, the wireless-device industry generated $113.5 billion in revenue as usage time exceeded 1.4 trillion minutes and the number of cell phone users in the United States surged to 208 million in 2005 from 340,213 in 1985, when wireless networks were first being built. Now, handheld devices hold computers, cameras, microphones, live video feeds, all in addition to a multifunction telephone. There is no evidence of adverse health effects, according to a wireless industry trade association. There is a general controversy about adverse effects of chemical exposures, but this is about exposure to sound waves. This industry has large profits, little regulation, and strong trade organization representation at the congressional level in Washington, D.C.

Wireless phones emit low levels of radio-frequency energy while in use and in stand-by mode. Radio-frequency energy can heat and damage body tissues. Despite industry assertions, there is uncertainty about the effects of frequency, intensity, proximity, and duration of exposure. This uncertainty finds its way into a legal battleground. A federal judge has refused to dismiss a lawsuit that charges manufacturers are making and selling cell phones with the knowledge that they are dangerous. The federal court ruled that because the federal Food and Drug Administration (FDA) has not issued standards regulating cell phones, the states are not preempted from regulating the devices and state courts are not barred from hearing lawsuits about the potential dangers of cell phones. As discussed earlier, the FDA's stance on the research showing potential harmful effects on the brain was to rigorously examine whether it needed more research but do nothing in terms of law or regulation. This lawsuit does not claim specific health problems but argues that cell phone manufacturers have been aware that cell phones emit radiation into users' brains and have done nothing to protect their customers. The suit argues that cell phones should have been sold with headsets as standard equipment to reduce consumers' exposure to potentially harmful and known emissions of radiation. Among other legal remedies requested by the victims, this lawsuit asks that cell phone manufacturers supply headsets to past and future customers and reimburse those who purchased headsets on their own. The cell phone industry argued that Congress had given the FDA the sole authority to oversee cell phone safety, and those courts and state legislatures had thus been preempted. They argued that this is true even if the FDA decides not to exert its authority, delegates it to the states, and engages in no activity except considering research potential. The court ruled that because the FDA has not issued standards, the courts and other jurisdictions are free to set up their own regulations. This greatly concerns the cell tower–based communications industry because the lack of uniformity makes it expensive for them. Each state, and sometimes municipalities within the state, depending on their state constitution and enabling legislation, could have different rules and procedures. The placement of the tall cell towers necessary for many U.S. cell phones in residential neighborhoods is also very controversial. They are considered ugly, may emit dangerous radiation, can potentially decrease property values, and are placed in locations with vulnerable human populations.

The legal battleground for this issue is a very large one. There are several class-action court cases being litigated at high levels, with many appeals. The U.S. Supreme Court has declined to dismiss a number of class-action lawsuits that challenge radiation emissions from cell phones. Class-action lawsuits are large lawsuits representing an entire class of victims. It is in the interest of judicial economy and decisional consistency that class actions are allowed, because otherwise there would be many cases with different decisions. Also, if there is gross manufacturing negligence that is actually harming people, it is more effective for the courts to handle it all at once to reach a more effective remedy. The problem with class action lawsuits is that if you are part of the class and do not opt out of it, your case is decided without you and you are prevented by the legal principle of *stare decisis* from refiling your case. This happened in the case of African American farmers and their lawsuit against the federal government for land loss. Also, class actions are not a regulatory tool per se but a way to protect the public health. With the then-new Chief Justice John Roberts presiding, the court declined to consider an appeal from cell phone manufacturers, who wanted the U.S. Supreme Court to overturn a lower court decision allowing a class action to stand. Class-action lawsuits currently pending in Maryland, New York, and Pennsylvania allege that cell phone radiation caused brain tumors. The suits allege the manufacturers are aware of the danger and have deliberately kept consumers without necessary knowledge to make informed consumer choices. The suits seek to force phone manufacturers to reduce the amount of radiation produced by phones and to warn users of the alleged health hazards. Some of the largest U.S. and international cell phone manufacturers are defendants named in the lawsuits. Respected neurologists are also named plaintiffs in some of these lawsuits. Whatever the course of all this litigation on this issue, the next usual battleground for all stakeholders in issues like these is the legislature.

POTENTIAL FOR FUTURE CONTROVERSY

This controversy is being waged on scientific fronts. Industry scientists say cell phones are safe; the U.S. government sides with industry, allowing phones to proliferate without testing or well-developed standards. Consumer groups here and abroad would rather cell phone usage be proven safe rather than just not deadly. What is the risk? In developing countries and rural areas of others, the mobile phone is replacing land-based wire systems. Cell phone towers are themselves a controversy in the United States, with some arguing they are a "taking" of private property. This controversy has just started to percolate up through the U.S. courts. It will require years of scientific research and legislative and administrative initiatives before the regulations catch up with current technology. Consumers have learned that safe does not always mean risk free. In this case, the jury is out as to what the risks are.

See also Children and Cancer; Ecological Risk Assessment; Human Health Risk Assessment

Web Resources

The Cell Phone Chip Store: Cell Phone Use Associated with Decline in Fertility. Available at http://www.thecellphonechipstore.com/lovelife.htm. Accessed January 20, 2008.
Cell Phone Index: Cancer. Available at www.mercola.com/article/cell_phones/index.htm. Accessed January 20, 2008.
Wave Guide: EMF/RFR Bioeffects and Public Policy. Available at www.wave-guide.org/. Accessed January 20, 2008.

Further Reading: Burgess, Adam. 2003. *Cellular Phones, Public Fears, and a Culture of Precaution.* Cambridge: Cambridge University Press; Kane, Robert C. 2001. *Cellular Telephone Russian Roulette: A Historical and Scientific Perspective.* New York: Vantage Press; Ma, Jian-Gou. 2004. *Third Generation Communication Systems: Future Developments and Advanced Topics.* New York: Springer Press.

CHILDHOOD ASTHMA AND THE ENVIRONMENT

Recent increases in childhood asthma have created controversies about the environmental causes. Children living in urban areas are especially vulnerable to asthma because of the high number of pollutants and allergens in their environment. Others argue that exposure to pesticides in the air and food helps cause it.

Asthma is a disease that affects breathing. It attacks and damages lungs and airways. Asthma is described as breathing through a straw, and it can be serious for any age group. Childhood asthma attracts attention because of its potential developmental consequences. Asthma is characterized by partially blocked airways. It can occur periodically or reactively, and attacks or events can range from mild to severe. The nose, sinuses, and throat can become constricted. Breathing becomes difficult and is accompanied more coughing and wheezing. During an asthma event, the muscles around the breathing passages constrict. The mucus lining of the airways becomes inflamed and swollen. This further constricts air passages. These episodes can last for hours or days. They can be terrifying events for parents and children. Childhood asthma and its disproportionate impact on vulnerable populations is one of the foundational issues of environmental justice in the United States.

CAUSES OF ASTHMA: FOUNDATIONS OF A CONTROVERSY

Asthma is a complex disease with many causes, some known, some contested, and some unknown. Each one presents its own battlegrounds. Environmental causes are controversial because they represent a broad, catchall category. Controversies about science, industry trade secrets, and unequal enforcement of environmental laws merge with a very high level of citizen concern. There is an emerging role for public health experts and advocates in urban environmental policies around childhood asthma. There is a greater incidence of asthma among children in U.S. inner cities. Asthma often accounts for a large number of emergency room visits, especially in poor areas underserved by medical insurance. Hospitals, health care administrators, and health insurance corporations are all very interested in the cause of asthma. Employers and educators know that a number of days in school or on the job are lost because of asthma. They also have an interest in understanding the cause. Some stakeholders may fear liability for causing asthma. They have a strong interest in not being named as the cause.

ENVIRONMENTAL TRIGGERS FOR ASTHMA

The policy question posed now is what triggers an asthma attack. Others, such as public health experts and advocates, ask what prevents it. They are concerned that focus on a trigger overlooks vectors of causality as opposed to last exposure.

INDOOR AIR CONTAMINATION

Dust mites, cockroach droppings, animal dander, and mold can be environmental conditions that cause asthma. Exposure to allergens could induce the onset of asthma itself. Exposure to secondhand tobacco smoke is also a contributor to childhood asthma. Certain insecticides may be triggers for asthma. Some researchers consider pesticides to be a preventable cause of asthma in children. Indoor air may be made worse by the increasing use of synthetic materials in homes in the form of carpets, carpet glues, curtains, and building materials. There is concern that as these materials age, they release potentially dangerous chemicals. This is a small battleground in this controversy. Manufacturers of these items strongly contest any conclusion that their products cause childhood asthma. However, concern about release of toxic synthetic materials has affected market trends in these products. Because many household products could cause asthma, the marketplace is a major battleground. Large big-box retailers like Wal-Mart are accommodating these consumer concerns about asthma causes that consumers can control, such as dust mites and animal dander.

OUTDOOR AIR POLLUTION

There is strong evidence from longitudinal studies that ambient air pollution acts as a trigger for asthma events among persons with it. Truck and automotive exhaust is a big part of the polluted air, especially in dense, urban areas. Combined with industrial and municipal emissions and waste treatment practices (such as incineration) the quality of the air becomes so degraded that the total polluted air load in some urban areas and nearby urban areas is a threat to the health of children. It is a threat that they can get asthma through longtime exposure, and a threat that they could have an attack at any time.

CHILDREN IN THE CITY

It is clear that the increasing severity of asthma in the United States is concentrated in cities among children who live in poor conditions. Children are especially vulnerable to air pollution and other exposure vectors compared to adults. Children have more skin surface relative to total body mass. According to Frederica Perera, director of the Columbia Center for Children's Environmental Health:

> They consume more water, more food, and more air per unit body weight than adults by far. Today's urban children are highly exposed to allergens, such as cockroach and rodent particles, and pollutants, such as diesel exhaust, lead and pesticides. And these elements affect them even before they are born. Preliminary evidence shows that increased risk of asthma may start as early as in the womb before birth.

The small particles of soot "are very easily breathed into your lungs, so they really exacerbate asthma," say Peggy Shepard, executive director of the West Harlem Environmental Action, Inc., adding that she believes these diesel particles also may play a role in cancer. Shepard says New York City is second in the nation when it comes to the amount of toxins released in the air, preceded only by Baltimore, Maryland.

When it comes to cockroach particles, they pose a problem for urban areas nationwide, says David Evans, who runs the Columbia Center's "Healthy Home, Healthy Child" intervention campaign. He says, "simple house cleaning won't solve the problem, because the cockroach residue tends to be present in many city neighborhoods." According to the Harlem Lung Center, childhood asthma rates have increased 78 percent between 1980 and 1993. And according to the Columbia Center, there are an estimated 8,400 new cases of childhood cancer each year nationwide.

DISPARITIES IN ASTHMA CARE

Access to health care is an important aspect of the asthma controversy. Many low-income groups do not have health insurance and tend to use the emergency room when necessary. An asthma attack often presents that necessity. Language and cultural differences can make a tense medical situation worse. Even with regular medical intervention differences in asthma treatment by race, gender, and class create a battleground out of this issue. It is not contested that disparities in the burden and treatment of African Americans and Puerto Ricans with asthma are well documented.

Among African Americans and Puerto Ricans the rates of asthma, hospitalizations, and deaths are higher when compared to whites. This is especially true among children. Different medicines are prescribed and used. Research shows that the use of long-term medications to control asthma is lower in African Americans and Puerto Ricans. Cost may be a factor, especially if there is no insurance coverage. Access to medical care is affected by many factors. There are shortages of primary care physicians in minority communities and issues of trust about the role and usefulness of medications.

COSTS OF ASTHMA

Asthma is a cause of death among U.S. children. There are 247 deaths each year due to childhood asthma. It is the leading cause of hospital admission for urban children. Asthma is also the leading cause of days of school missed. It is estimated that about 30 percent of acute episodes of childhood asthma are environmentally related.

Air pollution is considered a major cause of asthma, and asthma and public health are major regulatory justifications for clean air laws. The U.S. Environmental Protection Agency (EPA) has estimated the cost savings that resulted from the Clean Air Act. For the years 1970–1990, the EPA calculated that the annual monetary benefits of reductions in chronic bronchitis and other respiratory conditions was $3.5 billion. That is, these are health care costs that would have been incurred if there were no clean air regulations. There are other costs of course. Also, if there were no costs and if asthmatics could get free and accessible medical attention, the cost of human resources necessary to handle the scope of the problem could be large. Additional childhood asthma benefits are projected by the EPA to accrue over the years 1990 to 2010, assuming full implementation of the Clean Air Act Amendments of 1990.

POTENTIAL FOR FUTURE CONTROVERSY

This controversy is very salient among communities and public health professionals. Schools, hospitals, nursing homes, and other places where vulnerable people live hold strong views but lack resources. Emissions from traffic, industry, and heating and cooling systems are now part of the U.S. urban landscape. Environmentalists note that the law does not cover all the pollutants and is not enforced equally. Environmental justice advocates consider childhood asthma as proof of at least one disproportionate environmental impact. Asthma generally has resulted in a substantial increase in sales and profits of pharmaceutical companies. This controversy is structural in that it pits public health concerns against industrial emissions, and is therefore deep.

There will be many battlegrounds. The environmental controversies around childhood asthma will focus on air pollution and use other controversial methods such as ecosystem risk assessment or cumulative risk assessment. Childhood

FACTS ABOUT ASTHMA

Childhood asthma is an environmental controversy with much saliency in metropolitan areas. The following summary is from the Center of Children's Health and the Environment at the Mount Sinai School of Medicine:

- In 1993–1994, approximately four million children aged 0–14 reported asthma in the preceding 12 months.
- Self-reported prevalence rates for asthma among children ages 0–4 increased by 160 percent from 1980 to 1994; rates among children ages 5–14 increased by 74 percent.
- Asthma was selected as the underlying cause of death among 170 children aged 0–14 in 1993.
- Among children ages 5–14, the asthma death rate nearly doubled from 1980 to 1993.
- Over 160,000 children aged 0–14 are hospitalized for asthma annually.
- Among all age groups, children aged 0–4 had the highest hospitalization rate in 1993–1994 (49.7 hospitalizations per 10,000 persons).
- The cost of illness related to asthma in 1990 was estimated to be $6.2 billion, according to a 1992 study. Of that amount, inpatient hospital services were the largest medical expenditure for this condition, approaching $1.6 billion.

Many more suffer asthma without medical care. There may be rural, institutional, and pockets of urban populations that lack public health resources. Many citizens do not have health care coverage for asthma.

asthma is a big part of the new inclusion of cities by the EPA. In the early 1990s the visionary EPA administrator Carol Browner reduced the level of particulate matter allowed in urban air districts, effectively banning many diesel and leaded gas vehicles. She started an urban air toxics policy task force to help engage cities and the EPA, along with several other successful policy initiatives. Exxon and other oil companies responded with letters to their accounts about the new air pollution regulations. In an unusual step, the American Lung Association and other public health organizations responded in support of the EPA. As U.S. environmental policy matures into including all environments, including cities, a greater battleground in the areas of public health can be expected. Battle lines were drawn then and are much deeper now. Asthma is worse, there is greater consensus that air pollution not only triggers but also causes asthma, there are large documented environmental injustices by race and class, and there is a large overall push for sustainability.

See also Children and Cancer; Cumulative Emissions, Impacts, and Risks; Environmental Justice; Sustainability

Web Resources

American Family Physician. Management of Asthma in Children. Available at www.aafp.org/afp/20010401/1341.html. Accessed January 20, 2008.

American Lung Association: Childhood Asthma Overview. Available at www.lungusa.org/site/pp.asp?c=dvLUK9O0E&b=22782. Accessed January 20, 2008.

Beyond Pesticides. Asthma, Children, and Pesticides. Available at www.beyondpesticides.org/children/asthma/index.htm. Accessed January 20, 2008.

Further Reading: Bernstein, I. Leonard. 1999. *Asthma in the Workplace.* New York: Marcel Dekker; Cherni, Judith A. 2002. *Economic Growth versus the Environment: The Politics of Wealth, Health, and Air Pollution.* London: Palgrave Macmillan; Christie, Margaret J., and Davina French. 1994. *Assessment of Quality of Life in Childhood Asthma.* London: Taylor and Francis; Institute of Medicine. 2000. *Clearing the Air: Asthma and Indoor Air Exposures.* Washington, DC: National Academies Press; Naspitz, Charles K. 2001. *Pediatric Asthma: An International Perspective.* London: Taylor and Francis.

CHILDREN AND CANCER

A core controversy is whether environmental stressors, such as pollution and pesticides, are responsible for the increase of childhood cancers.

BACKGROUND

The cause of cancer is always a controversial topic. The rate of cancer among U.S. children has been rising since the 1970s. The mortality rate, however, has decreased since the 1980s. There are scientifically established causes of childhood cancer. Family history of cancer, radiation exposure, genetic abnormalities, and some chemicals used to treat cancer are known causes of childhood cancer. The plethora of new chemicals in food, air, water, clothing, carpets, and the soil is strongly suspected as being part of the cause of cancer in children. The scientific model of causality struggles with proof of the cause of childhood cancer, and engages fierce environmental controversy in the process. Some aspects of this controversy have moved into the courtroom. There science struggles both with causality by a certain chemical, and liability of a specific person (the defendant).

CHILDHOOD CANCER FACTS

According to the National Cancer Institute (http://www.cancer.gov/cancertopics/types/childhoodcancers), the following set of statistics measures the expanding parameters of childhood cancer.

A newborn child faces a risk of about 1 in 600 of getting cancer by 10 years of age. The rate of increase has amounted to almost 1 percent a year. From 1975 to 1995 the incidence of cancer increased from 130 to 150 cases per million children. During this time mortality due to cancer decreased from 50 to 30 deaths per million children. In the United States, cancer is diagnosed each year in about 8,000 children below age 15. Cancer is the most common form of fatal childhood disease. About 10 percent of all deaths in childhood are from cancer. There are big differences between types of cancer, and researchers investigate these differences because it may lead them to the environmental stressors. Leukemia was the major cancer in children from 1973 to 1996. About one-quarter of all childhood cancer cases were leukemia. Brain cancer, or glioma, increased nearly 40 percent from 1973 to 1994. The overall rate of central nervous system tumors increased from about 23 per million in 1973 to 29 per million children in 1996. These two forms of cancer account for most of the disease in children. Lymphomas are the third most diagnosed category of childhood cancer. Lymphomas are diagnosed in about 16 percent of cases. There are different kinds of lymphomas; for some categories childhood incidence rates have decreased and for others they have increased. (Non-Hodgkins lymphomas increased from 8.9 per million children in 1973 to 11 per million in 1996, for example.)

According the U.S. Environmental Protection Agency, Office of Children's Health protection (http://Yosemite.epa.gov/ochpweb.nsf/content/childhood_cancer.htm) there are substantial differences by age and type of cancer.

> Rates are highest among infants, decline until age 9, and then rise again with increasing age. Between 1986 and 1995, children under the age of 5 and those aged 15–19 experienced the highest incident rates of cancer at approximately 200 cases per million children. Children aged 5–9 and 10–14 had lower incidence rates at approximately 110 and 120 cases per million children.

The U.S. EPA also reports some ethnic differences in childhood cancer rates:

> Between 1992 and 1996, incidence rates of cancer were highest among whites at 160 per million. Hispanics were next highest at 150 per million. Asian and Pacific Islanders had an incidence rate of 140 per million. Black children had a rate of 120 per million, and Native Americans and Alaska Natives had the lowest at 80 per million.

Also, different types of cancer affect children at different ages. According to the U.S. EPA:

> Neuroblastomas, Wilm's tumors (tumors of the kidney) and retinoblastoma (tumors in the eyes) usually are found in very young children. Leukemias and nervous system cancers are most common through age 14; lymphomas, carcinomas, and germ cell and other gonadal tumors are more common in those 15–19 years old.

SCIENTIFIC MODEL: STRUGGLING TO KEEP UP WITH POLICY

The last century saw a drastic lowering of infectious disease rates due to strong public health measures and education. In the United States and other industrialized nations this has been accompanied by a general rise in systemic, whole-body or immune system breakdowns. Cancer is considered a possible result of a whole-body immune system breakdown. About 100,000 chemicals are released into the environment. Less than 2 percent of them are tested for public health impacts. The tests are done in constrained laboratory conditions, generally considering a given chemical safe if less than one-half or one-quarter of the mice exposed to it die. The scientific model requires the isolation of an extraneous possible cause or intervening variables. It ignores cumulative, synergistic, and antagonistic real-world chemical interactions that are the exposure vectors of chemicals for children. The actual biological vulnerability of the affected humans is not taken into account. A developing fetus is much more vulnerable to harm by cancer-causing chemicals. It takes a newborn child at least one year to develop an efficient blood-brain barrier. The blood-brain barrier works to protect the brain while the central nervous system develops. Before it begins fully functioning an infant could be exposed to whatever the mother is exposed to. There is research that indicates that children of people who work with dangerous chemicals have an increased frequency of childhood cancer. The problem of childhood cancer is a driving force behind many other environmental controversies. The real-world number of cancer cases in industrialized nations has increased overall, although it depends on demographics and type of cancer.

Environmental scientists, from government, industry, and environmental groups, have been laboring for many years to unravel some of the exposure vectors to children with cancer, and sometimes endocrine disruption. Many chemicals are much more dangerous when mixed with other chemicals. Children are especially vulnerable to many of the chemicals used around the house, such as cleaners and pesticides. Research found over twice the risk of brain cancer for children exposed to household insecticide. Some studies found even higher rates of risk. These early studies focus on just one type of cancer from a few known exposure vectors. The cumulative and synergistic emissions of the past are becoming the cancer risks of the present.

COSTS ARE VERY HIGH: WHO PAYS?

Health care in the United States is another controversy altogether. Access is difficult, and cancer treatments are very expensive. The annual overall incidence of cancer is 133.3 per million for children under 15 years old in the United States. There were 57.9 million children under 15 years of age in the United States in 1997. About 7,722 cases of childhood cancer are anticipated each year, which is very close to the 8,000 reported. Experts have estimated the cancer related costs for children to be about $4.8 billion. There are other costs. Psychological stress, transportation, time with medical staff and insurers, and time as a health care provider are all also costs.

The cost of treatment of childhood cancer is controversial in that it is generally too much for an average family to afford. This plays into other controversies about the health care system. If the family cannot afford it, or if their insurance company requires it, they file a lawsuit against the most likely cause of the cancer. The litigation hurdles of proof and the burden of proof are often insurmountable obstacles.

WHAT ENVIRONMENTAL STRESSORS CAUSE CHILDHOOD CANCERS?

The following brain cancer figures, from the American Cancer Society, show a disturbing trend in the number of cases being found:

1940: 1.95 per 100,000 population
1945: 2.25 per 100,000
1950: 2.90 per 100,000

1955: 3.40 per 100,000
1960: 3.70 per 100,000
1965: 3.85 per 100,000
1970: 4.10 per 100,000
1975: 4.25 per 100,000

These figures show a steady increase for all industrialized nations. To many public policy makers the cancer rates in these countries implicate chemicals used there. Similar increases are occurring in children. Many chemical manufacturing industries would contest this association, stating that in most cases the scientific evidence neither proves nor disproves causality.

CHEMICALS AND CHILDHOOD CANCERS

As discussed previously, a major form of childhood cancer is brain cancer. Which chemicals have been linked to brain cancers? Chemical workers are often the most exposed to a particular chemical. They make it, store it, and transport it. Sometimes they also use it. Epidemiologists follow the exposure vector to workers of various suspected chemicals. Brain cancer risks follow workers exposed to chemicals used in vinyl and rubber production, oil refineries, and chemical manufacturing plants. Another study by the National Cancer Institute of 3,827 Florida pest control operators found they had approximately twice the normal rate of brain cancer. Pesticide exposure increases risks for childhood cancer. Because adult workers had higher rates of brain cancer when exposed to these chemicals in their occupations, researchers surmise that because children are more vulnerable they may get more brain cancer when exposed to these chemicals.

Some chemicals used in pesticides concern public health officials more than others. Chlordane is one of high concern. Research on children who developed brain cancer after their homes were treated with pesticides led them to this chemical. The battleground for this chemical has moved to litigation in these cases. Chlordane is a high-risk chemical for brain cancer. It is a fat-soluble compound. Fat-soluble compounds are absorbed into the nervous system, which is developing rapidly in children from birth to age 5.

Legal chlordane use was stopped in the United States in April 1988. However, the law was and is poorly enforced. One reason it was made illegal was its long-term killing power, which also made it an effective pesticide. The degree to which a chemical persists in the environment is one measure of how dangerous it could be to the environment and to humans. Chlordane is such a persistent chemical that it is still being detected today. Tests of more than 1,000 homes performed by federal agencies found that approximately 75 percent of all homes built before 1988 show air contamination with chlordane. They also found that 6 to 7 percent are suspected of being over the maximum safe levels for chlordane exposure in residential housing set by the National Academy of Sciences, a limit that some have argued is too low.

Research into this controversial area has increased. Authors Julia Green Brady, Ann Aschengrau, Wendy McKelvey, Ruthann A. Rudel, Christopher H. Schwartz, and Theresa Kennedy from the Boston University School of Public

Health in Massachusetts published, "Family Pesticide Use Suspected of Causing Child Cancers, I" (*Archives of Environmental Contamination Toxicology* 24, no. 1 [1993]: 87–92). In this peer-reviewed article the relationship between family pesticide use and childhood brain cancer was closely examined. The researchers compared brain cancer rates for families using pesticides and those not using pesticides. They concluded that the chemicals did increase the risk of getting cancer. Significant positive associations with brain cancer rates were observed in families using regular household supplies and pest control chemicals. Bug sprays for different kinds of insects, pesticide bombs, hanging no-pest strips, some shampoos, flea collars on dogs and cats, diazinon in the garden or orchard, and herbicides to control weeds in the yard were all found by the authors to be part of the chemical vector increasing risk of brain cancer. These results are still being disputed. Some argue that the sample sizes are very small in some of these studies and the results may not be typical. Unanswered questions fueling the uncertainty that underlies this controversy concern the total range of effects of chemicals. What happens when they combine in water or sunlight over time? Are there possible generation-skipping effects? What happens to the typical child when exposed to these chemicals in their normal environment? What constitutes their regular environment? Does air pollution pose another cancer-causing vector for children? The evidence is fairly conclusive now that secondary tobacco smoke can cause health risks. Originally, tobacco smoking and chewing were considered good for your health. The danger they posed was a conclusion resisted tenaciously by the tobacco industry. Secondary smoke was highly controversial and remains contested when local land-use ordinances forbid tobacco products.

POTENTIAL FOR FUTURE CONTROVERSY

Childhood cancer is a traumatic event for all involved. The costs are very high. Right now it is difficult to overcome scientifically based burdens of proof in litigation. Families with children with cancer often seek legislative recourse to the incidents they believe caused the cancer. Children as growing beings naturally absorb more from the environment than adults. The increase in most childhood cancer rates is a cause for alarm for environmentalists and public health officials. Industry tries to cap environmental liabilities through legislation and internal agency advocacy. This all means that this controversy will intensify as more chemicals are linked with childhood cancers.

See also Cancer from Electromagnetic Radiation; Cumulative Emissions, Impacts, and Risks; Human Health Risk Assessment; Pesticides

Web Resources

National Cancer Institute. National Cancer Institute: Research on Childhood Cancers. Available at www.cancer.gov/cancertopics/factsheet/Sites-Types/childhood. Accessed January 20, 2008.

National Institutes of Environmental Health Sciences: Childhood Cancer Research. Available at http://www.niehs.nih.gov/health/topics/agents. www.niehs.nih.gov/external/resinits/ri-22.htm. Accessed January 20, 2008.

Natural Resources Defense Council. Children, Cancer, and the Environment. Available at www.nrdc.org/health/kids/kidscancer/kidscancer1.asp. Accessed January 20, 2008.

Pressinger, Richard W., and Wayne Sinclair. Environmental Causes of Cancers. Available at www.chem-tox.com/cancerchildren/. Accessed January 20, 2008.

Further Reading: Hayman, Laura L., ed. 2002. *Chronic Illness in Children: An Evidence-Based Approach.* New York: Springer Publishing; Pangman, Julie Klaas. 1994. *Guide to Environmental Issues.* Darby, PA: Diane Publishing; U.S. Environmental Protection Agency. 1998. *The EPA Children's Environmental Health Yearbook.* Darby, PA: Diane Publishing.

CITIZEN MONITORING OF ENVIRONMENTAL DECISIONS

In many communities citizens are monitoring the chemicals in the land, air, and water for pollutants, aided by the expansion of right-to-know laws. Currently, most citizen environmental complaints about pollution must go through the state and/or federal environmental agency that granted the permits to the industrial and municipal emitters in that region. Industry environmental compliance is self-reported by the industrial entity to the agency, not the local community. Controversies about lack of notice and involvement of the citizens, the actual measurement of the emissions, and enforcement flare up around these dynamics and push many citizens into organized monitoring. Many communities that suffer from cumulative and ongoing emissions may begin monitoring if cancer, miscarriage, asthma, or other public health degradation indicators show cause for concern.

Citizens and other residents monitor the environment every day. Farmers monitor weather to grow crops efficiently. Emergency response personnel monitor weather for natural disasters. Citizens are now monitoring the conditions of the land, air, and water around them, with and without the assistance of state or federal government agencies. Citizens are often in the best location to gather environmental knowledge and have the most at stake. Environmentalists strongly encourage citizen monitoring. Recent increases in citizen access to information about the Toxics Release Inventory encourage citizens to ask questions. Concern about cancer-causing chemicals lurks in the public's mind despite scientific assertions otherwise. Many science and health courses in elementary and secondary education now teach important monitoring skills. Citizen monitoring is going to continue, but the controversies will erupt around whether they should count in government regulation of development and industry in that area. The battleground can begin in both the land-use and industrial permitting arenas. From here they both can end up in court. Whether citizen monitoring counts by law or not, it is a strongly salient political force in all U.S. environmental policy. Community environmental controversies can fester long after judicial intervention.

WHAT DO PEOPLE DO WHEN THEY MONITOR
THE ENVIRONMENT?

Monitoring the environment is both a recreational and a scientific pursuit. People tend to monitor the environment as it relates to their own safety and that of their loved ones. Agribusiness monitors the environment for profit. Mothers with asthmatic children may monitor the air more closely for pollution. Kayakers and canoeists may monitor water discharges more closely, especially in controlled water flows. Citizen monitoring often springs from a concern about the environment without knowledge of the source of the pollution.

One function of citizen monitoring of the environment is to expand what the regulating state and federal environmental agencies know. The coverage of environmental regulation in the United States is thin and relatively new. From the perspective of direct enforcement of environmental law, citizens supply information necessary to help both government and industry ensure compliance with the law. This would also help alleviate problems with unequal enforcement of environmental law that have occurred as a result of whole communities being left out of the scope of environmental policy and law.

Many community-monitoring programs begin with dissatisfaction with local, state, or federal agency response to an environmental issue. They often will search the telephone book and Internet for industries near the site. Illegal waste sites are a large concern for many communities. They want more enforcement from environmental agencies and do not get it. They want the site cleaned up and put back into productive use and do not get it. Citizen monitoring can take the form of an inventory or a survey of the offending activities. They often seek the name of the owner of the illegal waste or property. Communities will often take photographs. Aerial photographs of suspicious sites are useful and now available on the Internet. Communities are powerful stakeholders in terms of information. They will interview past and present employees of industries suspected of pollution. Communities will take down license plates and note warnings and signage on the truck or delivery vehicle. At landfills this can indicate how much waste, the kind of waste, and origin of the waste. Communities resisting a landfill expansion, modification, or renewal, or the siting of a new one, find this information very helpful. Polluting industries are thoroughly examined by information-hungry communities. Industry tends to resist this. This struggle for accurate information affects other controversies like environmental impact statements and environmental audit privileges for industry. When researching a particular industry, information tools used by citizen monitoring groups include the Toxics Release Inventory, county deeds and records for the property owner, the corporation's articles of incorporation filed with the secretary of state in the corporation's home state, a Dunn and Bradstreet report to find out subsidiaries and any otherwise undisclosed financial relationships, environmental records in other states, and, for publicly traded corporations and, some utilities, their 10(K) annual report filed with the Securities and Exchange Commission in Washington, D.C.

Another way citizen monitoring can expand agency knowledge is by helping with sampling. Sampling programs are expensive. Sampling requires sample selection, collection, transportation, storage, analysis, and data interpretation. Most focus on sample collection. Citizens whose homes are next to a lake or stream or within its watershed are available to do routine sampling of water volume and quality. Citizens can and want to do extra monitoring during periods of concern. Storm event information for floods, hurricanes, tornadoes, desert flash floods, and other natural disasters can be easily missed by regulatory environmental agencies.

Citizen monitoring is a personal experience, whether in a controversial or collaborative policy, program, or project. Many citizens have a view of nature that was shaped by happy memories of childhood, of camping, and of recreating in nature. Most have childhood memories of a certain place in nature. This same type of personal interest develops in monitoring. Natural places are familiar, interesting, complex, and integral parts of the ecosystem.

CONCERNS ABOUT CITIZEN MONITORING

Ever since members of the Audubon Society first did annual bird counts at the turn of the 1900s, citizens have monitored the environment. Farmers, ranchers, indigenous people, pioneers, and others monitored the environment. As environmental controversies and land-use conflicts have increased so has the involvement of communities. With this growth in power, mainly through information from the Toxics Release Inventory and through organization, came more battlegrounds and controversy.

DO STATE AND FEDERAL ENVIRONMENTAL AGENCIES NOT DO ALL THE NECESSARY MONITORING?

The U.S. government does not have a complete monitoring system for the environment. The parts that are monitored, in part based on self-reported industry data, are generally not done by personnel who work with the public. Citizens and citizen monitoring are part of the agency's engagement with the public.

Agencies do monitor the chemicals within their statutory purview, and those that threaten public health if and when science requires them to do so. There are many chemicals in the United States that are not monitored because of a lack of legislation and science. Citizen monitoring moves to fill this void. There are some recent changes in the dynamics of this controversy.

THE EPA WILL NOW MONITOR UNREGULATED CHEMICALS IN WATER

A large issue has been whether the EPA should monitor water for chemicals that are not regulated. Because not all chemicals are regulated, and there is much scientific dispute on a chemical-specific basis about whether enough or too little of the chemical is covered. This battleground is tied to another controversy, the development of total maximum daily loads per chemical in water. This policy of ignoring unregulated water pollution has changed. Approximately 4,000 public

VOLUNTARY BIOLOGICAL MONITORING (VBM) IN THE UNITED KINGDOM

Voluntary biological monitoring activities involve volunteers, from specialist amateurs to typical members of the public, in collecting data about the spatial or temporal distribution of species and habitats. In the United States, the Audubon Society's bird counts have proven invaluable in developing environmental policy to protect their populations. In the United Kingdom, there are few, if any, species that are not counted. Population counts of indicator species, like many birds and fish, are an essential component of an accurate ecological risk assessment. These risk assessments form the basis for developing sustainable approaches to fishing, logging, and farming that do not degrade the natural environment or threaten public health.

Britain probably has the best-documented flora and fauna anywhere in the world. The rich tradition of natural history that flourished in Europe from the eighteenth century and the learned societies whose aim was to enhance knowledge of their adopted group of organisms evolved into a much more structured and policy-generated demand for data. The current VBM landscape in the United Kingdom consists of (in approximate order of historical appearance):

- recording schemes for different biological groups (such as plants, mosses, or dragonflies), focusing on data collection, usually with acknowledged experts appointed as vice county recorders;
- specific projects to produce atlases, such as the *New Atlas of the British Flora*, based on more than nine million records collected by 1,600 volunteers;
- wildlife trusts and other nature reserves and protected areas, which need data in order to plan and monitor management of such areas;
- the more environmentally oriented political NGOs that seek to involve sectors of the public in lobbying for greater security for nature. National public surveys are organized with high media publicity, for example, the Garden Birdwatch Scheme, or PlantLife's annual single-species survey (bluebells in 2003, poppies in 2004);
- the UK Biodiversity Action Plan, and its associated Local Biodiversity Action Plans, linked in various ways with the previous, and focusing on data required to monitor their individual species and habitat action plans.

water systems will monitor drinking water for up to 25 unregulated chemicals to inform the EPA about the frequency and levels at which these contaminants are found in drinking water systems in the United States. This is called the unregulated contaminant monitoring rule. According to the EPA, the rule is needed because it is through continuous monitoring and research that the EPA collects the information needed to make effective policy decisions.

The EPA currently regulates more than 90 contaminants. This does not mean it prevents their discharge into water, although it can. One criticism of the EPA is that they allow too much pollution from their regulated entities. The Safe Drinking Water Act requires the EPA to identify up to 30 contaminants for monitoring every five years. The first cycle was published in 1999 and covered 25 chemicals and one microorganism. The new rule requires systems to monitor for contaminants that are not regulated under existing law. It is a necessary challenge for the maturation of water quality regulation but will require massive amounts of sampling. When communities hear that the cost of sampling is preventing action on any environmental issue, they are often eager to volunteer. This does cause concern for some.

The EPA will monitor the new contaminants in the water through a process that includes a review of EPA's Contaminant Candidate List, which contains high-priority contaminants that are researched to make decisions about whether regulations are needed. The contaminants on the list are known or anticipated to occur in public water systems. However, they are unregulated by existing national drinking water regulations. Many communities have chemicals of concern. Additional contaminants of concern are based on current research about occurrence and various health risk factors. This is often a place where community involvement and participation can make a difference in prioritizing agency budgets.

Costs for five years will total about $44.3 million. The EPA will conduct and pay for the monitoring for those water systems serving 10,000 people or fewer at a cost of $9 million. Citizen monitoring of the environment will greatly assist the development of this program and the scope of coverage of U.S. water quality law.

IS THE CITIZEN DATA GOOD?

Citizens can come into a project with different competencies. There is concern that overzealous activists or industrial sabotage may unfairly interfere with business. Another concern is whether the citizen group and agency follow the same biologically based protocols regarding sampling. Citizen groups who sample would like to see some results from their efforts. When the policy does not change, citizens feel that their data collection is simply a keep-them-busy type of cooption. If the data collected by citizens cannot meet the same protocols in collection, sample selection, and other areas, then it tends to be ignored. Minimally, it is not comparable to agency data. Data quality is a contested area. Data can become information from which policy is constructed or not. Data integrity is increased by tested and verifiable protocols. These are the types of sampling protocols agencies claim to use. There are many procedures for collecting and analyzing samples. The procedure used determines the quality of the data collected. If less rigorous sampling procedures are used, the resultant data are of lesser quality. Agencies claim it is not comparable to their data set derived after protocol was established. Protocols can vary greatly and include aspects unrelated to the monitoring of environmental conditions. Monitoring and sampling procedures need to be very clear before any collaborative efforts with environmental agencies begin. Citizen groups interested in monitoring need to ask themselves:

- What is the purpose of your study?
- What do you want to know?
- What do you hope to accomplish?
- What would you like the final outcome to be?

- How will the data be used by you or your group?
- Will you want to present the data to an agency or decision makers?
- Who will interpret the data?
- What are you expecting from agencies or other organizations?
- What are you expecting from local politicians and decision makers?

It is exciting to collect data, but sampling control must be strictly implemented to ensure the credibility of the information.

Roles of citizen monitoring include:

- Education should serve as a significant goal.
- Guide legislative funding of resource protection and policy.
- Keep current on new protocols, protocol adjustments, issues, and action options.
- Increase the resolution of our information and knowledge.
- Citizens can serve as knowledgeable local eyes and ears for natural resources.
- The network should be a trusted source of quality information that is:

  - Usable for planning and decision making (at least higher-tier information)
  - Accomplished through standards certification and guides
  - Volunteer responsibility and QA/AC and volunteer follow-up

- Increase awareness of habitat restoration and land-use planning.
- Should serve as the basis for a state-of-the-environment Report.
- Serve as an early warning system.
- Help pull together historic data and make it usable, for example, by digitizing it.
- Citizen involvement is crucial and influential in the legislative process.

There are ample opportunities for citizens to get involved.

- Citizen lake monitoring provides quality testing information.
- Citizen monitoring decreases costs and increases the efficiency of the agency.
- Water quality is a key component in land-use planning.
- There is so much diversity that help is needed to cover all bases.
- Determine who the most credible interpreter will be based on the issue.
- Advocacy for use of data, for example, water quantity legislation bringing together shared interests.
- Role of monitors will be to collect, disseminate/utilize/evaluate data, educate, and protect.
- With partners, establish goals; set standards; provide training; gather, interpret, and disseminate information; and advise legislators and decision makers to provide education and protect and restore resources.

Ideas that have worked in other states or programs:

- Establish quality assurance criteria for each tier.
- Have program-specific training support.

- A service provider network, training, certifying training.
- Individual attention to groups that want to move up a tier.
- Tiered data entry system.
- Engage data users in program development.
- Bring friends and foes to the table.
- Data can be used for enforcement and regulation.
- Gear the program to meet abilities and needs of target constituencies.
- Quality, not quantity, in numbers of volunteers.
- Establish a concise model of the program's relationship with its constituency.
- Pilot programs.
- Make science the basis for what you do.
- Communicate frequently with volunteers, for example, through a Web page.
- Web-based data management system (input access and retrieval by anyone).
- Different parameters used for different systems.

COMMUNITY-BASED CONSERVATION MOVEMENT

As many as 2,200 lake-monitoring volunteers advocate for water quality in the United States. They coordinate citizen monitoring data with satellite imagery. They also coordinate citizen monitor training with colleges and tech schools. Agencies use volunteer monitoring to bridge data gaps (e.g., data in basin reports). The greatest water quality improvement can come from data that can be used for regulation and enforcement. Volunteers can get involved beyond data collection, for example, lobbying or advocacy to make something happen. Citizen monitors are often the citizens involved in other important environmental and community issues. As such, they become part of controversial environmental issues when decisions are made that affect them.

DEVELOP STATEWIDE CITIZEN MONITORING: ANOTHER BATTLEGROUND

State environmental agencies are under increasing pressure to know the physical environment they regulate, especially in the area of water quality. Credible information from any source may help complete an ecological picture of a place. This can enlarge the cast of stakeholders to include hunters, trappers, and anglers. They get access to data that is being collected by biology teachers and university instructors in repeated class projects. Data submitted by industrial and municipal permit applicants could be combined with citizen monitoring data. There is a strong basis for a collaborative approach in this controversy. Common data collection goals, common data quality standards, and an expanded range of low-cost environmental information all decrease the uncertainty characteristic of large and entrenched controversies. Lower tier (data quality) data can be used for a lot of management decisions. Last, citizen monitors could gather regulatory-quality data when the polluter pays for cleanup.

POTENTIAL FOR FUTURE CONTROVERSY

Monitoring alone is an innocuous act of observation. As these observations become information and knowledge, citizens become empowered to challenge agency regulation of industry as too weak and lacking enforcement. Environmentalists become empowered to file environmental lawsuits because they have better evidence. Industry, however, now seeks to hide information they collect about themselves in privileged audits. Citizens and environmentalists continue to challenge the reliance on self-reporting by industry of their own environmental compliance. The need for information drives citizen monitoring in environmental policy, and distrust of government and industry fuels this environmental controversy.

See also Collaboration in Environmental Decision Making; Cumulative Emissions, Impacts, and Risks; Environmental Impact Statements: United States; Public Participation/Involvement in Environmental Decisions; Total Maximum Daily Loads (TMDL) of Chemicals in Water; Toxics Release Inventory

Web Resources

A Citizen's Guide to Understanding and Monitoring Lakes and Streams. Available at www.ecy.wa.gov/programs/wq/plants/management/joysmanual/1advantages.html. Accessed January 20, 2008.

Ecologia. Citizen's Environmental Monitoring Network. Available at www.ecologia.org/about/programs/monitoring.html. Accessed January 20, 2008.

Further Reading: Borbone, Stephen A. 2000. *Seagrasses: Monitoring, Ecology, Physiology and Management.* Boca Raton, FL: CRC Press; Busch, David E., and Joel C. Trexler. 2003. *Monitoring Ecosystems: Interdisciplinary Approaches for Evaluating Ecoregional Initiatives.* Washington, DC: Island Press; Gibbs, James P., Malcolm L. Hunter, and Eleanor J. Sterling. 1998. *Problems in Conservation Biology and Wildlife Management: Exercises for Class, Field, and Laboratory.* Malden, MA: Blackwell Science; Spellerberg, Ian F. 2005. *Monitoring Ecological Change.* Cambridge: Cambridge University Press; Strangeways, Ian C. 2003. *Measuring the Natural Environment.* Cambridge: Cambridge University Press; Wiersma, Bruce. 2004. *Environmental Monitoring.* Boca Raton, FL: CRC Press.

CLIMATE CHANGE

The controversy that initially accompanied global warming has spilled over to climate change. The consequences of climate change are uncertain in terms of specific impacts for many stakeholders. This uncertainty heightens are everyone's concern. The general impacts, such as rising ocean levels, are known. Specific weather changes in regions are not yet known and could occur abruptly. Many scientific, legal, and international battlegrounds are emerging in this controversy.

WHAT IS CLIMATE?

Climate is the total of all weather conditions over time. Whether there are seasons such as winter and summer, and when they start and end, are also parts of the climate. Amount of rainfall, hours of sun, prevailing wind patterns, and temperature are all parts of the climate. The global warming controversy focuses

CAN TECHNOLOGY SOLVE ABRUPT CLIMATE CHANGE?

One part of an environmental controversy is people's faith in technology to solve environmental problems. Many environmentalists point out that the application of technology to industry created some of the pollution problems in the first place. Technology has increased the scale and type of human environmental, which may now affect the globe. Engineers' claim that is precisely the case for technological intervention, because it is possible for humans to affect the climate. While human climatic impacts are accidental up to now, it could be possible to purposefully intervene into climatic processes. Technology could be used to engineer the climate and slow down the rate of global warming and the abruptness of climate change. This is a type of engineering called geoengineering.

Scientists have come up with several ideas about how to do it, all of them untested and controversial. One idea is to put shields on the outside edge of Earth's atmosphere to block the sun. These shields would be very thin and could be controlled. Reflective balloons, reflective space dust, iron dust (which absorbs carbon dioxide), reflective roofs and roads, and reforestation are also technological fixes being discussed. At one point Australian engineers floated the idea of building a big pipe to pump ozone into the hole in the ozone that had developed above them. Another idea is to increase the reflective cloud cover by having oceangoing vessels spray saltwater mists into the air to form strong reflective ocean clouds, increasing the albedo. Many of these ideas are present and future battlegrounds in the controversy of climate change. Many of the engineering concepts suffer from cost issues, possible unintended impacts like pollution and acid rain, and lack of information and computational power. The response to these concerns is that cost is a reflection of priority and as climatic changes affect more of the earth it will become more and more of a priority. More research, monitoring, and analyses can diminish questions about lack of information and unintended impacts. Computational power is developing very rapidly. In 1993 the most powerful computer could process 597 billion floating point operations per second (597 gigaflops). In 2006 the most powerful computer processed the same at 280 teraflops. In 2012 supercomputers should reach 10,000 teraflops. Computers with this power will help other challenging issues like research, monitoring, analysis, and prediction of climate change.

All of these ideas have potential benefits and burdens, and a mistake could have climatic implications. Controlling the climate could mitigate some of the weather extremes predicted by scientists. Crop productivity could increase with predictable weather. Natural disasters, species extinction threats, and rising sea level threats could be mitigated. Cloud cover could reduce the threat of skin cancer for many. Some fear that climate control could also be an instrument of war.

on whether the climate change is too fast, and the extent that humans cause it and can remedy it.

When is climate change too fast, too abrupt? There is some controversy about this. According to the National Oceanic and Aeronautics Administration (NOAA, a federal agency), it refers to:

> the changes in average weather conditions that generally occur over long periods of time, usually centuries or longer. Occasionally these changes can occur more rapidly, in periods as short as decades. Such climate changes are characterized as "abrupt."

Other scientists tend to couch climate change in terms of effects on ecosystems and social systems and their ability to adapt to climatic changes.

SCIENCE AND FORECASTING

Underlying the battleground of most global controversies is contested science. The scale is so large that until the advent of computers the science of climate change was theoretical and controversial. Now there is consensus among scientists and engineers that climate change is driven by greenhouse gases. The sun provides about 344 watts of energy per square meter on average. Much of this energy comes in the part of the electromagnetic spectrum visible to humans. The sun drives the earth's weather and climate and heats the earth's water and land surfaces. The earth radiates energy back into space through our atmosphere. Much of this is reradiated energy. Water vapor, carbon dioxide, methane, particulate matter, and other gases are called atmospheric gases; they exist in a very delicate and dynamic balance. They act like glass windows, letting heat in and holding in some of it. This is why this is called the *greenhouse effect*. Scientists can now observe atmospheric conditions further in the past than recorded weather observations by examining deeply embedded ice cores from old ice. By doing this, scientists helped to isolate the effects of human development on the atmosphere. This research also helps to pinpoint large catastrophic natural events in Earth's history.

Since about 1750–1800, atmospheric concentrations of carbon dioxide have increased by almost 30 percent. Carbon dioxide is about .04 percent of the atmosphere. With more carbon dioxide and warmer air more moisture develops in the atmosphere and increases the warming trends. Methane concentrations have more than doubled in this period of time; this is discussed further on. Nitrous oxide concentrations have risen by about 15 percent. Increases of this scale in these gases have enhanced the heat-trapping capability of the earth's atmosphere. They have blurred the windows on the greenhouse. This has the effect of heating the planet up, melting ice caps, and raising ocean levels.

Human activities are the main reason for the increased concentration of carbon dioxide. Human impacts have affected the usual balance of plant respiration and the decomposition of organic matter release heavily on the side of CO_2. Fossil fuels are responsible for about 98 percent of U.S. carbon dioxide emissions, 24 percent of methane emissions, and 18 percent of nitrous oxide emissions. Increased agribusiness, deforestation, landfills, incinerators, industrial production, and mining also contribute a large share of emissions. In 1997, the United States emitted about one-fifth of total global greenhouse gases. This estimate is based on models and industry self-reporting.

Knowledge of past and present environmental emissions is limited. Most environmental regulation and monitoring are new, and most of the world is still unregulated and unmonitored. Even environmentally regulated countries still allow large amounts of chemicals into the land, air, and water without complete knowledge of short- or long-term ecological risks and impacts. It is difficult to assess these impacts because not all emissions into the environment from humans are regulated. Large amounts of unregulated industrial emissions, municipal emissions, agricultural emissions, and commercial and residential emissions all remain unregulated and a source of uncertainty. Each of these categories represents future stakeholders in a growing controversy. Because so much is unknown at the present regarding the scale and scope of these emissions, it is impossible to predict environmental impacts like synergistic and cumulative risks. Over time, with an increasing human population and more extensive climatic monitoring, the level of uncertainty as to effects of climate change will decrease. However, fear of liability under common polluter-pays principles may increase the resistance among the various stakeholders and thus increase the controversy. Uncertainty about the best policies to follow to mitigate climate changes will increase controversy. Most agree that it will require better knowledge about actual emissions to be developed over time. The policy need for this information and the stakeholder fear of liability and increased regulation will fuel the first fires of the climate change policy wars. Currently, the state of knowledge is highly dependent on modeling and weather data.

THE CASE OF METHANE

Methane emissions demand special attention in climate change scenarios. Methane remains in the atmosphere for approximately 9–15 years. Methane is more than 20 times more effective in trapping heat in the atmosphere than carbon dioxide. Former U.S. Vice President Al Gore and others conclude that there are large pockets of methane in the north and south poles and that as they warm more methane than currently predicted would be released, greatly increasing the rate of global warming and rate of climate change. Methane is a powerful gas in the atmosphere. It has both human and natural sources. It is estimated that 60 percent of global methane emissions are related to human activities. It has a strong industrial level of use all over the world. Methane is emitted during the drilling, refining, use, and transport of coal, natural gas, and oil. Methane emissions also result from the decomposition of organic wastes. Wetlands, gas hydrates, permafrost, termites, oceans, freshwater bodies, nonwetland soils, volcanic eruptions, and wildfires are natural sources of methane. Some sustainable dairy farms in

Vermont have defrayed expensive heating costs by collecting and burning the methane from the cow manure collected underneath the barn. It also collects in municipal solid waste landfills, which are near capacity. Sometimes landfills burn off the methane gas that can form in a landfill. Methane is a primary constituent of natural gas and an important energy source all over the world.

Other emissions may also be affecting climate change. As human populations and industrialization increase, these emissions will also increase.

Greenhouse controversies will escalate substantially. The climate changes that occur could be dramatic. Most experts generally expect

1. Land to increase in temperature more than oceans
2. A substantial retreat of northern hemisphere sea ice
3. Sea level to rise more than a meter over the next several hundred years, although there is some controversy over the rate, both generally and at specific locations
4. A sharp reduction in the overturning circulation of the North Atlantic ocean
5. Substantial reductions in midcontinental summer soil moisture (~25 percent)
6. Increases in the intensity of tropical hurricanes/typhoons, at least for those that tend to reach mature stages
7. Sharp increases in summertime heat index (a measure of the effective temperature level a body feels on a humid day) would be likely in moist subtropical areas (http://www.gfdl.noaa.gov/aboutus/article/aree_page9. html—Delworth)

There could be other impacts not yet known. There could be increases in natural disasters and in conflicts over power between nations.

CLIMATE DATA AVAILABILITY

Climate data provide the basics for our characterizations of the time-averaged climate states for various statistics for temperature, pressure, wind, water amounts, cloudiness, and precipitation as a function of geographical location, time, and altitude. Data provide invaluable information on the natural variability of climate, ranging from seasons to decades. These data sets have led to important understandings of how the climate system works. These data sets also provide valuable information on how ice ages and warm epochs interact with climatic changes. For meteorological purposes thousands of observation points collect information continuously for weather forecasting. Most of this information is also important research on longer-term climate change. Climate change data are more expansive than weather forecasting data sets. Some climate change data not usually included in weather data sets are vertical velocity, radiative heating/cooling, cloud characteristics (albedo), evaporation, and properties of critical trace species such as particles containing sulfate and carbon. Weather data sets do not provide information on the vegetative cover and its role in governing surface water evaporation. The ocean's currents, waves, jets, and vortices are important climatic measurements not included in usual weather data sets.

Weather often determines human settlement patterns, and as world population increases, the unstable and sometime contradictory computer modeling of climate changes lends itself to controversy. Weather forecasts can also determine financial lending patterns in agricultural areas, as well as economic development based on industrial manufacturing. This expands the role of industrial stakeholders from one of being regulated by various international and state governments to one of engagement with the accuracy of the models.

CLIMATE CHANGES AND THEIR EFFECTS ON ANIMALS

The Earth is warming and the climate is changing. One aspect of climate change is that the climate changes faster than an ecosystem. Some species in the food chain will be affected first and, unless they evolve or move, will become extinct. Abrupt climate change has dramatic and unknown effects on the environment. Robins have been sighted in the Arctic for the first time. Species may move to new territories and interbreed. Grizzlies and polar bears have done this in northern Canada. The polar bears lose the ability to get to the seals they hunt because climate change has melted the ice. They move inland in search of food. Other aspects of an ecosystem will be tested with rapid climate change.

WHERE ARE THE ANIMALS GOING? CURRENT RESEARCH

Vast ecosystem changes cause plants and animals to migrate. They can cause migrating animals to alter their genetically inbred routes of travel. In 2000, scientists from 17 nations examined 125,000 studies involving 561 species of animals around the globe. Spring was beginning on average six to eight days earlier than it did 30 years ago. Regions such as Spain saw the greatest increases in temperatures. (This contradicts some climate change models that forecast the greatest temperature changes at the north and south poles.) Spring season began up to two weeks earlier there. The onset of autumn has been delayed by an average of three days over the same period. Changes to the continent's climate are shifting the timing of the seasons. There is a direct link between rising temperatures and changes in plant and animal behavior.

Research this year examined 125,000 observational series of 542 plants and 19 animal species in 21 European countries from 1971 to 2000. The results showed that 78 percent of all leafing, flowering, and fruiting records are happening earlier in the year, while only 3 percent were significantly delayed. Species that are dependent on each other but changing at different rates could break down the food web. Current research is based only on indicator species not entire ecosystems.

POTENTIAL FOR FUTURE CONTROVERSY

Scientific and political controversies about climate change will increase. International environmental responsibilities and choices and rising local concern will raise some inconvenient environmental issues. Industrialization has had and continues to have a large environmental impact, perhaps affecting climate stability. Some of the nations most benefiting from industrialization are now debating policies about sustainability. They request poorer nations to refrain from using the same fuels that began their own economic development under free market capitalism. This is the so called north–south debate. Poorer nations want the quality-of-life improvements of free markets and do not like interference from richer nations. This aspect of the battleground is global.

Another continuing aspect of climate change is the rate of climate changes. In August 2007, a study by Professor James Hansen predicted that oceans could rise substantially more than predicted. Hansen is director of the National Aeronautics and Space Administration's (NASA) Goddard Institute for Space Studies in New York and a professor at Columbia University. Because of positive feedback loops in the atmosphere, global warming events could cause a spiraling effect, causing oceans to rise much more quickly then predicted. By 2100, oceans could rise hundreds of feet instead of the smaller predictions of two to four feet by conservative climate watch organizations. His prediction was disputed by other international climate change sources until recently. Recent ice quakes in Greenland, ice core samples from the poles indicating rates of melting, and rapid release of methane from thawing permafrost all give greater credibility to this still controversial prediction. There will be more battlegrounds on the accuracy of the climate changes themselves, their rate of change, and their environmental impacts.

It is also a local battlefield. As the push for sustainability rises to a policy level in richer nations, they confront an industrial past. Large programs of waste cleanup and assessments are begun. Environmental regulations are tightened to include all environmental impacts. Local communities begin to adopt environmental principles, like the precautionary principle, in their land-use laws. All these areas are controversial, and part of the bigger battlefield of climate change. Climate change concerns may be creating a greater environmental consciousness and in that way create a supportive environment for policies like sustainability and 100 percent waste cleanup.

See also Citizen Monitoring of Environmental Decisions; Cumulative Emissions, Impacts, and Risks; Global Warming; Sustainability

Web Resources

Global Warming: Early Warning Signs. Available at www.climatehotmap.org/. Accessed January 20, 2008.

Greenpeace. Stop Climate Change. Available at www.greenpeace.org/international/campaigns/climate-change. Accessed January 20, 2008.

U.S. Environmental Protection Agency. Climate Change. Available at www.epa.gov/climate change/. Accessed January 20, 2008.

Further Reading: Cox, John D. 2005. *Climate Crash: Discovering Rapid Climate Change and What It Means to Our Future.* New York: National Academies Press; Faure, Michael, and Joyeeta Gupta. 2003. *Climate Change and the Kyoto Protocol.* UK: Edward Elgar

Publishing; Jepma, Catrinus J., and Mohan Munasinghe. 1997. *Climate Change Policy: Facts, Issues and Analyses.* Cambridge: Cambridge University Press; Victor, David G. 2004. *Climate Change: Debating America's Policy Options.* Washington, DC: Council on Foreign Relations.

COLLABORATION IN ENVIRONMENTAL DECISION MAKING

The cost, unpredictability, rancor, and ill will generated by environmental litigation push nonadversarial problem-solving approaches such as collaboration to the fore. Collaboration in environmental decision making is encouraged by government but is new and lacks a solid basis in legislation. It is controversial because it is new, involves community stakeholders, and could be expensive and time consuming.

Within the federal government, the EPA has been a leader in the use of collaborative approaches to accomplish strategic goals and objectives. The ability to collaborate effectively, internally and externally, is becoming more important as environmental problems become more complex and combine land, air, water, and ecosystem approaches.

Collaboration is an attitude that prompts people to approach environmental decision making in the spirit of proactive cooperation and shared effort that leads to better, more creative outcomes. Taking collaboration to a new policy level will require this attitude to be developed in legislation, promulgated rules and regulations, and application to environmental decisions. This policy goal would be realized when:

- Environmental decisionmaking explicitly considers whether collaborative approaches should be used;
- EPA managers and staff are fully equipped with the skills, tools, and resources to effectively implement collaborative problem-solving projects across EPA programs, media, and organizations and with external stakeholders and the public;
- These collaborations achieve superior environmental outcomes;
- EPA accountability systems for air and water media are aligned with these new expectations; and
- Organizations, communities, and groups outside the EPA (government, nonprofit, and private) see this approach as a catalyst for other collaborative efforts to improve the environment.

Collaboration is a specific approach to working with stakeholders, in which participants develop a mutually agreeable process for joint learning and problem solving. Collaboration takes many forms and can be either formal or informal, but it is distinct from other forms of engaging stakeholders and the public, such as informing, consultation, involvement, or empowerment. The requisite degree of formality will depend on the purpose of a collaboration process; the number and diversity of stakeholders; the scale, scope, and complexity of the issue at hand; the duration of the process; and other factors. In some situations, an informal, ad hoc process may be appropriate and sufficient to solve an environmental problem; in other situations, it may be necessary to convene a formal advisory committee under the auspices of the Federal Advisory Committee Act (FACA). In all of its forms, collaborative approaches to environmental protection can foster superior environmental outcomes.

Collaborative processes, however, are not considered appropriate in all situations. Precollaboration situation assessments are a valuable tool that can help EPA managers and staff by providing an objective assessment to determine whether collaboration is appropriate in a given situation and, if so, what type of formal or informal collaborative approach would be most effective. Such assessments, which vary in their complexity and are often conducted by a neutral third party, entail a situation-specific analysis of an environmental problem's complexity and history, the needs and resources of interested and affected parties, and time constraints. They need to do essentially a *placestudy*. Placestudy assessment results often lead to a detailed plan that describes the goals, process, participants, timing, and structure of EPA-stakeholder interactions. As this is a new policy approach, there is not much controversy. When these approaches are mandated or otherwise required, and costs in time, capital, and lost opportunity affect a wide range of stakeholders, some controversies could arise. Currently, the environmental and community health data needs are not met to the standard this approach would require.

THE EPA'S ROLE

The EPA's role in collaborative environmental problem solving will vary. In many situations, the EPA's statutory responsibilities place it in a leadership role that requires convening the relevant parties and facilitating interaction. In other situations, the EPA will simply act as one of many interested parties in a collaborative problem-solving effort convened by another federal agency; a state, tribe, or local government; or a private-sector entity. In still other situations, the EPA may be the beneficiary of a collaborative problem-solving effort without actively participating in the collaboration itself. Collaboration cannot replace the core functions of a regulatory agency, standard setting, permitting, and enforcement and compliance assurance, nor compromise the EPA's decision-making responsibility. In general, however, the EPA looks to all affected stakeholders for ideas and innovative solutions and, where appropriate, incorporates stakeholder recommendations into policy and practice.

Where environmental problems require collaboration, some common themes in successful collaborative problem solving emerge. Collaborative efforts to solve environmental problems are more likely to succeed when these factors are present.

A SHARED ENVIRONMENTAL PROBLEM

Stakeholders are motivated to collaborate when all parties would benefit by solving a problem, but no single party has the capacity or incentive to do so. Collaborative responses to shared problems allow stakeholders to coordinate their activities, leverage resources, and enhance accountability. Without a shared problem, stakeholders have little reason to collaborate. In some cases, the intensity of a problem will have risen to the level where common pain brings the parties to a table. In other instances, a common sense of the opportunity for better outcomes through a collective process will be sufficient.

This aspect of successful collaboration suggests the following EPA practices:

- The EPA should target its collaborative efforts on high-priority shared problems. Collaborative processes are unlikely to sustain themselves in the absence of a motivating problem and are unlikely to resolve issues associated with fundamental rights or social values.
- In some instances, a shared problem may exist but not be recognized by affected stakeholders. In these situations, EPA can catalyze collaboration by raising awareness of an environmental problem and the need for a collaborative approach to develop and implement of solutions.

CONVENER WITH POWER

A convener with power can catalyze collaboration by legitimizing the process, encouraging stakeholder participation, and shouldering initial organizational costs to bring parties together to address a shared problem. A convener can help to sustain collaborative processes by reaffirming the process and the importance of all parties working together to solve a common problem.

This factor in successful collaboration suggests the following EPA practices:

- Convening a collaborative process takes time, energy, and financial resources, so the EPA must strategically decide when to serve as a convener. Shared problems of national significance that clearly fall within the statutory authority are particularly strong candidates for EPA investment.
- Where EPA lacks statutory authority or recognized credibility, it may facilitate collaboration by encouraging other agencies or organizations to convene stakeholders to solve a shared problem. In complex situations that involve multiple issues and stakeholders, it may be necessary for the EPA to instill confidence in a collaborative process by using a respected neutral party to convene the relevant stakeholders and facilitate the process.

COMMITTED LEADERSHIP

While a convener is necessary to bring a group together, committed leadership is necessary to craft an agreement among collaborating parties. When participants become disappointed or disillusioned, the committed leader—staff or manager—can sustain a group by reiterating the benefits of collective, coordinated action compared to the drawbacks of independent, uncoordinated action and by emphasizing the personal commitments that participants have made to each other and to the collaborative process.

The EPA and other governmental agencies can foster the committed leadership necessary to sustain collaboration.

- Giving the individuals responsible for leading a collaborative process the capacity (time, resources, skills) to do so will enhance the likelihood that these EPA representatives can serve as committed leaders.
- Successfully leading a collaborative process requires the authority to make decisions. In many instances, fostering committed leadership by EPA representatives will require conscious delegation of authority to the EPA from Congress.
- Assigning a dedicated, knowledgeable, and reputable person who is known to the other participants to represent EPA in a collaborative process can provide an initial reserve of trust and goodwill that enables effective leadership and facilitates collaboration.

STAKEHOLDERS OF SUBSTANCE

Successful collaborative problem solving requires direct involvement by representatives of substance, individuals who have sufficient authority to decide on behalf of, or sufficiently influence, their represented interest, and who, collectively, by virtue of prominence, role, or market share, can implement timely solutions to a given problem. These representatives must represent a critical mass of affected stakeholders; by bringing these stakeholders together, a collaborative process can foster the development and implementation of an effective policy. Excluding key stakeholders from collaborative processes, by contrast, frequently leads nonparticipants to reject resulting decisions, undermine timely and complete implementation, and inhibit subsequent efforts to develop collaborative solutions to environmental problems.

This part of successful collaboration suggests that the EPA should:

- Ensure that a critical mass of affected stakeholders participates in collaborative processes. EPA will need to identify and recruit a balanced group of representatives who represent the full range of interests. Principals familiar with collaborative processes and with long time horizons (i.e., expectations of long-term involvement with an activity or issue) are particularly good candidates to participate in collaborative processes because of their knowledge and expectations for the future.
- Collaborate processes are generally more likely to succeed when the number of participants is relatively small. While it may be tempting to reduce the complexity of a collaborative process by ignoring differences among stakeholder groups (e.g., geographic or cultural differences, different risk perceptions or exposures, etc.), differences should not be ignored in an attempt to facilitate collaboration by assembling like-minded participants.
- Stakeholders are much more likely to support (and participate in) collaborative processes if these processes provide ample opportunities for meaningful collaborator involvement. Early stakeholder involvement in collaborative processes enhances group ownership of both process and

outcomes. In some instances, the EPA may need to provide training and resources to ensure that stakeholders have the capacity to participate meaningfully in a collaborative problem-solving process.

- To ensure the legitimacy of collaborative decision-making processes, the EPA will need to foster the accountability of stakeholder representatives to their constituents. Stakeholder accountability can be fostered through both formal (e.g., elections) and informal (e.g., town hall meetings) mechanisms.
- Truly participatory decision making involves not only consulting with affected stakeholders but also enlisting them as partners in decision making.

CLEARLY DEFINED ENVIRONMENTAL PROBLEM

Collaborative efforts are more likely to succeed when groups develop a clearly defined purpose for themselves. This purpose should respond naturally to the collective problem that the group shares. An overly ambitious or misaligned purpose can frustrate groups, undermining both the collaboration process and the development of viable policy solutions. It may also affect the leadership abilities of involved stakeholders.

This suggests that collaborators should:

- Set boundaries around problems that are large or ill defined (in scale or scope) to facilitate successful collaboration by providing focused, manageable problems for stakeholders to address and achieve results. In some instances, however, allowing stakeholder groups to define the scope of a problem more broadly can create new opportunities for negotiation and/ or areas of common ground.
- Focus on shared problems and problem solving in order to help maintain group purpose, facilitate collaboration, and minimize conflict. In the course of a collaborative process, participants may lose sight of their purpose.

FORMAL CHARTER

Because collaboration is a complex and high-stakes process that often involves many individuals and issues, clearly and collectively articulated roles and responsibilities are critical to timely success. A formal written charter fosters successful collaborative problem solving by reducing the uncertainties and ambiguities among collaborating parties that can cause conflict and, thus, enhances participants' confidence in each other and the collaborative process as a whole. A formal charter can also help to ensure that decision-making processes are transparent and participatory, enhancing the legitimacy, accountability, and ownership of collaborative processes by allowing stakeholders to understand how decisions are made and to have a voice in decision making. Collective definition of purpose, roles, and procedures also enhances group ownership of both process and outcomes, enhancing the likelihood of successful collaboration.

This key to successful collaboration suggests that the EPA or other convening stakeholders should adopt the following measures into the formal charter.

- Timelines establish a framework that encourages decision making and re-sults, providing useful measures of success and instilling group confidence in the process and progress. In many cases, establishing important mile-stones at the beginning of a collaborative process can provide momentum for participants.
- Reliable measures help stakeholders to define the magnitude of a problem and to track progress toward its resolution. Measures also foster account-ability by tracking indicators of stakeholder behavior and performance.

Measures should address both procedural outcomes (e.g., development of a shared understanding of an issue) and environmental outcomes (e.g., increase in a population of organisms) because changes in public policy or practice may take time to have the desired impact on human health or the environment.

COMMON KNOWLEDGE BASE

A common information base enables collaborators to develop a shared un-derstanding of the problem and possible solutions, facilitating development of viable, legitimate policy solutions through information exchange and dialogue. Information asymmetries (where different actors hold different information) can exacerbate power inequalities and foster conflict among collaborators.

This key to successful collaboration suggests that:

- Participants should work jointly to identify key questions, assemble the relevant information, and determine how to address information gaps.
- Stakeholders should be sufficiently knowledgeable and able to understand necessary information, and the EPA should provide stakeholders with the capacity to obtain independent technical assistance where necessary.
- Information regarding both process and substance should be broadly dis-seminated in order to enhance the transparency of collaborative processes. Particular attention should be given to ensure that populations at risk and other marginalized groups have access to information.

POTENTIAL FOR FUTURE CONTROVERSY

As an untested policy, collaborative environmental decision making is criti-cized as too theoretical. Its foundations rest on many currently unresolved con-troversies. Accurate environmental information and transparent environmental transactions are still controversial. In the United States, many environmental disputes are fought in the courts. The EPA is an agency that must meet the spe-cific statutory missions of most national environmental legislation. The EPA is a young and dynamic agency that always has a policy development or reinvention office with important and timely ongoing projects.

Within the EPA, the Office of Environmental Justice has taken the lead in de-veloping case studies of collaborative environmental problem solving. In many ways, collaborative environmental policy represents the new foundation for a generation of new environmental policies that are based on ecosystems and focused on multiple stakeholders.

However, a very strong push for sustainability globally, nationally, and locally has brought a groundswell of support for collaborative methods. While there are many controversial hurdles to overcome with this approach, it offers a cheaper, cleaner, and smarter way to achieve politically salient environmental policies like sustainability. Resistance, and some controversy, comes from the cost to each stakeholder in terms of time, money, and lost opportunity. Businesses claim to lose money, many community people do not get paid if they do not work and therefore lose money by participating, and many environmentalists prefer the traditional lawsuit.

While collaborative environmental policy approaches may evoke controversy as they are rolled out from research and development into real public policy, their focus on environmental outcomes and public engagement meets many of the demands of a maturing environmental policy.

See also Different Standards of Enforcement of Environmental Law; Litigation of Environmental Disputes; Sustainability

Web Resources

Cires. Collaboration Features/Success Stories. Available at cires.colorado.edu/collaboration/features/ Accessed January 20, 2008.

Collaboration in Environmental Enforcement: Experiences with the Build-Up of a Coordinated Enforcement Structure. Available at http://www.inece.org/3rdvol1/pdf/tindem.pdf. Accessed January 20, 2008.

Collaborative Efforts to Restore a Contaminated Reservoir. Available at http://www.epa.gov/region6/6xa/collaboration.html. Accessed January 20, 2008.

Oregon Solutions. Community Collaboration for Sustainability. Available at www.orsolutions.org/. Accessed January 20, 2008.

Further Reading: Carmin, JoAnn, Toddi A. Steelman, Craig W. Thomas, Cassandra Moseley, Katrina Smith Korfmacher, and Tomas M. Koontz. 2004. *Collaborative Environmental Management: What Roles for Government?* Washington, DC: Resources for the Future; Collin, Robert. 2006. *The U.S. Environmental Protection Agency: Cleaning Up America's Act.* Westport, CT: Greenwood; Durant, Robert F., Daniel J. Fiorino, and Rosemary O'Leary. 2004. *Environmental Governance Reconsidered: Challenges, Choices, and Opportunities.* Cambridge, MA: MIT Press; Marshall, Graham R. 2005. *Economics for Collaborative Environmental Management: Renegotiating the Commons.* London: James and James/Earthscan; Stern, Alissa J. 2000. *The Process of Business/Environmental Collaborations: Partnering for Sustainability.* Westport, CT: Quorum/Greenwood; Wondolleck, Julia Marie, and Steven Lewis Yaffee. 2000. *Making Collaboration Work: Lessons from Innovation in Natural Resource Management.* Washington, DC: Island Press.

COMMUNITY-BASED ENVIRONMENTAL PLANNING

Community-based environmental planning (CBEP) is new in the United States. Controversies around the capacity of the community to plan for natural resources and environmental protection, the failure of state and federal agencies to support it, and industry concern that the community will know too much of its environmental information form the current, dynamic boundaries of this battleground.

Communities are very aware of environmental conditions around them, and are growing more so with the Toxics Release Inventory, accumulating industrial pollution, and environmental justice activism. The EPA, the federal agency responsible for environmental protection, has not included cities as places for environmental protection until recently. Their view of CBEP has not been developed as a serious environmental policy until very recently in test cases. One such placestudy is described here.

CAN COMMUNITIES REALLY PLAN FOR THE ENVIRONMENT?

Some environmentalists are concerned that communities may not be able to preserve their environment or resources. A logging community may vote to cut down all the protected old growth trees, for example. A very poor and desperate community may agree to not enforce or monitor environmental conditions to accommodate any type of job creation. The U.S. history of giving tax abatements to industry in exchange for jobs and in the hopes of economic development shows that as soon as the tax abatement is over, if not sooner the industry leaves. Most of the time the promises of job creation were illusory. Because property taxes are used to fund education in many places, giving up taxes effectively reduces the education budget in that community. When the industry leaves or fails to meet its promises, there is little recourse unless there is an enforceable Good Neighbor Agreement, which is new and rare. The community ends up with very little economic development and often a more polluted environment.

COMMUNITY-BASED ENVIRONMENTAL PLANNING AND THE EPA

Community-based environmental planning refers to an integrated, place-based, participatory approach to managing the environment that simultaneously considers environmental, social, and economic concerns. In its CBEP framework document, the agency describes CBEP as a process that "brings together public and private stakeholders within a place or community to identify environmental concerns, set priorities, and implement comprehensive solutions. Often called a place-based, or ecosystem approach, CBEP considers environmental protection along with human social needs, works toward achieving long-term ecosystem health, and fosters linkages between economic prosperity and environmental well-being." The agency has identified several key attributes that characterize CBEP, including:

- collaboration through a range of stakeholders;
- assessments that cut across environmental media;
- integration of environmental, economic, and social objectives;
- use of regulatory and nonregulatory tools;

- monitoring to allow adaptive management;
- and a focus on a geographic area.

The EPA facilitates CBEP efforts by coordinating traditional regulatory programs to support CBEP; providing tools to communities pursuing CBEP activities; and collaborating directly with stakeholders. The Office of Policy, Economics, and Innovation (OPEI) coordinates the agency's CBEP efforts.

PLACESTUDY: CHARLESTOWN, SOUTH CAROLINA

The EPA's involvement in CBEP is new. Community-based environmental planning's distinctly urban focus is not yet an integral part of current U.S. environmental policy. Nonetheless, some urban environments are next to important ecosystems and highly contaminated with industrial pollution. Community residents, environmentalists, churches, state and federal agencies, and national environmental justice organizations all advocated for environmental protection.

The Charleston/North Charleston CBEP project focuses on the 17-square-mile neck area of the Charleston, South Carolina, peninsula that is bordered on the west by the Ashley River and on the east by the Cooper River. The area consists of more than 20 neighborhoods in the cities of Charleston and North Charleston with more than 40,000 inhabitants, of whom roughly 70 percent are minority and 40 percent live at or below the poverty level.

An industrial corridor in close proximity to the residential population as well as to the peninsula's abundance of tidal creeks, marshes, and rivers characterizes this place. Heavily industrialized since the 1800s, the neck area faces a complex set of environmental problems, including historical releases of hazardous waste and former and active industrial and commercial sites. Environmental contamination at one of these industrial properties, the site of a former wood treating facility, brought EPA Region 4's Superfund program to the Charleston/North Charleston area in the mid-1990s. As part of the program, EPA provided a grant for hiring a community technical advisor to meet with area residents and respond to questions about the site cleanup. Based on environmental justice and other concerns raised by several of the area neighborhoods, EPA began exploring the value of helping to organize a CBEP project.

The EPA held sessions with the South Carolina Department of Health and Environmental Control (DHEC) and other partners and, in the spring of 1997, assisted in the formation of a multistakeholder group to guide the CBEP project. The EPA suggested that a community advisory group (CAG) could provide an effective vehicle for the community to develop and guide its community-based environmental protection project. The resulting CAG consisted of representatives from neighborhoods and businesses in the CBEP area, local environmental and social advocacy organizations, and local, state, and federal agencies. The EPA provided funding to the Medical University of South Carolina (MUSC) to support the organization of the CAG. Through a detailed organizational process, a 25-member self-nominated group emerged, complete with a chairperson and other elected officers serving two-year terms, a mission and a vision statement, and a comprehensive set of bylaws. The CAG consisted of voting community and business representatives and nonvoting ex officio members, including MUSC and the other founding partners. The CAG also established subcommittees (e.g., a group addressing business/industry issues) to solidify its operation.

Once organized, the CAG confronted a complicated, overlapping set of human health, socioeconomic, environmental, and other quality of-life issues in the Charleston neck area. The environmental concerns cut across all media, including air, surface water, groundwater, sediments, and soil. Residents had long-standing concerns about cancer rates, childhood lead poisoning, and other health problems in their communities and the potential for links to chemical releases, contamination, and other effects of improper environmental compliance and management. Although the original idea for the project arose because of concerns expressed by a handful of central neck-area neighborhoods, the CAG set the project boundaries to cover the seven-square-mile area described previously, which encompasses the historical industrial corridor and also approximates the boundary lines of Charleston's Enterprise Community (now the Greater Charleston Empowerment Corporation), a distressed area targeted for economic and cultural revitalization.

GOALS AND OBJECTIVES

The long-term goal of the Charleston/North Charleston Community Project is to improve the quality of the land, air, water, and living resources to ensure human health, ecological, social, and economic benefits. To achieve the multiple aspects of this goal, project managers have established many short-term objectives through partnerships with citizens, industry, conservation groups, and other stakeholders.

The CAG developed its own mission and vision statements to guide it in its activities. Its stated mission is "to address environmental quality programs and concerns as they relate to the community's well-being and that of the environment. It exists to increase environmental awareness through education and effective collaboration with diverse groups and to promote and cultivate cooperation with industry and government. Finally, the group exists to empower, create, and sustain a healthy, livable community that will positively impact residents' quality of life."

Both the CAG and the overall CBEP project have environmental improvements and human health concerns as long-term goals as well as ecological, social, and economic well-being. To accomplish these overall goals, CAG members have established the following short-term objectives:

- To develop a baseline for environmental conditions
- To reduce both lead contamination of soil and childhood lead poisoning
- To identify and remediate locations with elevated indoor radon levels
- To minimize the effects of environmental contamination from former phosphate/fertilizer facilities
- To provide targeted compliance assistance and pollution prevention information for small businesses

In developing and carrying out efforts to address these objectives, the CAG has drawn on several partnerships with industry, government, academic institutions, and other stakeholders. Numerous activities and indicators have been developed to facilitate progress toward these objectives.

PROJECT ACTIVITIES

The first activities undertaken by the CAG were the development of the previous objectives, which emerged from its neighborhood research. To begin to address all of the challenges facing the more than 20 neighborhoods in the targeted area, the CAG and its partners embarked on outreach, research, environmental remediation, and other activities. Through monthly gatherings, public forums, and subcommittee meetings, the CAG developed several short-term and long-term initiatives to help in the achievement of its goals. The short-term activities, the full set of which is beyond the scope of this evaluation, have included river cleanup events, Earth Day fairs, and other outreach events aimed at increasing understanding of community-based environmental protection and environmental awareness in general. Long-term initiatives led by or associated with the CBEP project include the following.

CHARACTERIZATION OF COMMUNITY CONCERNS

The priority concerns as determined by the CAG are the ones addressed by the activities described further on. Other issues identified among residents relate to crime, excessive noise, poor air quality, the need for economic development, a lack of safe playgrounds and open spaces, improper drainage and flooding, contamination of open ditches and associated safety risks, environmental justice concerns, and poor environmental compliance among local commercial and industrial facilities.

Baseline Environmental Data Compilation: CAG partners undertook an extensive effort to assemble data about regulated industrial facilities, chemical releases, water quality, and other environmental conditions to meet their first objective of a baseline environmental characterization of the CBEP area. The collected data were to represent baseline conditions for the CBEP project. The CAG also intended to complete an outreach effort to make the information available to residents in the surrounding communities.

Lead Poisoning Prevention: The purpose of this effort was to provide education to new and expectant mothers to meet the objective of reducing childhood lead poisoning. Much of the housing within the neck area dates from the early and mid-1900s, when lead paint was still used widely. With the help of EPA grant money, MUSC provided training to community members (termed *advisors*) hired to conduct outreach with new and expectant mothers and other family members about how to protect their children from lead exposure in homes and other locations. The introduction of lead exposure tracking will provide indicator data for the success of the initiative.

Testing for and Mitigation of Elevated Indoor Radon Levels: Because of past phosphate mining (a factor in the presence of elevated radium levels in soil), the CBEP area is considered to be at risk for elevated indoor radon levels. CAG members began a radon testing survey and a related educational outreach effort and will provide mitigation in homes where elevated levels are discovered. These efforts address both radon reduction objectives and broader goals of community involvement.

Assessment and Remediation of Former Phosphate/Fertilizer Facility Sites: The goal of this initiative is to evaluate the contamination present at nine former phosphate/fertilizer facilities. Where unacceptable risk is found, CAG partners will ensure that an adequately protective site management strategy is implemented.

SMALL BUSINESS COMPLIANCE ASSISTANCE

In light of the number of industrial and commercial facilities, including many small businesses, two CAG partners, EPA Region 4 and DHEC, have collaborated to address compliance issues. This initiative focuses on providing targeted compliance assistance to two industries, dry cleaners and auto paint and body shops, which appear to present the greatest potential for environmental impacts in the CBEP area. Researchers are using behavioral changes, compliance records, and environmental and human health improvements as indicators of success in meeting the compliance objective. With EPA involvement in this project, the community had to incorporate aspects of EPA agenda into the project.

Environmentally Friendly Small Business/Pollution Prevention Initiative: Focusing mostly on auto paint and body shops, CAG partners undertook an outreach effort to inform small businesses of pollution prevention opportunities. Outreach team members conducted site visits and provided small business owners with information on environmental performance beyond that relating to regulatory compliance. This initiative will ensure that environmental gains are sustained and enhanced in the future and that small businesses are part of the process.

While several of these initiatives are still ongoing, the CAG and its partners are currently evaluating the results of the CBEP efforts thus far and determining next steps. One of the most significant developments since the CBEP project's inception is the decision to incorporate the CAG as an environmental subcommittee of the Greater Charleston Empowerment Corporation to take advantage of issue and organizational overlap.

Like the initial CAG formation process, the majority of CBEP activities have been fully funded by the EPA. The lead poisoning prevention, radon reduction, and small business pollution prevention projects were all funded by the EPA through the RGI. The project has also leveraged in-kind contributions and other resources from a variety of sources, including MUSC; the USGS; DHEC; other local, state, and federal health agencies; Youth Build and other local nonprofit organizations; and businesses, such as Lowe's and Home Depot. Part of the rationale for making the CAG part of the Greater Charleston Empowerment Corporation is to leverage resources between efforts with similar sustainable development goals.

EMERGING APPLICATION OF THE EPA'S ROLE IN CBEP

According to everyone involved, the EPA has acted as the driving force within the Charleston/North Charleston CBEP project from the beginning. The Charleston site became a major EPA project when it was listed on the Superfund National Priorities List (NPL). The agency has supplied specialized information, facilitation support, and sources of funding to launch and carry out all of the activities detailed previously. At the same time, the key role played by EPA has had both positive and negative implications, as viewed from the perspectives of different CAG members and project stakeholders.

IMPACT OF OPERATIONAL DIFFERENCES BETWEEN THE EPA AND OTHER STAKEHOLDERS

Some participants felt that the project has been influenced by differences in expectations and approach between the EPA (as well as other institutional members) and community members. Although the priority of everyone involved has always been to improve the area's quality of life, some residents expected more immediate results (e.g., health screenings, repair work to address risks posed by drainage ditches). Some feel that the EPA and others have been overly concerned with developing the project itself, such as formation of CAG procedures, use of resources to publicize the project, and so on. Some participants noted, for example, that the communities had previously voiced their priority issues, so they felt that the effort to record resident concerns was not the most efficient use of time and resources. For some participants, EPA-facilitated developmental process was perceived as only further bureaucracy rather than a process to build credibility and trust, and added to the cynicism of residents who viewed previous partnership efforts as failing to deliver concrete results. However, some CBEP participants viewed the structured CAG process as an asset. In fact, these participants credit the CAG structure with gathering different community view points at the table and keeping participants engaged when differences of opinion arose.

CONTROVERSY ABOUT EPA SUPPORT AND FACILITATION

In this project the EPA claims it has always expressed the desire that the Charleston/North Charleston efforts be community-led and thus has encouraged operational mechanisms such as the CAG. From the perspective of some participants, however, the project has been neither community-directed nor responsive to community voices. This sentiment originates from perceptions about a lengthy CAG formation process dominated by the EPA and other institutional partners, which may have helped lead to a subsequent lack of involvement from residents (e.g., lack of public attendance at CAG-sponsored meetings and events). Participants holding this view would have preferred that the EPA provide less overall facilitation in exchange for more up-front support for existing community priorities (e.g., technical assistance for targeted health screenings, repair of drainage ditch hazards, etc.). Some participants also suggested performance tracking and evaluation as a valuable niche role for the EPA within CBEP projects.

Confronted with a complex set of environmental problems and other challenges that are very characteristic of the environment where most people live, work, and play, the EPA and its partners established an ambitious agenda of objectives and strategies for the Charleston/North Charleston CBEP project. Tracking of some of the project's completed initiatives remains unfinished, and other efforts are still ongoing.

Measuring progress toward the project's overall goals of improving the environmental quality to ensure human health and ecological, social, and economic benefits is a long-term process. Nevertheless, participants can point to several environmental and other accomplishments to characterize the project's progress in meeting the previously stated objectives.

In the summer of 1999, CAG partners finished the environmental data compilation effort to meet their objective of determining the baseline environmental conditions. They released a draft document titled *Summary of the Environmental Information Collected for the Charleston/North Charleston Community-Based Environmental Protection Program*. The document contains more than 20 maps and tables with data ranging from a summary of area Toxics Release Inventory (TRI) releases to the location of facilities with NPDES permits for discharges. The CAG has provided comments on the document as well as recommendations for the next phase of the effort. Based on these recommendations, the CAG is making plans to use the information to assess certain environmental conditions, create maps showing the data points on a neighborhood-specific level, and develop a user-friendly system to enable community access to the data.

To address the lead poisoning prevention goal, MUSC trained eight area residents who were hired to be community educators or advisors. The purpose of the outreach was to inform new and expectant mothers and other family members about childhood lead poisoning and preventative behavioral measures (e.g., frequent washing of hands). By the summer of 2000, the advisors had reached more than 900 community members in interactions that ranged from brief one-on-one conversations to group meetings in residents' homes. To the surprise of the advisors and their CBEP partners, a large percentage of young mothers were unaware of lead poisoning risks and reported that their children were not being screened at their regular medical check-ups. As a result of the outreach efforts, many families have reported taking their children in for lead level screening. In addition, DHEC CBEP participants are investigating the adequacy of regular lead level screenings within the Charleston area.

Identification of homes with elevated radon levels is under way. Thus far, testing is complete at 200 out of a targeted 2,000 residences for which test kits have been obtained. CAG members have secured support from the Southern Regional Radon Training Center, which will provide training to the local Youth Build program to complete the mitigation work, and Home Depot and Lowe's have offered to contribute mitigation materials.

The minimization of impacts from former industrial sites is under way. Preliminary environmental assessments are now complete at the nine former fertilizer/phosphate facilities targeted by CAG partners. Additional results to date under this initiative include a removal action at one site, a remedial investigation at another site, a Superfund NPL designation and subsequent remediation plan at one site, and voluntary cleanup agreements with several responsible parties.

WAS CBEP EFFECTIVE AS ENVIRONMENTAL POLICY?

The Charleston/North Charleston project showed how an effective CBEP has resources to complete assessments, remediation, and other environmental outcomes; increase capacity-building within the community (e.g., lead-poisoning prevention training); and nurture multistakeholder partnerships (e.g., through the CAG). Although in some ways the CAG represents the most controversial aspect of the project, its continued operation is perhaps the strongest demonstration of the effectiveness of the CBEP process. Despite the group's difficulties, many local organizations have participated in the CAG (with some requesting to join following its initial formation). In fact, several participants noted that the CAG represents a significant first in terms of bringing diverse community viewpoints to the table to discuss environmental issues. They noted that without the unique collaborative, comprehensive nature of the CBEP approach, this enlarged discussion could not have occurred. Although some project participants questioned the extent to which community voices are represented on the CAG, the group's membership includes the leadership of diverse organizations, most of which are new CBEP recruits. DHEC, for instance, which had no previous CBEP experience, has maintained active CAG participation all along and has implemented changes suggested by the group (e.g., providing better public access to an environmental release log within its offices).

Although many of the project's objectives were either accomplished or are in progress, frustrations with the initial stages of the CBEP process were still evident. The EPA is still assisting the community in the CBEP process; for example, in early 2002, the EPA organized and delivered a workshop for planning boards and citizens on the planning process and methods for encouraging public participation.

POTENTIAL FOR FUTURE CONTROVERSY

Community-based environmental planning is taking place with or without the EPA or state environmental agency involvement. Citizen monitoring of environmental decisions, the rise of Good Neighbor Agreements and collaborative approaches to environmental decision making, and a groundswell of support for sustainability all push U.S. environmental policy in this direction.

Many of the environmental controversies that face state and federal environmental agencies could soon face communities. Some of these very controversies, such as Superfund cleanups, can mobilize a community into environmental action. Others may not. Some challenges to CBEP are controversies now because of their newness, because of a slight increase in environmental empowerment of a community, and because of underlying controversies based on race and cultural differences.

In the rush of many new voices describing the ecology where they live, work, and play; deliberating about local environmental problems in inclusive ways; and planning for ways to solve them, many controversies will arise. CBEP is a controversial way to handle the grassroots end of environmental policy. This is usually the most controversial part of any policy, especially new environmental policies. However, an environmental policy with inadequate implementation and enforcement is none at all. The community in any given ecosystem is one of the best witnesses to implementation and enforcement. Their testimony presents many inconvenient and controversial truths about the real state of the environment in many places.

See also Citizen Monitoring of Environmental Decisions; Collaboration in Environmental Decision Making; Community-Based Science; Cumulative Emissions, Impacts, and Risks; Ecological Risk Management Decisions at Superfund Sites; Good Neighbor Agreements; Public Participation/Involvement in Environmental Decisions

Web Resources

Check Your Success: A Community Guide to Developing Indicators. Available at www.uap. vt.edu/checkyoursuccess/. Accessed January 20, 2008.

EPA: Community-Based Environmental Protection: A Resource Book for Protecting Ecosystems and Communities. Available at http://www.epa.gov/care/library/howto.pdf. Accessed January 20, 2008.

Operations Manual for Hispanic Community-Based Organization. Available at http://www.epa.gov/care/library/hispanic_community-sed_orgs.pdf. Accessed January 20, 2008.

Further Reading: Bowen, William Milton. 2001. *Environmental Justice through Research-Based Decision-Making.* New York: Routledge; Boyce, James K., and Barry G. Shelley, eds. 2003. *Natural Assets: Democratizing Environmental Ownership.* Washington, DC: Island Press; de Roo, Gert, and Donald Miller. 2004. *Integrating City Planning and Environmental Improvement.* Aldershot, UK: Ashgate Publishing; Honachefsky, William B. 2000. *Ecologically Based Municipal Land Use Planning.* Singapore: CRC Press; Randolph, John. 2004. *Environmental Land Use Planning and Management.* Washington, DC: Island Press; Riddell, Robert. 2004. *Sustainable Urban Planning: Tipping the Balance.* London: Blackwell Publishing; Stoll-Kleemann, Susanne, and Timothy O'Riordan. 2002. *Biodiversity, Sustainability and Human Communities: Protecting beyond the Protected.* Cambridge: Cambridge University Press.

COMMUNITY-BASED SCIENCE

Community-based scientific research is controversial because it is not considered valid, may misuse the community, and provides information that challenges government and industry environmental decision makers. In the United States, universities can provide science to communities, but this is just developing.

BACKGROUND: SCIENTIFIC RESEARCH AND COMMUNITIES

Most research occurs at research institutions, such as universities, private laboratories, and some federal government agencies. It is assumed that scientific research is always done by scientists. Even when communities are studied in the social sciences or public health professions, communities are not consulted except as research subjects. Part of the problem is the definition of community, which can create a battleground. Definitions of community can determine who can participate in the research. Many communities do not like to be research subjects, but even when they agree to do so, the results can be used against them. In the case of the Hanford Nuclear Power Plant in the state of Washington, a community group known as the Downwinders learned this. This group was organized around concerns about nuclear emissions and waste from the nuclear power plant upwind of them. One large area of concern was the health effects on pregnant woman and the rate of miscarriage there. When the power plant, via its consultant, offered free medical consultation and screening for all pregnant women downwind from the plant most of the woman agreed. When the same consulting firm, still working for the power plant, performed a survey of all those downwind from the power plant, they eliminated all woman who had been to the screening as subjects in the survey on downwind health effects because they had already been subjects. According to them their inclusion would have biased the sample. Their survey found very few health effects downwind of the power plant but did increase community organization and interest in science. Another problem is that some communities are asked to scientifically prove that a given land use or industrial expansion is harmful to them while other communities can exercise political power and resist pollution-causing land uses. This is especially a battleground in environmental justice communities and raises controversies about the unequal enforcement of environmental laws. Most often, industry and/or government scientists are trying to persuade the community that everything is safe and that the risk is minimal. Another major issue, and potential battleground, is the question of exactly who represents the community. How many from the community and who from the community are logistical and practical questions that can prevent scientists from community engagement at all. However, many environmentally significant research questions can only be addressed in a community setting with the active assistance of community members.

POLITICAL INFLUENCES ON GOVERNMENT SCIENTISTS

Testimony of Rick Piltz , Director, Climate Science Watch Government Accountability Project Washington, D.C.

Before the Committee on Oversight and Government Reform U.S. House of Representatives Hearing on Allegations of Political Interference with the Work of Government Climate Change Scientists, January 30, 2007.

Despite the utility of the National Assessment, beginning in 2001, and more aggressively from the second half of 2002 onward, the Bush Administration acted to hide the National Assessment. . . . The Administration failed to consider and utilize the National Assessment in the *Strategic Plan for the U.S. Climate Change Science Program* issued in July 2003. From my experience, observation, analysis of documentation, and personal communications with others in the program, I believe it is clear that the reasons for this were essentially political, and not based on scientific considerations. I believe this is generally understood within the program.

I was directed by the White House Office of Science and Technology Policy to delete the section of the draft report on the National Assessment. No documented explanation was provided to the program leadership and the program office as to why this alteration was necessary and appropriate. However, I was given to understand that the directive from OSTP was related to the Administration's intention to settle a lawsuit that had been brought by *Competitive Enterprise Institute et al. v. George W. Bush et al.*, seeking to suppress the distribution of the National Assessment. Specifically, that CEI et al. would withdraw the lawsuit in return for an assurance by Administration officials that the Administration would, in effect, disown the National Assessment. (CEI is an industry-funded policy organization that has aggressively promoted the position of denying that global warming is a significant problem calling for a significant policy response strategy.) The Administration was uncomfortable with the mainstream scientifically based communications suggesting the reality of human-induced climate change and the likelihood of adverse consequences. The administration had adopted a policy on climate change that rejected regulatory limits on emissions of greenhouse gases, and cited scientific uncertainty about climate change as one of its justifications for the policy. Straightforward acknowledgement of the growing body of climate research and assessment suggesting likely adverse consequences could potentially lead to stronger public support for controls on emissions and could be used to criticize the administration for not embracing a stronger climate change response strategy. Administration political officials appeared increasingly to take an interest in managing the flow of communications pertaining to climate change in such a way as to minimize the perception that scientifically based communications might be seen as conflicting with the Administration's political message on climate change policy.

Immediately prior to taking the position of CEQ Chief of Staff, Cooney had been employed as a lawyer-lobbyist at the American Petroleum Institute (API), the primary trade association for corporations associated with the petroleum industry. He was the climate team leader at API, leading the oil industry's fight against limits on greenhouse gas emissions. CEI also had a close relationship with the oil industry, having reportedly received $2 million in funding between 1998 and 2005 from ExxonMobil.

ENVIRONMENTAL ADVOCACY VS. SCIENTIFIC INQUIRY

Community-based researchers must negotiate the difficult, often blurred boundaries between advocacy and inquiry. Although the research may serve a larger advocacy function, its claims must be able to withstand public and scientific scrutiny.

The research work is complex, involving diverse skills and decisions. Community members are more appropriately part of some aspects of research than others. For instance, although it is both appropriate and valuable to consult with community members on the design of a survey instrument, they will probably not have a role to play in the statistical interpretation of its structure. Clear boundaries should be stated at the outset. Community members are much more likely to let us do our job if we make it clear that we are going to let them do theirs. Finally, although consultation with community representatives prior to publication is important, they should be aware that no individual or group will have the authority to prevent publication.

Good community-based research is a difficult social achievement. Poorly handled projects can, and do, leave damaging trails of mistrust. Yet contributing to the life of state and regional communities through research is central to the mission of all land grant colleges and universities. When handled well, community-based projects enhance public support for research universities.

PRINCIPLES OF COMMUNITY PARTNERSHIP

Researchers and government agencies, along with many nonprofit groups that work with communities, have struggled to attain some principled way to collaborate. According the U.S. Environmental Protection Agency, some of these basic principles include:

1. Community partners should be involved at the earliest stages of the project, helping to define research objectives and having input into how the project will be organized.

This is often controversial because academic requirements may differ from community-defined research objectives. Also, the use of human subjects in research requires an ethics committee approval from the university. News that they are subjects in research is not usually well received by the studied community.

2. Community partners should have real influence on project direction and at least enough power to ensure that the original goals, mission, and methods of the project are adhered to.

This can be controversial because the learning needs of the students may require the professor to prioritize her students.

3. Research processes and outcomes should benefit the community. Community members should be hired and trained whenever possible and appropriate, and the research should help build and enhance community assets.

Communities often feel used when they interact with a community science project and are left with no product in hand afterwards.

4. Community members should be part of the analysis and interpretation of data and should have input into how the results are distributed. This does not imply censorship of data or of publication, but rather the opportunity

to make clear the community's views about the interpretation prior to final publication.

Academic and scientific models of causality face significant challenges when viewed through the lens of the studied community. The usual scientific explanation for lack of no causality or of causality (the null hypothesis) usually strikes communities as validating the status quo and discounting their known and shared experience. If the sample size is too small to be statistically valid, as is often the case with a community science project that is designed around a course, then the community may view the exercise as a waste of time.

5. Productive partnerships between researchers and community members should be encouraged to last beyond the life of the project. This will make it more likely that research findings will be incorporated into ongoing community programs and therefore provide the greatest possible benefit to the community from research.

6. Community members should be empowered to initiate their own research projects that address needs they identify themselves.

POTENTIAL FOR FUTURE CONTROVERSY

As citizens become frustrated with governmental environmental regulation that seems to ignore cities and cater to industry's self-reported style of enforcement, they are seeking scientists to help understand their local environmental concerns. As scientists from industry or government seldom help communities some community leaders turn to universities in their neighborhood for scientific help. Community-based science is a controversy that fills in the void left when communities have no access to government or industry scientists. There is usually a felt need for science in environmental controversies when a community feels threatened and does not trust the information from industry and government stakeholders.

Universities are often an excellent resource for communities to find scientists. However, universities are stakeholders too. The students expect to receive a safe and valuable educational experience. Universities are also responsible for environmental impacts for the operation of their physical plant and for the chemicals used in research. In most places, colleges and universities are exempt from right-to-know laws. They produce toxic, hazardous, and radioactive waste in close proximity to young, child-rearing adults. Some universities do a large amount of research for the military in these areas. Tension and controversy between the university and host community-form part of the battleground of community based, university-affiliated science shops.

In one sense community-based science laboratories are an extension of citizen-based environmental monitoring. As in citizen monitoring, any evidence of environmental permit violations can only come through the permit holder (unlikely) or the permit grantor (usually a state environmental agency or regional office of the Environmental Protection Agency). Communities, students, faculty, and scientists are likely to become frustrated by the lack of power scientific information about the environment has on government decision making.

See also Citizen Monitoring of Environmental Decisions; Collaborative Environmental Decision Making; Environmental Impact Statements; Public Participation/Involvement in Environmental Decisions

Web Resources

Science and Environmental Health Network: Community Research. *The Networker* 3(3). Available at www.sehn.org/Volume_3-3.html. Accessed January 20, 2008.

Street Science: Community Knowledge and Environmental Health. Available at www.pubmedcentral.nih.gov/articlerender.fcgi?artid = 1280385. Accessed January 20, 2008.

U.S. Environmental Protection Agency. National Estuary Program: Implementing a Community Based Watershed Program. Available at www.epa.gov/neplessons/. Accessed March 2, 2008.

Further Reading: Agrawal, Clark C. 2001. *Communities and the Environment: Ethnicity, Gender, and the State in Community-Based Conservation.* Piscataway, NJ: Rutgers University Press; Horgan, John. 1997. *The End of Science: Facing the Limits of Knowledge in the Twilight of the Scientific Age.* New York: Broadway Books; Hynes, H. Patricia, and Doug Brugge. 2005. *Community Research in Environmental Health: Lessons in Science Advocacy and Ethics.* Aldershot, UK: Ashgate Publishing; Randolph, John. 2004. *Environmental Land Use Planning and Management.* Washington, DC: Island Press; Spellerberg, Ian F. 2005. *Monitoring Ecological Change.* Cambridge: Cambridge University Press.

COMMUNITY RIGHT-TO-KNOW LAWS

Many community residents assume they have the right to know about industrial emissions and environmental conditions generally, especially if they pose risks. Often they do not. This extremely contentious issue resulted in federal, state, and local laws that give communities some rights to know about some of the emissions from industry and some of the environmental conditions of their permits.

The reporting of environmental impacts is always contentious. Industry resents the cost, the exposure to environmental liability, and the potential loss of trade secrets. Firefighters, police, and sanitation workers were among the first exposed to the hazards of interacting with unknown chemicals. Toxic spills are difficult to handle without knowledge of their contents. They can spread and affect a large area, sometimes for a long time. Residential communities along major roads and rail lines began to get concerned about the contents of these spills, especially as they might taint water supplies. The city of Eugene, Oregon, was among the first cities to enact right-to-know laws, basically copying part of the federal law discussed further on. There are still many battles in this battleground. Many industries are still not included under right-to-know law coverage, such as universities, defense agency research sites and agribusiness. The direction of more inclusion of all industries and all environmental impacts is gaining rapid momentum, spurred on by a new wave of sustainability proponents and policies. The reporting requirements established under community right-to-know laws provide the public with important information on chemicals that can be hazardous in their communities. Hazardous chemicals and materials are a part of everyday life. They range from common household products such as cleaning solvents to industrial substances used in businesses surrounding your neighborhood. Being aware of chemical hazards within a community will help facilitate environmental planning, land-use decisions, and emergency planning and help reduce the effects of chemical incidents.

Industry concerns about public knowledge of their emissions include fear of litigation, partially driven by citizen suit provisions. Sometimes, concerns about protected trade secrets emerge and are challenged by communities and environmentalists. Most industries are concerned with profit and perceive environmental compliance as reducing profit. They are concerned about liability for waste they did not produce and other contingent environmental liabilities.

FEDERAL LAW

On October 17, 1986, the president of the United States signed into law the "Superfund Amendments and Reauthorization Act of 1986" (SARA). This act amended the already-existing law titled "Comprehensive Environmental Response, Compensation and Liability Act of 1980" (CERCLA), which was also known as *Superfund*. The Emergency Planning and Community Right-to-Know Act (EPCRA) was part of this legislation. States and some municipalities have used this law as a model for their own right-to-know laws.

BIRTHPLACE OF THE TOXICS RELEASE INVENTORY (TRI)

This inventory was established under the Emergency Planning and Community Right-to-Know Act of 1986 (EPCRA). It was later expanded by the Pollution Prevention Act of 1990.

EPCRA's primary purpose is to inform communities and citizens of chemical hazards in their areas. Sections 311 and 312 of EPCRA require businesses to report the locations and quantities of chemicals stored on-site to state and local governments in order to help communities prepare to respond to chemical spills and similar emergencies. EPCRA Section 313 requires the EPA and the states to annually collect data on releases and transfers of certain toxic chemicals from industrial facilities and make the data available to the public in the Toxics Release Inventory (TRI). In 1990 Congress passed the Pollution Prevention Act that required that additional data on waste management and source reduction activities be reported under TRI. The goal of TRI is to empower citizens, through information, to hold companies and local governments accountable in terms of how toxic chemicals are managed.

EPCRA does not place any limits on which chemicals can be stored, used, released, disposed of, or transferred at a facility. It simply requires accurate environmental information. The law applies different requirements, has different deadlines, and covers a different group of chemicals, which is a reporting cost borne by industry now. These specific requirements are contained in the following sections of EPCRA:

- Emergency Planning (Sections 301–303)
- Emergency Release Notification (Section 304)
- Community Right-to-Know Reporting Requirements (Sections 311–312)
- Toxic Chemical Release Inventory Reporting (Section 313)

Each one of these sections contains specific and detailed procedures and processes. They themselves are not controversial. What is controversial is their level of enforcement. In many states the protocol is not followed. Communities now want to know more about their environment, especially anything potentially dangerous. Environment justice advocates want to measure the benefit and burden of past and present environmental decisions. Sustainability advocates want to know all past and present environmental impacts, usually at the ecosystem level. As the coverage of right-to-know laws is rapidly expanding, so too is the level of enforcement. Citizen monitoring of environmental decisions, environmental litigations, and political organizing are part of the backdrop to this battleground. As they increase so does the controversy about right-to-know coverage and enforcement.

POTENTIAL FOR FUTURE CONFLICT

As populations grow and cumulative risks rise, communities will insist on completely transparent environmental transactions. The tendency for right-to-know provisions to grow from a few standard industrial classifications to more and more comprehensive categories is very strong. Accurate information about cumulative emissions requires knowledge about all environmental impacts. Most environmental information is self-reported. The battleground here is defined by high levels of distrust, fear and actuality of contentious environmental and toxic tort litigation, and awkward relations between state environmental agencies and local communities.

Today, communities that can do so seek sustainable industries. In the late 1990s a wave of cities began enacting municipal right-to-know ordinances. They need to know all environmental impacts. City employees that deal with fire, law and order, and emergency services all need to know what chemicals they may be forced to work and interact with. This concern is spreading to public schools and impacts on children, especially in schools built on landfills. Parents need to know all the environmental impacts on their children. The need to know environmental impacts is often greatest at the actual point of impact. Air, water, and land vectors of exposure are often of great interest to public health agencies. Many communities that relied on military bases that have since closed also need to know about any remaining environmental risks. Sometimes, the wing de-icers used on military landing strips can contaminate local groundwater supplies, for example.

The security concerns raised by the 9/11 attack on New Yorks' Twin Towers have slowed the rapid growth of right-to-know coverage, as discussed previously. One of the last proposals to move forward in the waning days of the Bush administration's EPA was to change the reporting requirements from once a year to once every two years. Communities and environmentalists strongly protested this attempt. In rural states, the battleground of right to know is the area of agricultural emissions. Agricultural fertilizer can be and was used for bombs, so post-9/11 concerns about where it now goes have increased. However, whether the community has a right to know about agricultural uses and emissions is a developing area in many state legislatures now. Farmworkers and their exposures are also part of this push for more information.

Despite consistent concerns about the cost of these requirements by complying industries and state agencies and local environmental agencies, right-to-know laws show no signs of slowing down. There are still categories of uncovered environmental impacts, such as low emitters and universities. The new push for sustainability will also push open these categories. Right-to-know laws may remain controversial because they open the door to other unresolved and currently unknown environmental controversies.

See also Citizen Monitoring of Environmental Decisions; Toxics Release Inventory

Web Resources

Right to Know Network. Available at www.rtknet.org/. Accessed March 2, 2008.
South Coast AQMD. Cumulative Impacts Working Group Maps. Available at www.aqmd.gov/rules/ciwg/maps.html. Accessed March 2, 2008.
U.S. Environmental Protection Agency. Local Emergency Planning Committee (LEPC) Database. Available at yosemite.epa.gov/oswer/lepcdb.nsf/HomePage?openForm. Accessed March 2, 2008.

Further Reading: Baram, Michael S., Patricia S. Dillon, and Betsy Ruffle. 1992. *Managing Chemical Risks: Corporate Response to SARA Title III.* Boca Raton, FL: CRC Press; Boyce, James K., and Barry G. Shelley. 2003. *Natural Assets: Democratizing Ownership of Nature.* Washington, DC: Island Press; Gray, Peter L. 2002. *Epcra.* Washington, DC: American Bar Association; Legator, Marvin S. 1993. *Chemical Alert: A Community Action Handbook.* University of Texas Press; Rainey, David L. 2006. *Sustainable Business Development.* Cambridge: Cambridge University Press; West, Bernadette M. 2003. *The Reporter's Environmental Handbook.* Piscataway, NJ: Rutgers University Press.

CONSERVATION IN THE WORLD

World conservation efforts focus on the most important areas of the environment in terms of biological diversity. Controversies occur when these efforts clash with cultural and developmental interests in these places. Marine world conservation issues are just beginning.

WHAT DO WORLD CONSERVATIONISTS CONSERVE?

Conservationists have national and international organizations. They focus on conserving, and sometimes protecting, natural resources and the environment. They are a traditional environmental advocacy group. One prominent example of such a group is the World Wildlife Fund (WWF). They try to conserve nature and natural systems. This often becomes the basis for large government parks or animal protection zones.

Conservationists now try to protect endangered species and habitats. Many view their mission as combating the extinction of species. World conservationists tend to view the planet in terms of ecoregions and look closely to save endangered species' habitat. They try to avoid political entanglements in countries but are very practical when it comes to getting and protecting habitat. Environmentalists, cooperating scientists and governments, and universities have examined many species and habitats. To this end, they have mapped out almost 900 ecoregions of the world. For a variety of reasons, some controversial, they have so far found 238 of them to be of great importance for biological diversity. This type of conservationist ethic also resonates with sustainability proponents, who tend to view biological diversity as an important ecosystem survival mechanism over time. According to the WWF, which ran the main project, the definition of an ecoregion is:

> An ecoregion is a relatively large unit of land or water containing a geographically distinct assemblage of species, natural communities, and environmental conditions. Ecoregion conservation aims to address the fundamental causes of biodiversity loss by looking across whole regions to identify the actions needed to secure long-term conservation and results that are ecologically, socially and economically sustainable.

World conservationists now focus on preservation of these areas while working with local populations. Issues of plant and animal poaching, fair pay to local workers, the emerging role of nongovernmental organizations (NGOs), and the never-ending goal of discovering all the species can all present battlegrounds.

WORLD CONSERVATION AND HUMAN RELATIONSHIPS

Protected ecoregions may occur on someone's land. While private property will form the battleground for conservation efforts in Western countries, other countries have different land ethics and practices. About 18 percent of U.S. land is protected while 0.4 percent of U.S. waters are protected. Battlegrounds may be tribal, community based, unknown, and rapidly developing. Research indicates that areas of high environmental biodiversity are also of high cultural diversity. Because many indigenous people are in oppressive and complex relationships with the current nation-state or country, conservation intervention in these areas is very controversial. This battleground can be complex. In many instances, indigenous people have accumulated important ecological knowledge. This knowledge is important to other people. Strong interests in potential medicines, patents, lumber, minerals, water, as well as ecoregion protection for biodiversity all lay claim to land that is someone else's. The right of indigenous peoples to self-development is a serious, ongoing battleground.

When indigenous people or poor nations want to economically develop and achieve a standard of living comparable to the United States, but occupy high-priority ecoregions, controversy may ensue from the imposition of conservation policies. Logging, mining, low-level nuclear waste sites, and other similar activities bring in needed capital but at a huge environmental price. A large problem is that the benefit of world conservation policies is focused on the world population but the burden of such policies rests predominantly on poor people with low life expectancies and the desire to use their country to achieve economic prosperity. Some environmental groups have worked hard to fill this void by offering to help with sustainable trade in these areas. For many environmentalists, this has meant increasing their level of cultural competency because of the cultural diversity often associated with areas of environmental biodiversity. The world is a very diverse place, both culturally and environmentally. In total, 6,867 ethnolinguistic groups in about 2,252 ecoregions are known.

WORLD CONSERVATION EFFORTS WITH OCEANS

Oceans represent the cutting edge of world conservation efforts. Marine areas are among the most highly prioritized ecoregions. Oceans cover most of the Earth. They are intrinsically part of the climate and life cycle. Ocean conservation efforts still comprise only a small part of total conservation efforts.

The deep ocean floor is unseen by most people, so the severe environmental impacts caused by mining, bottom trawling, and dredging equipment go unnoticed. Pollution affects oceans differently than it does land masses. As noted by the World Watch Institute:

> Once pollution enters the sea, currents and tides can move and mix them far from the original source. Or they may be consumed by a species and move up the food chain, becoming more concentrated as they go. Both pollutants and species continually migrate across boundaries and interact, complicating protection efforts.

Pollutants created and dispersed by one nation may have an impact on oceans and wildlife far from its borders. Persistent organic pollutants (POPs) synthetic chemicals do not degrade easily. Over time they tend to circulate toward colder environments such as the Arctic. There they accumulate in the fatty tissues of fish, then move up the food chain to predators at a more concentrated level. POPs have been implicated in a wide range of animal and human health controversies.

Perhaps the greatest difficulty in protecting the world's oceans is that although ocean waters are used by many nations, no nation owns them. A nation has sovereignty over its lands and territorial sea, the small coastal strip adjacent to its shores, but no nation has sovereignty over the high seas. Conserving open-ocean resources thus requires concerted international cooperation, a more complicated effort than is involved in national land conservation efforts, which can generally be coordinated within one nation. Prior to the mid-twentieth century, this international cooperation was lacking; in fact, the use of ocean resources was guided by the freedom-of-the-seas principle.

OCEAN LAW AND WORLD CONSERVATION

In 1958, however, a number of factors prompted the United Nations to develop an international law to govern the oceans. Some coastal nations had unilaterally claimed different parts of the oceans as their own and conflicts between nations

LOSS OF REEFS

The loss of ocean reefs is a great concern for many environmentalists:

- Twenty percent of the world's coral reefs have been effectively destroyed and show no immediate prospects of recovery;
- Of the 16 percent of the world's reefs that were seriously damaged in 1998 approximately 40 percent are either recovering well or have recovered;
- Twenty-four percent of the world's reefs are under imminent risk of collapse through human pressures; and a further 26 percent are under a longer-term threat of collapse;

Reefs are useful to the environment and to people because they

- Protect shores from the impact of waves and from storms
- Provide benefits to humans in the form of food and medicine
- Provide economic benefits to local communities from tourism

The reef damage is global, but early indicators suggest that some areas are worse hit than others:

- Coral reefs of Southeast Asia, the most species-rich on earth, are the most threatened at more than 80 percent, primarily from coastal development and fishing-related pressures.
- Most U.S. reefs are threatened. Almost all the reefs off the Florida coast are at risk from runoff of fertilizers and pollutants from farms and coastal development. Close to half of Hawaii's reefs are threatened, while virtually all of Puerto Rico's reefs are at risk.
- Nearly two thirds of Caribbean reefs are in jeopardy. Most of the reefs on the Antilles chain, including the islands of Jamaica, Barbados, Dominica, and other vacation favorites, are at high risk. Reefs off Jamaica, for example, have been ravaged as a result of overfishing and pollution. Many resemble graveyards, algae-covered and depleted of fish.

Some actions that can mitigate the loss of reefs are creating marine parks; treating sewage before it reaches reefs; and eliminating costly government subsidies of reef-destroying activities.

However, polluted runoff from land into the ocean remains largely uncontrolled. Ocean dumping of all kinds of toxic and hazardous waste and materials continues unabated. Ships incinerate waste at sea. The environmental effects of this waste could be as devastating to the ocean as air pollution is to the climate. The reefs are among the first signs that there may be deep-ocean environmental impacts.

began to arise. In addition, ocean pollution had begun to threaten coastal regions and wildlife worldwide. In response, the United Nations held a series of conventions between 1958 and 1982 that resulted in the United Nations Convention on the Law of the Sea (UNCLOS), which went into effect in 1994 and, as of February 2002, has been ratified by 138 nations. Although the United States has not yet ratified the convention, it has accepted it in principle. UNCLOS covers a wide range of ocean issues, including protection of the marine environment and the conservation and management of its resources. UNCLOS also approved a twelve-nautical-mile territorial limit for coastal nations and a two-hundred-nautical-mile exclusive economic zone, in which the adjacent nation may control fishing rights, marine environmental protection, and scientific research. The provisions of UNCLOS, some claim, are insufficient to protect the marine environment. The scarce language in UNCLOS regarding the conservation of marine biodiversity is far more aspirational than operational. Treaties such as UNCLOS, they argue, operate using unsuccessful marine policies, based on the outdated maxim, "Take as much as can be taken and pollute as much as can be polluted until a problem arises." These conservationists argue that a precautionary principle, preventing damage before it occurs, must be incorporated into conservation treaties.

For many, marine conservation efforts continue to lag behind those on land, and identifying how best to meet the challenges of conserving ocean and coastal resources remains the subject of ongoing controversy.

POTENTIAL FOR FUTURE CONTROVERSY

The rapid advancement in knowledge about the world's biodiversity fuels the power of the world conservation movement. There is a strong emphasis on

SEA LIONS VS. SALMON

California sea lions, many kinds of salmon, and Pacific Harbor seals on the West Coast are protected species. Sea lions in particular have thrived, with their numbers increasing at an annual rate of about 5 to 7 percent, tripling since the 1970s.

Sea lions and harbor seals have increased in population and pursue food sources that now include piers, docks and marinas, and other public spaces. That raises a battleground of public safety concerns. Some of the sea lions can be aggressive and large. An environmental battleground develops when the sea lions pursue returning salmon and other fish. Sea lions are smart and inquisitive, especially in search of food. They follow the salmon up fish ladders when the salmon return from the ocean to spawn. Salmon numbers then decrease, which reduces the allowable fishing for licensed anglers and sports fisherman. Fishermen want to shoot and kill the sea lions, still a protected species. Federal and state wildlife agencies have tried to deter them with loud noises, bean-bag bullets, and their presence. Sea lions quickly learn any weaknesses in these deterrents and are very persistent. This creates a fundamental conservation question.

NOAA's official position is that in cases where seals or sea lions are causing repeated, serious conflict with human activity, state or federal managers should be authorized to lethally remove identified problem marine mammals after individual animals fail to respond to repeated attempts to deter them. Killing problem sea lions has diminished salmon predation problems in some areas. Wildlife researchers are quick to point out that there are many other factors affecting diminished fish populations, such as pollution, overfishing, fishery reductions, and poaching.

monitoring of environmental conditions in land, air, and water. Global warming and climate change controversies have also sparked renewed interest in world conservation. Species extinction, indigenous cultural integrity, increasing world population and consumption, and more accurate environmental information will make this controversy continue for a long time.

See also Climate Change; Cultural vs. Animal Rights; Ecotourism as a Basis for Protection of Biodiversity; Endangered Species; Precautionary Principle; Rain Forests; Sustainability

Web Resources

United Nations Environmental Programme—Earth Monitor for World Conservation. Available at www.earthwatch.unep.ch/observation/index.php. Accessed January 20, 2008.

World Conservation Union. Available at www.iucn.org/en/news/archive/2005/newjune05. htm. Accessed January 20, 2008.

World Conservation Problems: Factsheet. Available at www.yptenc.org.uk/docs/factsheets/ env_facts/world_con_probs.html. Accessed January 20, 2008.

Further Reading: Jenkins, Martin D., and Brian Groombridge. 2002. *World Atlas of Biodiversity: Earth's Living Resources in the 21st Century.* Berkeley: University of California Press; Levitt, James N. 2002. *Conservation in the Internet Age: Threats and Opportunities.* Washington, DC: Island Press; McDuff, Mallory D., Martha C. Monroe, and Susan Kay Jacobson. 2006. *Conservation Education and Outreach Techniques: A Handbook of Techniques.* Oxford: Oxford University Press; Nierenberg, Danielle. 2006. *State of the World 2006: A Worldwatch Institute Report on Progress toward a Sustainable Society.* Washington, DC: Worldwatch Institute; Ray, G. Carleton. 2003. *Coastal Marine Conservation: Science and Policy.* London: Blackwell Publishing; Sayer, Jeffrey A., and Bruce Morgan Campbell. 2004. *The Science of Sustainable Development: Local Livelihoods and the Global Environment.* Cambridge: Cambridge University Press.

CULTURAL VS. ANIMAL RIGHTS: THE MAKAH TRIBE AND WHALING

Whaling is an ancient industry that has defied meaningful environmental regulation despite the near extinction of some whales. Whaling is one of the cultural and aboriginal rights of the Makah people. The whales they hunt are endangered and environmentalists protest the killing of these whales.

THE MAKAH

The Makah Tribal Reservation is centered on Neah Bay (on the northwest tip of the Olympic Peninsula in the state of Washington), an area that is far from large population centers and difficult to find. The tribe has about 2,500 members, about 1,300–1,500 of whom live on the reservation in Neah Bay itself. The Makah tribe are a coastal people with a close relationship with the sea. Jobs are hard to find locally. There is one small town, and few health facilities.

The Makah tribe were whalers and traded whale products. In 1856 they sold $8,000 worth of whale oil. The tribe stopped whaling around 1926 because California gray whales were almost extinct. Makah people resumed the whale hunt recently, taking a gray whale on May 17, 1999. Environmentalists strongly protested, including physical confrontation.

WHAT GOOD ARE WHALES?

Whales provided the sustenance for the Makah. All parts were used, and none wasted. Whales gave oil, meat, bone, sinew, and gut. According to the Makah, whaling was approached ceremoniously.

> To get ready for the hunt, whalers went off by themselves to pray, fast and bathe ceremonially. Each man had his own place, followed his own ritual and sought his own power. Weeks or months went into this special preparation beginning in winter and whalers devoted their whole lives to spiritual readiness. Spring was the time for the hunt. Men waited for favorable weather then paddled out, eight in a canoe, timing arrival on the whaling grounds for daybreak. (http://makah.com)

Tribal whaling methods contrasted sharply with modern whaling. Greater efficiency in ocean fishing used more boats and explosive harpoons. Whales tend to have lower population replacement rates. With the rapid overfishing of the whale stocks, the whale population plummeted.

THE INTERNATIONAL WHALING COMMISSION (IWC)

The International Whaling Commission (IWC) was created in 1948. The whaling was self-regulated even after the formation of the IWC. The IWC had no enforcement authority to penalize wrongdoers or to correct inaccurate reporting of the numbers of whales killed.

A MORATORIUM ON COMMERCIAL WHALING

The United Nations (UN) held its first conference on the environment in Stockholm, Sweden, in 1972. Whales were high on the agenda. At this meeting whale protection advocates rallied against the IWC. The UN delegates present unanimously adopted a resolution recommending a 10-year moratorium on commercial whaling. A few weeks later at the IWC meeting in London, whalers soundly defeated the call for a moratorium. The moratorium remains in place now only because the pro-whaling nations of Japan and Norway cannot get the three-fourths majority of votes necessary to overturn it.

The Makah tribe maintains that they have a right to hunt whales, as evidenced by a signed treaty. The Humane Society protests the inhumane killing of all animals. They sought to stop the Makah from hunting whales by suing them in federal court.

THE TREATY OF NEAH BAY

The tribe maintains that its 1855 treaty, the Treaty of Neah Bay, expressly preserves its right to whale. However, the treaty does so "in common with all citizens of the United States," the appeals court judges held, quoting from the treaty itself. They further wrote:

> The Tribe has no unrestricted treaty right to pursue whaling in the face of the MMPA [Marine Mammal Protection Act]. Instead, having concluded that the MMPA is applicable to regulate the Tribe's whaling because the MMPA's application is necessary to effectuate the conservation purpose of the statute, and because such application is consistent with the language of the Neah Bay Treaty, we conclude that the issuance by NOAA [the National Oceanic and Atmospheric Administration] of a gray whale quota to the Tribe, absent compliance with the MMPA, violates federal law.

In short, while the court did not rule that the Makah lacked a treaty right to hunt whales, it held that the Makah must comply with the MMPA if they wish to pursue any such right. Under the MMPA, entities may apply for waivers to the law's prohibitions. The court has ordered the Makah to follow the MMPA's waiver procedures before proceeding with their hunt. These procedures do not guarantee the cultural and subsistence rights the Makah thought were protected by treaties.

The Makah have stated that this ruling will have damaging implications for treaty rights across the board, that is, will be used to negate treaty rights. The Humane Society points out that the MMPA has specific procedures that allow tribes to exercise their treaty rights to hunt marine mammals. According to them, this ruling in no way negates those rights. They feel the federal court opinion ensures that strong conservation principles will govern the exercising of those rights.

The Humane Society has opposed the Makah hunt from the beginning, not because they oppose native treaty rights or even the aboriginal subsistence hunting of marine mammals. The HSUS does not oppose the Makah or Makah culture. They do oppose any killing of marine mammals when it is done in an inhumane manner. They also oppose marine mammal hunts when they do not meet clear subsistence needs. They oppose all whaling that is not in compliance with international treaty obligations.

The HSUS has worked for decades at the international level to improve the humaneness of aboriginal subsistence whaling and to minimize the quotas needed to meet subsistence needs.

The ban on commercial whaling does not affect aboriginal subsistence whaling, which is permitted by Denmark, the Russian Federation, St. Vincent and the Grenadines, and the United States. Nor does this ban cover small whales, as Japan and a handful of other nations refuse to accept the IWC jurisdiction over small whales and dolphins.

HOW THE MAKAH HUNT WHALES

One of the concerns of environmentalists is how the Makah people actually hunt and kill the whales. Here, in their own words, is how the Makah hunt the whale.

Paddling silently, whalers studied the breathing pattern of their quarry. They knew from experience what to expect. As the whale finished spouting and returned underwater, the leader of the hunt directed the crew to where it would next surface. There the men waited.

When the whale rose, the paddlers held the canoe just to its left matching the speed of their quarry. As the back broke the surface, the harpooner struck and the crew instantly paddled backward, putting all possible distance between the canoe and the wounded prey so as to avoid the thrashing tail flukes. A hit in the shoulder blade interfered with use of the flippers and slowed the whale. Floats of sealskin blown up like huge balloons and attached to the harpoon line slowed the whale. Harpoons weren't intended to kill the whale, but to secure the sealskin floats to them until they tired themselves and could be lanced fatally. Shafts of yew wood measured 12 to 18 feet long—heavy wood to add to the harpooner's thrust and help the blade pierce deeply. Splices in the shaft deadened the springiness and furthered the penetration. They also let the shaft break rather than hit the canoe repeatedly if the whale rolled. Furthermore, they allowed a clean break rather than a splintering. This aided repair.

Shafts fell away once the harpoon head had been set. In a whale, the head turned partly sideways. Barbs of elk antler helped to keep it from pulling out. They fit one on each side of the blade, which was mussel shell. Spruce pitch, at Ozette still pungent after 500 years within the earth, fared and smoothed the head. Whale sinew plied into rope and bound with wild cherry bark attached the harpoon head to as much as 40 fathoms of additional rope. This line, which was of twisted cedar boughs, was carried coiled within the baskets so that it would play out easily and wouldn't entangle the canoe's occupants.

A telltale float at the end of the line acted as a marker so that the whalers could follow the wounded whale, setting additional harpoons and staying out over night if need be.

Eventually the time came for the final kill of the whale which was done using a special lance. The next step was to tow the whale home—a distance of only a few miles if its spirit had heeded prayers to swim for the beach, perhaps 10 miles or more if not. As a precaution against the whale's sinking, a diver generally laced the mouth shut. This kept water from flooding into the stomach, weighing the carcass down and interfering with the tow. . . .

Songs welcomed the whale to the village, welcomed the returning hunters and praised the power that made it all possible. http://makah.com.

POTENTIAL FOR FUTURE CONTROVERSY

As populations increase and ocean hunting technology improves to the point of being harvesting, the whaling controversy will increase. There is a large market advantage for those countries that do not comply with the law. The information is self-reported in a self-regulated industry. Aboriginal people, such as the Makah exert cultural rights in traditional ways. Many indigenous groups around the world have very strong traditions and subsistence needs tied to ocean fishing. They too suffer when large floating whale-harvesting factories deplete the species beyond a sustainable level.

The battleground for this controversy is the ocean, the habitat of the whale. The lack of identified ownership by people or states and the uncharted vastnesses of ocean depths contribute to the inability to solve this particular environmental controversy. Strong political advocacy by environmental groups like Greenpeace have increased public awareness of this issue.

See also Endangered Species; Environmental Impact Statement: Tribal; Sustainability

Web Resources

Federal Government Reaffirms Makah Treaty Rights Regarding Whaling. Available at http://www.publicaffairs.noaa.gov/releases2001/jul01/noaa01r121.html. Accessed January 20, 2008.
Humane Society of the United States. Federal Appeals Court Bars Makah Tribe from Whaling. Available at www.hsus.org/marine_mammals/marine_mammals_news/federal_appeals_court_bars_makah_tribe_from_whaling.html. Accessed January 20, 2008.
Makah Tribe. Whaling. Available at www.makah.com/whaling.htm. Accessed January 20, 2008.

Further Reading: Erikson, Patricia Pierce, Helma Ward, and Kirk Wachendorf. 2002. *Voices of a Thousand People: The Makah Cultural and Research Center.* Lincoln: University of Nebraska Press; Heyning, John. 1995. *Masters of the Ocean Realm: Whales, Dolphins and Porpoises.* Vancouver: University of British Colombia Press; Kieran, Mulvaney. 2003. *The Whaling Season: An Inside Account of the Struggle to Stop Commercial Whaling.* Washington, DC: Island Press; Martello, Marybeth Long, and Sheila Jasanoff, eds. 2004. *Earthly Politics: Local and Global in Environmental Governance.* Cambridge, MA: MIT Press; Stoett, Peter John. 1997. *The International Politics of Whaling.* Vancouver: University of British Columbia Press; Tweedie, Ann M. 2002. *Drawing Back Culture: The Makah Struggle for Repatriation.* Seattle: University of Washington Press.

CUMULATIVE EMISSIONS, IMPACTS, AND RISKS

Since the mid-1850s the results of the industrial revolution have polluted the environment. When industrial manufacturing processes garner raw materials, they produce a product and by-products. These by-products are often wastes and chemicals. They have grown enormously since then. In many urban areas industrial emissions have accumulated for 150 years. These emissions are mixed with other waste streams as they percolate through soil or volatize into the air. This can result in accumulating impacts to the environment, almost all negative. Emissions that impact the environment can bioaccumulate in all species, including humans. Bioaccumulation of some chemicals, such as metals, is known to be very harmful, and therefore risky, to humans. Emissions, impacts, and risks fall under the collective label of cumulative effects. No one industry wants to be liable for the emissions of others. Many communities are concerned about the eroding health of their families. Environmentalists want cumulative effects to be accounted for in environmental impact statements. Policy development is weak, yet every day these cumulative effects increase. This is a young controversy that is growing and will drive and divide many other environmental policies.

BACKGROUND

Several reports have highlighted the importance of understanding the accumulation of risks from multiple environmental stressors. These reports, as well as legislation such as the Food Quality Protection Act of 1996 (FQPA), urged the EPA to move beyond single chemical assessments and to focus, in part, on the cumulative effects of chemical exposures occurring simultaneously. In 1999, the EPA's Risk Assessment Forum began development of EPA-wide cumulative risk assessment guidance.

CUMULATIVE EFFECTS

Cumulative risk means the combined risks from aggregate exposures to multiple agents or stressors. Several key points can be derived from this definition of cumulative risk. First, cumulative risk involves multiple agents or stressors, which means that assessments involving a single chemical or stressor are not cumulative risk assessments under this definition. Second, there is no limitation that the agents or stressors be only chemicals. They may be, but they may also be biological or physical agents or an activity that, directly or indirectly, alters or causes the loss of a necessity such as habitat. Third, this definition requires that the risks from multiple agents or stressors be combined. This does not necessarily mean that the risks should be added, but rather that some analysis should be conducted to determine how the risks from the various agents or stressors interact. It also means that an assessment that covers a number of chemicals or other stressors but merely lists each chemical with corresponding risk without consideration of the other chemicals present is not an assessment of cumulative risk under this definition. *Cumulative risk assessment* in this report means "an analysis, characterization, and possible quantification of the combined risks to health or the environment from multiple agents or stressors." One key aspect of this definition is that a cumulative risk assessment need not necessarily be quantitative, so long as it meets the other requirements. The framework itself is conceptually similar to the approach used in both human health and ecological assessments, but it is distinctive in several areas. First, its focus on the combined effects of more than one agent or stressor distinguishes it from many assessments conducted today, in which, if multiple stressors are evaluated, they are usually evaluated individually and presented as if the others were not present. Second, because multiple stressors are affecting the same population, there is increased focus on the specific populations potentially affected rather than on hypothetical receptors. Third, consideration of cumulative risk may generate interest in a wider variety of nonchemical stressors than do traditional risk assessments.

THE EPA APPROACH

The term *cumulative risk assessment* covers a wide variety of risks. Currently, EPA assessments describe and, where possible, quantify the risks of adverse health and ecological effects from synthetic chemicals, radiation, and biological stressors. As part of planning an integrated risk assessment, risk assessors must define dimensions of the assessment, including the characteristics of the population at risk. These include individuals or sensitive subgroups that may be highly susceptible to risks from stressors or groups of stressors due to their age (e.g., risks to infants and children), gender, disease history, size, or developmental stage. There are other risk issues, dimensions, and concerns that the EPA does not address. This broader set of concerns, recognized as potentially important by many participants in the risk assessment process, relate to social, economic, behavioral, or psychological stressors that contribute to adverse health effects. These stressors may include existing health conditions, anxiety, nutritional status, crime, and congestion. On the important topic of special subpopulations, the EPA and others are giving more emphasis to the sensitivities of children and to gender-related differences in susceptibility and exposure to environmental stressors. The EPA's stated focus is on risk assessments that integrate risks of adverse health and ecological effects from the narrower set of environmental stressors. There is a great deal of controversy about what specifically is an adverse impact. The EPA is engaged in several activities that involve working with stakeholders. However, the EPA still resists regularly incorporating cumulative risk concerns in most applied policy areas such as environmental impact statements.

HOW DO YOU AGGREGATE RISKS?

This is a very controversial area on many levels. In the scientific community there is a diverse body of opinion and perspective. Most would like to have a model of some kind that reliably predicts an effect from a given cause. Without a model many scientists do not believe it is possible to have a policy. On the community level, controversies surround what goes into the pool of risks that are aggregated. For example, low levels of fire and police protection may make an area more risky then one with high levels of fire and police protection. Environmental advocacy groups want to make cumulative effects part of the requirements for an environmental impact assessment.

Due to the current state of the practice, strongly vested stakeholder positions, and limited data, the aggregation of risks may often be based on a default assumption of additivity in the United States. This simply adds the risk per chemical for a sum total of risk. It also ignores antagonism, when chemicals mitigate the risk from one another. In many western European markets, synergized risk and risk to vulnerable populations determine entry into commerce. Some emerging cumulative risk approaches in Canada and western Europe may help set up data development approaches in the United States. However, U.S. approaches to emission control still leave many sources completely unregulated, and those that are regulated emit millions of pounds of chemicals per year. For an accurate cumulative risk assessment, all past and present emissions must be counted.

FLINT, MICHIGAN, ENVIRONMENTAL JUSTICE, AND CUMULATIVE LEAD EXPOSURE

Flint, Michigan, was the site of an early legal challenge based in part on cumulative impacts of lead on African American people, primarily African American children. The industrial plant was built by the Genesee Power Company. It was located in a predominantly African American residential neighborhood in Flint. The lawsuit (*NAACP-Flint Chapter et al v. Engler et al.,* No. 95–38228-CZ [Circuit Court, Genesee County, filed 7/22/95]), filed by two community groups—United for Action and the NAACP-Flint Chapter—and several African American women, challenged the state of Michigan's decision to grant a construction permit to the power company on environmental and environmental justice grounds. The complaint alleged that granting this permit would allow that facility to emit over two tons of lead per year into an African American community that already had very high levels of lead exposure and contamination.

The Maurice and Jane Sugar Law Center for Economic and Social Justice, a Detroit-based national civil rights organization represented the community.

Census data showed that a substantial disparity exists in the racial composition of the population around the proposed site. Then, the population within a one-mile radius surrounding the incinerator was 55.8 percent African American; in contrast, African Americans comprise 19.6 percent of those living in Genesee County and 13.9 percent of Michigan's population. There was evidence in the form of a risk assessment demonstrating that African Americans living in Flint constituted the population that would be most affected by the emissions from this incinerator. Health data included public reports, studies from scientific journals, and privately commissioned studies.

It Is Worse for Children

Health information specific to children was also very important in this case. Children under six are especially vulnerable to lead's negative effects, because children absorb more lead in proportion to their weight than do adults. Of children ages six months to five years living in the Flint metropolitan area 49.2 percent already have elevated blood lead levels. Lead exposure at an early age has been linked to attention-deficit disorder, problems with anger control and management, and other behavioral changes.

Lead Exposure: It Is Black and White in Flint

A much higher percentage of African American (22.5 percent) than white (8.1 percent) children living in cities similar in size to Flint have blood lead levels exceeding 10 pg/dl. These levels of exposure come from many sources, some controversial. The power plant in Flint was an incinerator used to burn old buildings. Old buildings built or renovated before1980 have lead paint that can enter the air after incineration. Flint was a community that saw the demise of a manufacturing base that left it a wasteland.

Flint has many sources of pollution, which comes in the form of unregulated junkyards that burn garbage and tires, bulk-storage gas and chemical tanks, an asphalt company and cement factory, and a fenced-off holding pond containing sludge and other liquid waste northwest of the industrial park. It also takes the form of vehicle emissions and leaching industrial waste sites. The question for the court here was whether this new proposed use, a power plant that is really an incinerator that burns and emits even more lead, is an acceptable cumulative risk for an already lead-poisoned community.

The lower state court issued an injunction stopping Michigan from granting any air permits for six months. Appeals ensued. They were granted and allowed the incinerator to go into the African American community in Flint.

BEGINNINGS

U.S. environmental policy is relatively new, with the EPA forming in 1970. U.S. policy is just beginning to study cumulative effects. The research is emerging slowly, and no one is anxious to hear the news. Cumulative effects often represent the environmental impacts of humans when there were no environmental rules or regulations. They can be significant, and represent large cleanup costs. If cumulative effects are an issue in a typical environmental impact statement, then a finding of significant impact on the environment is made and a larger-scale environmental impact analysis is required. This too is expensive. Cleanup costs and the cost of environmental impact assessments are usually borne by industry. Industry strongly resists assuming responsibility for what they did not cause, based on a weak model of cumulative effects to date. Many of these cleanup costs could affect the profitability of any single corporation in these industries. Currently, most corporations listed in the stock exchange place these types of environmental issues in a 10B5 Securities Exchange Commission reporting statement under "contingent liabilities." Nonetheless, communities are very concerned about any emissions, especially as they accumulate in their midst. Public accessibility has increased knowledge about emissions generally and locally, and they become easier to detect as they accumulate over time. As some legislation now contains some cumulative effects provisions, some federal agencies are beginning new policies. The first policy experiments are important in terms of lessons learned. Decisions made now about cumulative environmental and human effects in public policy will have a direct bearing on the future health of communities, the future profitability of corporations, and the place in government that resolves the hard parts of implementing this type of policy. Right now, data and information are being developed through pilot programs. Here are some of them.

- Cumulative acute and subchronic health risk to field workers' infants and toddlers in farm communities as a result of organophosphate pesticide exposure (that is, through respiratory, dermal, dietary, and nondietary ingestion) resulting from agricultural and residential uses in light of the nutritional status of field-worker families.
- Cumulative ecological risk to the survival and reproduction of populations of blue crabs or striped bass in the Chesapeake Bay resulting from water and air emissions from both urban and agricultural sources.
- Cumulative risk under the Food Quality Protection Act may be defined using terms such as aggregate exposure (that is, the exposure of consumers, manufacturers, applicators, and other workers to pesticide chemical residues with common mechanisms of toxicity through ingestion, skin contact, or inhalation from occupational, dietary, and nonoccupational sources) or cumulative effects (that is, the sum of all effects from pesticide chemical residues with the same mechanism of toxicity).

EXAMPLES OF CUMULATIVE RISK ASSESSMENT ACTIVITIES WITHIN THE EPA IN 2002

The U.S. Environmental Protection Agency is engaged in several cumulative risk activities. The Superfund program has updated its guidelines on risk assessment to include planning and scoping cumulative risk assessment and problem formulation for ecological risk assessments. The plan for the Office of Solid Waste's Surface Impoundment Study includes both a conceptual model and an analytical plan, per the agency guidance on planning and scoping for cumulative risk.

The Office of Water is planning a watershed-scale risk assessment involving multiple ecological stressors. This approach was developed through collaboration with external scientists and is now being evaluated in the field.

Several regional offices are evaluating cumulative hazards, exposures, and effects of toxic contaminants in urban environments. In Chicago (Region 5), citizens are concerned about the contribution of environmental stressors to endpoints such as asthma and blood lead levels. In Baltimore a regional/ Office of Prevention, Pesticides, and Toxic Substances/community partnership tried to address the long-term environmental and economic concerns in three neighborhoods that are adjacent to industrial facilities and tank farms. Dallas is developing a geographic information system approach for planning for and evaluating cumulative risks. The Food Quality Protection Act of 1996 requires that the EPA consider the cumulative effects to human health that can result from exposure to pesticides and other substances that have a common mechanism of toxicity. The Office of Pesticide Programs has developed guidelines for conducting cumulative risk assessments for pesticides and has prepared a preliminary cumulative risk assessment for organophosphorous pesticides.

The Office of Air and Radiation's (OAR's) air toxics program has a cumulative risk focus. Under the Integrated Urban Air Toxics Strategy, OAR will be considering cumulative risks presented by exposures to air emissions of hazardous pollutants from sources in the aggregate. Assessments will be performed at both the national scale (a national-scale assessment for base year 1996 was completed in 2002) and at the urban or neighborhood scale. In partnership with the Office of Research and Development (ORD) and the National Exposure Research Laboratory, the Office of Air Quality Planning and Standards is developing the total risk integrated methodology (TRIM), a modular modeling system for use in single- or multimedia, single- or multipathway human health and ecological risk assessments of hazardous and criteria air pollutants at the neighborhood or city scale.

ORD's National Center for Environmental Assessment (NCEA) has completed ecological risk assessment guidelines that support the cumulative risk assessment guidance. Five watershed case studies are being assessed to demonstrate the guidelines approach. Each of these cases deals with cumulative impacts of stressors (chemical, biological, and, in some cases, physical). In addition, federal agencies have prepared a draft reassessment of dioxin and related compounds.

As emissions, impacts, and effects continue to accumulate in the environment, more chemicals will be reevaluated for their contribution to environmental degradation and public health impacts. This is not happening fast enough for many environmentalists and communities.

POTENTIAL FOR FUTURE CONTROVERSY

The public is exposed to multiple contaminants from a variety of sources, and tools are needed to understand the resulting combined risks. The stakes are very high and getting higher every day. It is likely the first set of U.S. tools will eventually be tested in the courts. This controversy is not there yet. Cumulative effects are receiving much study here and are being implemented as policy abroad. With global warming and climate change developing into treaties and U.S. municipal ordinances, this controversy could flare up rapidly in the United States.

See also Environmental Impact Statements: United States; Permitting Industrial Emissions: Air; Permitting Industrial Emissions: Water

Web Resources

Assessment of Cumulative Environmental Effects, a Selected Bibliography (November 1995). Available at www.ec.gc.ca/ea-ee/eaprocesses/bibliography_1995_e.asp. Accessed January 20, 2008.

Considering Cumulative Effects under the National Environmental Policy Act (NEPA). Available at www.nepa.gov/nepa/ccenepa/ccenepa.htm. Accessed January 20, 2008.

Cumulative Effects Assessment Practitioners Guide. Available at www.ceaa-acee.gc.ca/ 013/0001/0004/index_e.htm. Accessed January 20, 2008.

Further Reading: Lawrence, David Peter. 2003. *Environmental Impact Assessment: Practical Solutions to Recurrent Problems.* Hoboken, NJ: John Wiley and Sons; National Academies Press. 2003. *Cumulative Environmental Effects of Oil and Gas Activities on Alaska's North Slope.* Washington, DC: National Research Council; Simon, Thomas P. 1999. *Assessing the Sustainability and Biological Integrity of Water Resources Using Fish Communities.* London: CRC Press; Social Learning Group. 2001. *Learning to Manage Global Environmental Risks.* Cambridge, MA: MIT Press.

D

DIFFERENT STANDARDS OF ENFORCEMENT
OF ENVIRONMENTAL LAW

The enforcement of environmental law often disappoints citizens. It is rare, difficult, and different depending on the community. Fines for the same environmental offense are lower in African American communities than in white communities.

Environmental cleanup efforts disproportionately benefit white Americans over people of color. A landmark 1992 study uncovered glaring inequities in the way the U.S. Environmental Protection Agency (EPA) enforced environmental laws. The study found a racial divide in the way the U.S. government cleans up toxic waste sites and punishes polluters. White communities see faster action, better results, and stiffer penalties than communities where African Americans, Hispanics, and other minorities live. This unequal protection often occurs whether the community is wealthy or poor. A recent study in Massachusetts found that communities where people of color compose 25 percent or more of the population face nearly nine times higher cumulative rates of exposure to hazardous materials than predominantly white communities.

Most U.S. environmental laws allow enforcement of pollution-control standards by state government officials and private citizens as well as federal officials. Beginning in the early 1980s, the EPA delegated enforcement responsibility to individual states while retaining general oversight authority. State environmental laws often are similar to federal statutes. One reason for the unequal enforcement of environmental law is that state approaches to enforcement vary widely. Some states provide the minimal amount of enforcement necessary for minimal EPA funding. Other states seek the maximum enforcement and acquire the maximum federal support and resources possible. Still other states supplement the federal environmental enforcement money with their own. Large and medium-sized cities now have environmental departments.

One battleground in this controversy is at the state environmental agency level. Environmentalists, citizens, some local governments, and others all expect that environmental laws will be equally enforced. If they are not, the fear is that industry will shop around for the least environmentally regulated community. States and cities sometimes seek economic development with lax environmental enforcement for desired industries. When states and cities compete with one another to have the least environmental regulation and enforcement, this is called the race to the bottom. By retaining oversight authority, the federal government mandates a minimum level of environmental protection. If the state does not do it, the federal government will step in and do it themselves. This was recently the case in Alaska. Alaska declined to regulate older hazardous waste facilities. The EPA stepped in and now runs that particular program. States usually seek to keep the federal government from deciding local issues.

State agencies are between the federal government and the citizens. They have to balance economic development with environmental enforcement. If they close an industry due to environmental regulations, then economic development could suffer. That is why the EPA, and virtually every state environmental agency, accommodates compliance policies, not punishment policies.

CITIZEN SUITS

Many U.S. environmental laws authorize citizen suits, lawsuits brought by any citizen against another party thought to be in violation of a pollution-control standard. These citizen suits are generally barred if federal or state agencies are diligently prosecuting an action against the same defendant for the same violation. Citizen suits must be pursued with respect to environmental rule violations that are ongoing, not alleged violations of pollution laws that occurred in the past.

ENFORCEMENT CONTROVERSY

One aspect of the law is that there is an expectation that it will be enforced, and enforced fairly. That generally means that like behavior is treated the same way. That is not the case in environmental enforcement. A core controversy that riddles this battleground is the best approach to enforcing environmental laws in the United States. Government works hard to facilitate compliance with environmental rules. One view is deterrent enforcement. Wrongdoers are punished for their acts. The punishment should be enough to deter, or prevent, them from benefiting from their actions or repeating them. The EPA will go through a complex process to determine how much to fine a polluting industry without putting it out of business.

The prevailing enforcement approach is the compliance model. Here, regulatory agencies advise regulated industries on how to come into compliance. They include compliance incentives that encourage regulated industries to police themselves by engaging in environmental audits and self-correction of violations. Sometimes, one part of the deal is that, by adopting cleanup measures that go above and beyond what the law requires, industry gets enforcement forgiveness or flexibility. Compliance programs also incorporate compliance assistance efforts. Environmental agencies, state or federal, assist regulated industries and communities to comply with environmental laws. This can include technical assistance and grants.

POTENTIAL FOR FUTURE CONTROVERSY

The difference in legal expectations from one group of stakeholders (citizens, environmentalists, environmental justice advocates, and sustainability proponents) to another (industry, state and federal environmental agencies, and courts) is very large. This is the battleground of environmental law enforcement generally. This particular controversy of unequal or poor enforcement of environmental laws shows that some of them can be enforced. As U.S. environmental policy begins to include the urban areas, a historical pattern of exclusion will emerge and cause conflict. A different standard of enforcement is interpreted by some to indicate where it is acceptable to pollute. Cities are where most people of color reside, where most immigrants arrive, and where the most pollution exists. Different standards of enforcement of environmental law between city and suburb will cause difficulties as concern about cumulative effects and sustainable policy development mounts.

See also Citizen Monitoring of Environmental Decisions; Environmental Audits and Environmental Audit Privileges; Environmental Justice; Litigation of Environmental Disputes; Supplemental Environmental Projects

Web Resources

Neal, Ruth, and April Allen. Environmental Justice: An Annotated Bibliography. Available at www.ejrc.cau.edu/annbib.html. Accessed January 20, 2008.

Environmental Protection Agency. Enforcement. Available at www.epa.ie/OfficeofEnvironmentalEnforcement/. Accessed January 20, 2008.

Further Reading: Hawkins, Keith, and John M. Thomas, eds. 1984. *Enforcing Regulation.* Boston: Kluwer Nijhoff; Landy, Marc K., Marc J. Roberts, and Stephen R. Thomas. 1990. *The Environmental Protection Agency: Asking the Wrong Questions.* New York: Oxford University Press; Mintz, Joel A. 1985. *Enforcement at the EPA: High Stakes and Hard Choices.* Austin: University of Texas Press.

DROUGHT

Droughts are environmental controversies in the United States in terms of both monitoring and governmental response. They are integrally related to the water cycle and have deep environmental impacts. On the international level, droughts can cause mass movements of people in search of food—environmental refugees. The environmental impacts of drought can be severe and long lasting. Climate change may cause drought to spread.

WHAT IS A DROUGHT?

Droughts have been the bane of human existence since they were one of the seven plagues in biblical references. Droughts highlight the role of water in ecosystems and in agribusiness. Irrigation can mitigate the impact of droughts on crops but not on ecosystems. Droughts can also increase the concentration of pollutants, such as fertilizer and pesticide runoff, for downstream water users. Droughts can cause low wells to run dry, and junior water rights holders to be completely without water. In this way droughts can pull out other controversies around water pollution, water rights, and environmental impacts. A pragmatic definition is that a drought is a period of unusually dry weather that lasts long enough to cause serious problems for important stakeholders. These problems include crop damage, water supply shortages, and risk to ecosystems. The severity of the drought depends on the degree of moisture deficiency, the duration, and the size of the affected area.

DROUGHT CHARACTERISTICS

Drought differs from other natural hazards in several ways. The effects of drought accumulate slowly over a time. Serious problems from drought may fester for years. Drought impacts are less apparent and spread over a larger geographic area than are damages that result from other natural hazards. The measurement of environmental impacts and the need for disaster relief are difficult with drought disasters. A big battleground is the definition of drought. How much drought is natural? How much is due to climate change? Can human intervention mitigate drought? When does the government intervene to assist drought-stricken communities? What kind of intervention best assists drought victims? Should there be government-assisted drought insurance similar to flood insurance? Are droughts an act of God, and insurable? The lack of a precise and objective definition has been an obstacle to drought policy development. Population increases and industrialization have increased the demand for water. Increased demand for water is increasing societal vulnerability to drought conditions. Future droughts could create greater impacts because of the increased social vulnerability.

Environmental impacts of drought include:

- Damage to animal species
- Reduction and degradation of fish and wildlife habitat
- Lack of feed and drinking water
- Greater wildlife mortality due to increased contact with agricultural producers, as animals seek food from farms and producers are less tolerant of the intrusion
- Disease
- Increased vulnerability to predation (for species concentrated near water)
- Migration and concentration (loss of wildlife in some areas and too many wildlife in other areas)
- Increased stress to endangered species

- Loss of biodiversity
- Hydrological effects
- Lower water levels in reservoirs, lakes, and ponds
- Reduced flow from springs
- Reduced stream flow
- Loss of wetlands
- Estuarine impacts (e.g., changes in salinity levels)
- Increased groundwater depletion, land subsidence, reduced recharge
- Water quality effects (e.g., salt concentration, increased water temperature, pH, dissolved oxygen, turbidity)
- Damage to plant communities
- Loss of biodiversity
- Loss of trees from urban landscapes, parks, and wooded conservation areas
- Increased number and severity of fires; wind and water erosion of soils; reduced soil quality; air quality effects (e.g., dust, pollutants); and visual and landscape quality (e.g., dust, vegetative cover, etc.)

CRISIS MANAGEMENT VS. RISK MANAGEMENT APPROACH TO DROUGHT MANAGEMENT

There is a major controversy in policy approaches to drought and its environmental impacts. The policy of government has been to react to severe drought through the provision of food, water, and shelter or other emergency assistance. Some argue that government intervention in this manner serves as a disincentive for sustainability because it reinforces poor drought management practices. Technological and social change is improving the ability to effectively manage water during periods of drought. Scientists claim that an improved understanding of complex atmospheric-oceanic systems and the development of new computer models have improved drought-forecasting skills.

DROUGHT IN THE UNITED STATES

Drought is a normal, recurrent feature of the climate of virtually all portions of the United States. Because of the country's size and the wide range of climatic systems present, it is rare for drought not to exist somewhere in the country each year. Drought frequently affects more than 10 percent of the United States, and sometimes more than 30 percent of the nation is affected. Other regions of the world can go for years with very little rain, and then for years experience a high amount of precipitation. Some areas have had drought so long they are considered deserts. Drought can be spotty in some areas, leaving only parts completely dry.

Drought-stricken areas are difficult to live in. Without water, many plants and animals suffer. What water resources remain can become quickly contaminated. Drought is devastating to certain industries, especially farming and ranching. On an international level, drought can cause the mass migration of people in search of food. The United Nations calls these people *environmental refugees*. An area's sensitivity to drought ultimately determines the level of financial investment and risk that banks and other financial institutions are willing to consider.

HOW CAN A POLICY MITIGATE A DROUGHT?

Should we mitigate all the impacts of all the droughts? Which ones should be mitigated, if possible? *Mitigation* is an environmental term that means activities that reduce the degree of long-term risk to human life and the environment. This definition is awkward in application to drought because there is often no direct loss of human life and measurements of long-term risk to the environment are unknown and debatable. States are given flexibility to define mitigation as including actions or activities that they felt were appropriate. Mitigation is a battleground in many environmental controversies, including this one. Sometimes, the mitigation activity itself is damaging, and there is often no obligation to mitigate the effects of mitigation attempts. Water conservation and storage at the home and

---

AUSTRALIAN NATIONAL DROUGHT POLICY: A MODEL FOR THE UNITED STATES?

Australia is a country with a harsh, drought-prone interior. Much of the settlement of the interior of Australia has been a struggle between civil engineering and the forces of nature, primarily water. The national government of Australia has been engaged in drought policy since European settlement.

The Australian drought policy is focused on agricultural drought, thus focusing on animal and crop impacts. According to Australian policy, drought is considered to be an integral part of a highly variable climate. People should expect periods of drought in certain regions of Australia (known there as the Big Dry). Drought is considered to be one of many weather risks that agribusinesses face in managing farm operations. The Australian federal government under this policy assists agribusiness in coping with drought. They do so through the provision of better and timelier information about potential drought conditions.

The objectives of Australian drought policy are:

1. to encourage primary producers and other segments of rural Australia to adopt self-reliant approaches in planning for climatic variation,
2. to facilitate the maintenance and protection of Australia's agricultural and resource base during periods of increasing climatic stress, and
3. to facilitate the early recovery of agricultural and rural industries to levels consistent with long-term sustainable production.

The long-term goals of this policy are to increase productivity, improve the allocation of resources, and enhance self-reliance. Given that previous attempts to mitigate drought in the United States have been unsuccessful, policy changes must occur to adequately address the drought management problems that exist today.

One big problem in applying the Australian approach to the United States is private property ownership.

---

local level may help mitigate some of the effects. More efficient use of water in agribusiness would also do so. When the government makes mandatory rules in both these areas, new battlegrounds develop about enforcement.

Conflicts between water users increase during droughts. In the western United States, water rights are often divided into senior and junior water rights to the same water. If a drought occurs, only senior water rights holders will get water. That can mean one farm goes bankrupt while the next farm prospers. Enormous controversies are brewing between municipal and agricultural water users. If a drought occurs, a city generally will get the water over agribusiness.

POTENTIAL FOR FUTURE CONTROVERSY

Water is an increasingly scarce resource in the United States, and human and industrial water use is increasing. The economic, social, and environmental impacts of drought are significant. Government continues to deal with drought in a reactive, rather than planning, mode. States have developed and implemented a wide range of mitigation measures, but the shift from crisis management to risk management continues to be controversial.

Drought shows the vulnerability of economic, social, political, and environmental systems to a variable climate. Projected changes in climate suggest a possible increase in the frequency and intensity of severe drought in the future. The battleground for much of this controversy lies within government agencies. These agencies are just developing the early parts of a comprehensive environmental policy around drought. All that will not necessarily stop a drought, or mitigate all the impacts if there is not enough water for everyone. A bigger battlefield for this controversy will occur when some of the policies require actions from private landowners that threaten water rights.

See also Climate Change; Ecosystem Risk Assessment; Environmental Vulnerability of Urban Areas; Global Warming; Sustainability; "Takings" of Private Property under the U.S. Constitution

Web References

National Oceanic and Atmospheric Administration. Drought Information Center. Available at www.drought.noaa.gov/. Accessed January 20, 2008.

Palmer Drought Index Graphic. Available at www.cpc.ncep.noaa.gov/products/analysis_monitoring/regional_monitoring/palmer.gif. Accessed January 20, 2008.

Further Reading: Alvarez, Joaquin Andreu. 2005. *Drought Management and Planning for Water Resources.* Singapore: CRC Press; MacGuire, Bill, Ian M. Mason, and R. Kilburn Christopher. 2002. *Natural Hazards and Environmental Change.* Oxford: Oxford University Press; Whitmore, Joan S. 2000. *Drought Management of Farmland.* MO: Springer; Wilhite, Donald A. 1993. *Drought Assessment, Management, and Planning: Theory and Case Studies.* New York: Springer.

E

ECOLOGICAL RISK MANAGEMENT DECISIONS AT SUPERFUND SITES

Risk assessment and ecological risk assessment are themselves controversial. Superfund sites listed on the National Priorities List (NPL) apply ecological risk assessment in very important and controversial legally mandated cleanups of dangerous, toxic, and polluted areas. Although very technical and pioneering at this juncture, every step of this infant policy will engender environmental controversy.

ECOLOGICAL RISK ASSESSMENT

Ecological risk assessment involves a holistic perspective of how the land, air, and water interact. It is new, expensive, and time-consuming. Assessment and evaluation often uncover pollution from years of unregulated industrial activity. Industry does not want to be liable for waste it did not create. However, the cumulative impacts increase daily, and the amount of cleanup required to achieve minimal levels of public safety can be unmanageable for individual corporations. Ecological risk assessment is much more comprehensive and uncovers much more pollution. It also provides important environmental information necessary for new policies like sustainability, emissions trading, and impact assessment. Superfund cleanups represent the cutting edge of ecological risk assessment applied to important environmental issues. It is both a pioneering and rudimentary policy approach from a relatively young federal agency. Both characteristics spark intense controversy that finds expression in courts and legislatures. If the EPA's ever-changing model survives lawsuits and legislative attacks, and it usually does, then many states and some cities may adopt their approach and policy.

ECOLOGICAL RISK ASSESSMENT: HOW THE EPA DOES IT NOW

Ecosystem and ecological risk assessment emerged from human health-based risk assessment. During the 1980s, risk assessment was a foundational part of environmental policy and law. The use of ecological information for decision making expanded slowly through the 1980s. In March 1989, the Environmental Protection Agency published *Risk Management Guidance for Superfund, Volume 2: Environmental Evaluation Manual,* which was among the first documents to address ecological risk (EPA540-/1–89/001). In 1992, the EPA published the *Framework for Ecological Risk Assessment* (EPA/63-R-92/001) as the first statement of principles for ecological risk assessments. In April 1998, the agency published the final Guidelines for Ecological Risk Assessment (EPA/630/R-95/002F).

Ecological risk assessments are most often conducted by the EPA during the remedial investigation/feasibility study (RI/FS) phase of the Superfund response process. They are used to evaluate the likelihood of adverse ecological effects occurring as a result of exposure to physical or chemical stressors. Adverse impacts is a technical term with multiple meanings. Here, it is defined as any physical, chemical, or biological entities that can induce adverse responses at a given site. Exposure profiles are developed to identify ecological receptors, habitats, and pathways of exposure. The sources and distribution of pollutants in the environment are also examined. Other information contained in ecological risk assessments can include evaluations of individual species, populations of species, wildlife communities, habitat types, ecosystems, landslide risk, or landscapes. All this information fuels many other environmental controversies. In this way the EPA's implementation of an ecological risk assessment becomes part of many later battlegrounds.

In October 1999 the EPA published *Ecological Risk Assessment and Risk Management Principles for Superfund Sites* in order to begin policy implementation. This policy describes six principles to consider when making ecological risk management decisions. The policy specifically requires that all ecological risk assessments be performed according to the eight-step process described in *Ecological Risk Assessment Guidance for Superfund: Process for Designing and Conducting Ecological Risk Assessments.* The EPA wants its risk management decision makers to be able to present a clear basis for their ecological risk management actions. They want them to be able to present them to the public in the proposed plan and the record of decision steps of the Superfund response process. This emphasis on public involvement facilitates the use of ecological risk assessments in other areas of controversy. Most cleanup sites listed on the Superfund National Priorities List have been the subject of community and environmental controversy anyway. They are risks to people and to the environment. Many of the pollutants at a site migrate from the site via the water, air, and land so that immediate cleanup is necessary to prevent ecosystem contamination. One premise of the use of ecosystem risk assessment here is that the policy goal is to prevent ecosystem contamination.

PRINCIPLED ENVIRONMENTAL DECISION MAKING

The policy requires EPA risk managers to adhere to the six principles summarized further on. Many community groups and environmental organizations find some of these processes incomplete in law, and especially incomplete in practice. This creates local controversies that simmer for years. Nonetheless, it is a U.S. starting point for beginning a long, expensive, necessary process to clean up the results of processes that contributed to our present-day affluence. These principles are taken directly from the previously described public documents and can be viewed in greater detail in the references and Web sites listed at the end of this entry.

> Principle 1: Superfund's Goal Is to Reduce Ecological Risks to Levels That Will Result in the Recovery and Maintenance of Healthy Local Populations and Communities of Biota
>
> Principle 2: Coordinate with Federal, Tribal, and State Natural Resource Trustees
>
> Principle 3: Use Site-Specific Ecological Risk Data to Support Cleanup Decisions

Site-specific data should be collected and used, wherever practical, to determine whether or not site releases present unacceptable risks and to develop quantitative cleanup levels that are protective. Site-specific data include plant and animal tissue residue data, bioavailability factors, and population or community-level effect studies.

> Principle 4: Characterize Site Risks

When evaluating ecological risks and the potential for response alternatives to achieve acceptable levels of protection, Superfund risk managers should characterize risk in terms of

1. magnitude,
2. severity,
3. distribution, and
4. the potential for recovery of the affected receptors.

> Principle 5: Communicate Risks to the Public

Clearly communicate to the public the scientific basis and ecological relevance of the assessment endpoints used in the site risk assessment.

> Principle 6: Remediate Unacceptable Ecological Risks

Superfund's goal is to eliminate unacceptable risks due to any release or threatened release. Contaminated media that may affect the ability of local populations of plants or animals to recover and maintain themselves in a healthy state at or near the site should be remediated to an acceptable level.

Acceptable levels can vary depending on land use. If it is industrially zoned land, then a lower level of cleanup is necessary, which is much less expensive. However, this partially polluted land remains in the land base for the community, and since it is not clean, it is more difficult to develop for residential and commercial uses, legally. This decreases the overall economic wealth of a community as well as suppressing real estate values and increasing public health risks. Communities want sites cleaned up to residential standards, which is more expensive but safer.

SUPERFUND CLEANUPS: NEW, EXPENSIVE, AND CONFRONTATIONAL

The context for the application of ecological risk assessment is Superfund cleanups. Superfund is a federal EPA policy and law that mandates the cleanup of polluted sites. The 1,200 or so worst sites are placed on the National Priorities List, or NPL. The liability for cleanup is complex and litigated frequently. At any given site, the producers of the waste, then the shippers, then the storage providers, and ultimately the owner of the property are liable for the cost of the cleanup. These are known as primary responsible parties, or PRPs. If they do not clean it up, then the EPA can do so and go after the PRPs for the costs. It often strikes many property owners as unfair because they can be liable for the costs of cleaning up waste they did not cause. To complicate matters further, PRPs can go against other PRPs for costs. Some PRPs go bankrupt before, during, and after a mandated EPA cleanup. This causes more technical, protracted, and complicated litigation that decreases the value of the real property. Municipal economic development prefers to have high-value property because this increases the tax base. The tax base, in turn, pays for municipal services like education, the fire department, police, and sanitation. Their position is complicated because if the site does not get cleaned up, it does nothing for the tax base, but if they seek strict enforcement, then they could suppress the value of the property.

Of fundamental importance is the level of cleanup required. Cleanup to residential standards can be very expensive. Some developers have complained that it has to be clean enough that the children can eat the dirt, in a case when there was lead contamination of the soil in a housing project. If the cleanup standard is much lower, for industrial purposes, then the costs tend to be lower. However, industrially zoned land has many dangerous and toxic impacts on the surroundings. It is much more difficult to increase the value of that land as opposed to residentially zoned land. Many subsequent land uses of that property, such as nursing homes, schools, and hospitals, would be put at risk if the site were never cleaned up to residential standards. Human health risk assessment, generally focusing on a single chemical via a single modality, is less expensive then an ecosystem risk assessment.

POTENTIAL FOR FUTURE CONTROVERSY

Litigation, level of cleanup, and emerging urban environmentalism are the policy and political context for the application of ecosystem risk assessment. Every stage of the emerging risk assessment process can be fraught with controversy. Who pays for it? Who pays for the cleanup? How long will it take to be accurate? Cleanup of polluted sites is the type of environmental policy that becomes more important if it is ignored. Pollution increases over time and with population increases; the longer it takes to clean up polluted areas the harder it will be to do so. This means that controversial issues will continue to flare as this pioneering and necessary policy evolves.

The unfolding EPA policy on ecosystem risk assessment requires communication to the public of the scientific basis and ecological relevance of the assessment endpoints used in the site risk assessment. This particular principle is subject to wide interpretation and much controversy. Some argue for as much community participation as necessary, taking all the time required to bring communities up to a capacity to understand and engage environmental issues. Others argue that a small legal notice in one of the newspapers or a copy of the ecological risk assessment in city hall fulfills the policy requirement. There are practical aspects to this battleground, but some questions will need to be answered. Who gets invited? Does it matter what they think? What if science is inconclusive? Do some communities have to use scientific levels of proof while other communities simply politically resist it? What does *adverse* mean? Will these public meeting requirements be the basis of a lawsuit? Industrial and environmental stakeholders have strong and opposing positions in this battleground. Environmental justice communities and sustainability proponents clearly want to be fully apprised of all environmental issues where they live, work, and play. Environmental justice analysts are interested in the benefit and burden of environmental decisions, and that is often answered by an ecosystem risk assessment. Sustainability proponents consider ecosystem risk assessments part of sound environmental policy. Government considers them expensive and controversial. Industry sees them as an unnecessary expense that uses money that could go toward actually cleaning up the site. Stakeholders have strong positions, but the trend is to do more ecosystem risk assessment.

See also Cumulative Emissions, Impacts, and Risks; Ecosystem Risk Assessment; Environmental Impact Statements: United States; Environmental Justice; Sustainability

Web Resources

U.S. Environmental Protection Agency Superfund. Cleaning Up the Nation's Hazardous Waste Sites. Available at www.epa.gov/superfund/. Accessed January 20, 2008.

U.S. EPA Compliance and Enforcement Portal. Available at www.epa.gov/compliance/. Accessed January 20, 2008.

U.S. EPA Cleanup Enforcement: Information Resources. Available at www.epa.gov/compliance/resources/cleanup.html. Accessed January 20, 2008.

Further Reading: Calabrese, Edward James, Linda A. Baldwin, and Lawrence H. Keith. 1993. *Performing Ecological Risk Assessments.* Stockport, MI: Lewis Publishers; Pastorok, Robert A. 2001. *Ecological Modeling in Risk Assessment: Chemical Effects on Populations, Ecosystems, and Landscapes.* Boca Raton, FL: CRC Press; Sunahara, Geoffrey Isao. 2002. *Environmental Analysis of Contaminated Sites.* Hoboken, NJ: John Wiley and Sons.

ECOSYSTEM RISK ASSESSMENT

Chemical manufacturers and others maintain that the chemicals they make are safe for the environment while environmentalists and others maintain that they pose risks to the ecosystem. Critics maintain that ecosystem risk assessments are unscientific and costly. Communities and environmentalists would like to see more of them included in environmental impact statements. Sustainability advocates want them performed every time there is a significant impact on the environment.

BACKGROUND

An essential part of environmental preservation is preserving the whole habitat for the entire food web. Extinction is one of the most extreme examples of failed ecosystems. There are many other ecosystem effects from pollution, natural disasters, and land development. This range of ecosystem effects is itself a battleground. The controversy escalates because many sustainability proponents are advocating ecosystem preservation and resiliency. The main concern is with ecosystems that are damaged for the foreseeable future. Cities, waste sites, roads, airports, mountain ecotones, and other large institutional uses (military bases, college campuses) are often put in this category. The battle commences when concern about the cumulative impacts on ecosystems in these areas affects growth and development.

An ecosystem is an area where the land, air, and water share a common place and interact to support life. Some scientists have more precise definitions, but there is some controversy in that community about its application. Within the life sciences community, some sciences challenge conclusions about cause and effect when risk assessments are applied to an entire ecosystem. According to the U.S. Environmental Protection Agency (EPA) an ecological risk assessment evaluates the likelihood that exposure to one or more chemicals may cause harmful ecological effects. Effects can be direct or indirect and short or long term. There is a battleground for every effect on the risk to the ecosystem. For example, in policy application, ecological risk assessment is used to determine whether a pesticide meets the requirements for registration with the EPA. In this process many controversies arise.

EPA AND ECOSYSTEM RISK ASSESSMENT

An important matter related to risk assessment becomes law and policy when pesticides are registered with the EPA. By their very nature, pesticides, herbicides, rodenticides, fungicides, and others are designed to kill. One issue is often how long they persist in the environment because they may kill, or negatively affect, parts of the ecosystem beyond their intended usage. This is where ecosystem risk assessments are applied. Pesticides are highly regulated. To propose a pesticide for registration in the United States, the manufacturer must conduct scientific studies according to the EPA's requirements. The manufacturer then submits the data to the EPA, where it is reviewed and the pesticide's potential to cause problems is determined.

This is only one application of ecosystem risk assessment in a controversial area. In April 1998, the EPA established Guidelines for Ecological Risk Assessment that describe the risk evaluation process in other applications.

SCREENING LEVELS OF ECOLOGICAL RISK IN PESTICIDES

The Office of Pesticide Programs (OPP) is the main government agency currently screening ecotoxicity data. After reviewing an individual toxicity or ecological effects study for a pesticide, EPA scientists develop a data evaluation record (DER) for the study. A DER summarizes the toxicity to certain species groups that are expected to be exposed to the pesticide.

The conclusions from all the individual ecotoxicity DERs are then integrated and summarized in a stressor-response profile, the final product of the ecological effects characterization. The profile presents the effects for various animals and plants and an interpretation of available incidents information and monitoring data. Then the OPP compares the stressor-response profile with potential exposure levels to determine the risk of exposure-related effects.

ECOLOGICAL RISK ASSESSMENT: OVERVIEW

An ecological risk assessment tells what happens to a bird, fish, plant, or other nonhuman organism when it is exposed to a stressor, such as a pesticide. In scientific terms, an ecological risk assessment "evaluates the likelihood that adverse ecological effects may occur or are occurring as a result of exposure to one or more stressors." Undesirable events can include injury, death, or a decrease in the mass or productivity of aquatic animals, terrestrial animals, plants, or other nontarget organisms. This does include endangered and threatened species. This assessment process combines all the information from the toxicity tests, the exposure information, assumptions, and uncertainties.

An ecological risk does not exist unless an exposure has the ability to cause an adverse effect. Further, that exposure must contact an ecological component to elicit the identified adverse effect. There is much dispute over whether a certain effect is adverse. If not enough is known about a given chemical and its interaction with mammals, then the effect is not considered adverse. Another set of questions around controversies of what is adverse are, to whom is it adverse, and at what stage in their development. Infants and the elderly tend to be more vulnerable and have lower tolerances to exposure.

ADVERSE IMPACTS: WOULD YOU KNOW THEM IF YOU FELT THEM?

The definition of adverse impacts is very important because it defines the contours of public health policy. It is a definition that can shift in meaning from legislation, to rule making, to judicial decisions. Adverse impacts are those that are proven by science to be medically harmful. Many in the public distrust scientists and science as it applies to them. Many communities simply do not want the threat of a chemical emission in their neighborhood, and they can find out about it on the Toxics Release Inventory. When the government permits an industry based on science, this controversy takes shape in policy. There are many medical conditions that are not considered adverse but can erode the quality of life, decrease worker productivity, and be unpleasant. This limited definition of adverse impacts makes such impacts seem authorized in the name of science. The government therefore issues more industrial permits allowing more emissions because they are not adverse. This is a scientific battlefield in this controversy.

AIR

The general definition of adverse respiratory health effects is "medically significant physiologic or pathologic changes generally evidenced by one or more of the following: (1) interference with the normal activity of the affected person or persons, (2) episodic respiratory illness, (3) incapacitating illness, (4) permanent respiratory injury, and/or (5) progressive respiratory dysfunction." All changes are not adverse, citing the example of carboxyhemoglobin. The level of carboxyhemoglobin, beyond that from endogenous production, is indicative of exposure but it is not predictive of adverse effects until it reaches threshold levels, depending on the effect and the susceptibility of the exposed person.

ANOTHER RECENT DEFINITION OF ADVERSE IMPACTS

The events of September 11, 2001, highlighted the need to enhance the security of the U.S. food supply. Congress responded by passing the Public Health Security and Bioterrorism Preparedness and Response Act of 2002 (the Bioterrorism Act), which was signed into law on June 12, 2002. The Bioterrorism Act contains the phrase *serious adverse health consequences* to describe the standard relating to the exercise of many of the new powers. Together with the final rules implementing sections 303, 306, and 307 of the Bioterrorism Act, and the other sections of the act incorporating the serious adverse health consequences phrase, a definition of the phrase will further enable the FDA to act quickly and consistently in responding to a threatened or actual terrorist attack on the U.S. food supply or to other food-related public health emergencies. A definition of the serious adverse health consequences phrase will promote uniformity and consistency across the FDA in the understanding of the term and determination of an appropriate response. In addition, a definition of the term will inform the public and stakeholders about what the FDA considers to be a serious adverse health consequence under the Bioterrorism Act.

The proposed rule would define the phrase serious adverse health consequences for purposes of the Public Health Security and Bioterrorism Preparedness and Response Act of 2002 (the Bioterrorism Act) and any implementing regulations and guidance.

In the interests of quickly providing the agency's interpretation of serious adverse health consequences to the public, the FDA considered explaining the term in guidance. The agency concluded, however, that this option is neither effective nor efficient because guidance does not have the force and effect of law. If the definition or its application is ever challenged, guidance will receive less deference than if the definition were specified in a regulation.

The FDA also considered explaining the term in guidance followed by a regulation at a later date. This option was considered because it offers the advantage of rapidly informing the public about the agency's position while the agency gathers more information and experience in applying the definition. The agency concluded that guidance followed by a regulation was undesirable. First, as to the initial guidance, the FDA would encounter the same problems described previously for the guidance-only option. Second, this option creates a burdensome process for the FDA by doubling the agency's responsibilities—first, to publish guidance, and, second, to engage in notice-and-comment rule making. FDA resources will be conserved by avoiding this two-step process. Further, there is the possibility that once guidance is published, a regulation might not follow. As a result, the definition might never have the force and effect of law.

The FDA also considered defining or explaining serious adverse health consequences in preambles to rules promulgated under the Bioterrorism Act. However, implementing regulations are not required for all sections of the Bioterrorism Act that incorporate the term. Thus, the term would not be publicly addressed in the context of all of the applicable sections of the Bioterrorism Act. In addition, because preambles are not codified and incorporated into the Code of Federal Regulations, the context and interpretation of the term eventually may become disassociated from the codified regulations.

The FDA also considered adopting one of the two similar definitions for serious adverse health consequences or the definition for *serious injury* in medical devices regulations to promote consistency within the agency and avoid confusion. (In the medical devices reporting regulations, the preamble to the final rule states that "the agency intends for 'serious adverse health consequences' to have the same meaning as 'serious injury' under the [Medical Device Reporting] rule.") This option could promote greater consistency within the agency, avoid confusion, and also save time. However, the agency believes that a broader definition must be used for foods and feeds in order to satisfy congressional intent.

Specifically, it must be clear that the definition of serious adverse health consequences, for purposes of the Bioterrorism Act, (1) expressly includes vulnerable populations, and (2) expressly applies to food for humans and animals. In addition, there are terms incorporating the concept of *serious* in regulations. The definitions of these terms are not entirely consistent because they are tailored to the needs of each center and apply only to specific portions of the applicable regulations, that is, they have specific uses and contexts. Thus, a specific definition for serious adverse health consequences under the Bioterrorism Act is necessary in order to avoid confusion with differing definitions of serious, serious injury, or serious adverse health consequences in other regulations, and the context in which these terms are defined and applied. The proposed definition would apply to (1) all foods and feeds in bioterrorist events and other public health emergencies; and (2) all populations, vulnerable or healthy, effectively having very wide applicability in a wide variety of emergency situations. Finally, the FDA considered leaving the term undefined, thereby providing maximum flexibility for determining what constitutes serious adverse health consequences on a case-by-case basis. By not defining, the agency could avoid the potential consequences of a definition that is either too broad or too narrow. However, leaving the term undefined could cause confusion and inconsistency in implementation.

The impact of this proposed rule will depend on how the FDA decides to define the phrase serious adverse health consequences, which is used as a standard for taking action under the administrative detention, record-keeping, and prior notice provisions of the Bioterrorism Act. The broader the definition, the greater the costs and benefits associated with it. For example, if serious adverse health consequences were defined to include any case of food-borne illness, then foods would be administratively detained more often than if the definition were limited to cases resulting in death. A broader definition will mean the term is used more frequently in conjunction with the provisions of the Bioterrorism Act; therefore, there would be more costs, but also more benefits.

POTENTIAL FOR FUTURE CONTROVERSY

The definition of *ecosystem* is context- and policy-specific. What an ecosystem is must be defined and bounded by the policy or management question being assessed. This is controversial now in the context of national parks. As ecosystem risk analysis develops more standard protocols and is applied to complex environmental problems, more battlegrounds will emerge. Because ecosystem risk assessment is seen as a process and answer to a specific debate about environmental impact, it will also be part of other controversies.

Risk assessment at the ecosystem level has always been a long-term goal of applied ecology. However, lack of detailed data on the interactions among species, coupled with the complexity of these interactions over time, challenges ecosystem risk assessment methods. There are enormous gaps of knowledge about species' interaction. The effects of global warming and climate change on ecosystems are still unknown. There are many battlegrounds on this point from the community level to the courtroom. For example, how does the risk assessor incorporate scientifically derived data versus observational information from long-term residents?

With environmental knowledge expanding rapidly to all sectors of society, and with ecosystem management a core value, it is likely that the controversies around ecosystem risk assessment will assist its growth and promote its robust and vigorous use by more environmental decision makers. This will increase the controversy surrounding ecosystem risk assessment and help refine it for better application. Ecosystem risk assessment is a favorite tool of sustainability advocates and is most frequently applied in controversial environmental situations, like hazardous waste cleanups. Concern about cumulative emissions, impacts, and effects also focuses more attention on ecosystem risk assessments. Controversies and policies about ecosystem risk assessments and cumulative risk assessment could merge as U.S. environmental policy incorporates the urban environment.

See also Climate Change; Cumulative Emissions, Impacts, and Risks; Human Health Risk Assessment; Pesticides; Toxics Release Inventory

Web Resources

Hawaiian Ecosystems at Risk Project. Available at www.hear.org/. Accessed January 20, 2008.

South Florida Information Access. An Evaluation of Contaminant Exposures and Potential Effects on Health and Endocrine Status for Alligators in the Greater Everglades Ecosystem. Available at sofia.usgs.gov/projects/eco_risk/endogtr_geer03abs.html. Accessed January 20, 2008.

U.S. Environmental Protection Agency. Pesticides: Environmental Effects. Available at www.epa.gov/pesticides/ecosystem/ecorisk.htm. Accessed January 20, 2008.

Further Reading: Efroymson, Rebecca A., Bradley E. Sample, and Glenn W. Suter. 2000. *Ecological Risk Assessment for Contaminated Sites.* Boca Raton, FL: CRC Press; EPA. 1997. *Ecological Risk Assessment Guidance for Superfund: Process for Designing and Conducting Ecological Risk Assessments.* Washington, DC: U.S. Environmental Protection Agency; Mazaika, Rosemary, Robert T. Lackey, and Stephen L. Friant, eds. 1995. *Ecological Risk Assessment: Use, Abuse, and Alternatives.* Amherst, MA: Amherst Scientific Publishers; Pastorok, Robert A., Steven M. Bartell, Scott Ferson, and Lev R. Ginzburg. 2001. *Ecological Modeling in Risk Assessment: Chemical Effects on Populations, Ecosystems and Landscapes.* Boca Raton, FL: CRC Press; Steingaber, Sandra. 1997. *Living Downstream: A Scientist's Personal Investigation of Cancer and the Environment.* New York: Random House.

ECOTOURISM AS A BASIS FOR PROTECTION OF BIODIVERSITY

Ecotourism is controversial because it directly challenges modern tourism concepts that tend to have many environmental consequences. Traditional tourism may displace indigenous and other local people from their homes and communities. The income from tourism can be used to support the protection of biodiverse areas.

BACKGROUND

Touring environmental wonders is not new or controversial. When countries such as Costa Rica began to develop tours of their rain forests, cloud forests, deep biodiversity, and other natural wonders, both as industry and environmentalists took note. Ecotours advertise a soft impact on the environment, generally rustic surroundings in nature, and sometimes wildlife viewing. What if they include a zipline ride in an endangered tree? What if the hotel built to accommodate ecotours displaces a necessary wetland? What if the wildlife is enticed to be viewed and has its natural habitat disturbed for the sake of a photograph? The environmental impacts of most aggressive tourist development generally include land development (with golf courses), accommodations for emissions-producing vehicles, and acres of paving. Ecotours generally do have less environmental impact than traditional western tourism models, however.

Another aspect of ecotours is the hope that they will increase understanding of the global environment and inspire a desire to protect it. Rain forests and tropical biodiversity are a traditional part of ecotours.

WHAT IS BIODIVERSITY?

Biodiversity includes the world's species with their unique evolutionary histories. It also includes genetic variability within and among populations of species and the distribution of species across local habitats, ecosystems, landscapes, and whole continents or oceans. The word *biodiversity* is used in many ways. In the population sciences it means "the variety and variability of biological organisms." The Convention on Biological Diversity defines biodiversity as the "variability among living organisms from all sources."

The operational definition of biological diversity was a topic of the 2002 United Nations conference on ecotourism. This is one battleground in this controversy. The year 2002 was declared the International Year of Ecotourism by the United Nations in 1998. The event included 15 preparatory conferences and seminars held in as many different countries on ecotourism. Many of the issues and environmental controversies surrounding ecotourism were discussed.

According to the United Nations, ecotourism represents between 2 and 4 percent of all international travel expenditure. Ecotour consumers are higher spenders than mass tourists, usually demanding smaller groups, special attention, and accommodation of customers. Many developing countries do not have convertible currency, and tourism is a good way to bring in foreign investment.

NONGOVERNMENTAL ORGANIZATIONS' CONCERNS

Many nongovernmental organizations (NGOs) are concerned that a sudden growth in ecotourism in previously undeveloped areas of the world may have negative cultural ramifications. Indigenous groups who met in the UN presummit preparatory conferences have issues with ecotours and cultural preservation. There is no international consensus on what constitutes ecotourism, and there is concern that it may simply be regular tourism under a marketing name and that it continues to degrade the environment.

Most conference attendees felt that there is the need to strengthen the NGOs engaged in advocacy for cultural preservation. UN participants felt that funds should be mobilized to support indigenous ventures designed to create a balance between the environment and the eradication of poverty. Some environmental groups, such as the World Wildlife Fund, do assist in these efforts.

Ecotourism still has many definitions, most controversial. There is an uneasy consensus that it should minimize the industry's negative environmental and cultural impacts and actively promote the conservation of biodiversity. Revenue from ecotourists may be the only resource available to pay for the protection of the environment and the cultural impact of development on populations living in or close to them.

HAWAII: A PLACESTUDY OF ECOTOURISM AND BIODIVERSITY PRESERVATION

Of the roughly 138 islands that make up the Hawaiian chain, only about eight are known to be inhabited. It is far from the mainland and the site of a large military base—Pearl Harbor. Several of the small islands and a few of the major volcanoes were shooting ranges and may still have ordnance.

Today, more than any other state, tourism is Hawaii's main industry. It has an economic impact per year of about nine billion dollars, supporting whole communities. Hawaii has marketed its beautiful tropical and diverse environment to tourism. Because of this, many would rather encourage ecotourism and therefore develop tourism that maintains Hawaii's attractive environment.

Hawaii's unique native flora and fauna have many levels of biodiversity. The waters surrounding Hawaii are teeming with fish. Geologically very active, Hawaii's big island, Hilo, offers ever-changing views of volcanic activity forming new land. Other islands have fought against development by restricting road development and have emphasized a garden island approach, punctuated with sustainable ecosystem practices. One example of this emerging form of ecotourism is Limahuli Garden, a national tropical botanical garden located in Ha'ena, Kaua'i.

Nonetheless, tourism has environmental impacts. The more tourists there are, the greater the environmental impacts. Water and air pollution increase. Emissions from vehicles and airplanes increase, adding to any volcanic emissions. Recreational users like hikers, surfers, and others have increased, and so increased human encroachment on a sensitive ecosystem. The naval base at Pearl Harbor has also been a consistent source of emissions into air and water. Increased pollution, accumulating emissions, and human encroachment can have powerful impacts on the environment.

The impacts of humans on Hawaii are far-reaching. The original Hawaiians cleared lowland forests to cultivate plants. They brought in small pigs, which led to the extinction of at least 35 species of birds. Many of the birds had no natural enemies and laid their eggs on the ground. Pigs and nonnative cats ate the eggs. Pigs also like to root around in dirt and dig up grubs and roots. Much of the flora and fauna of Hawaii cannot survive being uprooted by pigs. Western Europeans later introduced cattle, goats, and sheep. The extinction of at least 27 species of birds since their arrival is one example of their tremendous environmental impact, well before tourism. Hawaii is no stranger to extinctions.

Hawaii has 75 percent of the historically documented extinctions of plant and bird species. Of all U.S. bird species currently listed as endangered, 40 percent are Hawaiian species. Thirty-one percent of the 271 endangered U.S. plant species are from Hawaii. The Hawaii Natural Heritage Program tracks 30 vertebrates, 102 invertebrates, and 515 plants that are considered to be "critically imperiled globally" (1–5 occurrences and/or fewer than 1,000 individuals remaining, or more abundant but facing extremely serious threats range-wide) or "imperiled globally" (6–20 occurrences and/or 1,000–3,000 individuals remaining, or more abundant but facing serious threats range-wide). For comparison, in New Jersey, which is approximately the size of Hawaii, the Natural Heritage Program tracks 3 vertebrates, 14 invertebrates, and 21 plants that are critically imperiled globally or imperiled globally.

Tourism shows no sign of abating in Hawaii. It accounts for about a third of the state's economy. Hawaii's beautiful environment is used to beckon tourists, but the impact of overdevelopment may imperil a sensitive island ecosystem. More people are now engaged in observing environmental impacts, which continue to threaten species survival.

Ecotourism is an international business. The International Year of Ecotourism served to assist the replication of best practices among governments and private companies.

ECOTOURISM: WORLD TRADE ORGANIZATION (WTO) RECOMMENDATIONS

In line with the first three objectives, WTO recommended that its 139 member states, in September 2000, undertake activities at the national and local levels, such as:

1. define, strengthen, and disseminate, as appropriate, a national strategy and specific programs for the sustainable development and management of ecotourism;
2. provide technical, financial, and promotional support for, and facilitate the creation and operation of, small and medium-size firms;
3. set up compulsory and/or voluntary regulations regarding ecotourism activities, particularly in reference to environmental and sociocultural sustainability;
4. establish national and/or local committees for the celebration of the International Year of Ecotourism, involving all the stakeholders relevant to this activity; and
5. inform the WTO secretariat of the activities planned for 2002 requesting, if appropriate, whatever support they deem necessary.

POTENTIAL FOR FUTURE CONTROVERSY

Ecotourism is a growth industry. It is expanding its reach into some of the most inaccessible places on the planet. Many environmentalists and proponents of sustainability applaud this development because ecotours instill a global approach to the environment. With the increase in tourism come environmental and cultural impacts, however. These can be very controversial. Ecotourism focused on biodiversity will face special challenges. There is no real measurement in the area of successful intervention, what data are needed, and what the impacts of ecotourism are on biodiversity. Given that much scientific measurement and analysis must be done to establish baseline measures of biodiversity, that the need for species preservation is urgent, and that ecotourism is growing, it is likely that controversies will continue.

See also Conservation in the World; Rain Forests; Sustainability; Wild Animal Reintroduction

Web Resources

Ecotourism Resources. Available at www.tourisminsight.com/reports/ecotourismresources/. Accessed January 20, 2008.

United Nations Report of the Economic and Social Council Assessment of the Results Achieved in Realizing Aims and Objectives of the International Year of Ecotourism. Available at http://pdj.sagepub.com/cgi/contentabstract/6/6/146. Accessed January 20, 2008.

Further Reading: Buckley, Ralf. 2003. *Case Studies in Ecotourism.* New York: CABI Publishing; Buckley, Ralf. 2004. *Environmental Impacts of Ecotourism.* New York: CABI Publishing; Burgess, Bonnie B. 2003. *Fate of the Wild: The Endangered Species Act and the Future of Biodiversity.* Athens: University of Georgia Press; Honey, Martha. 1999. *Ecotourism and Sustainable Development: Who Owns Paradise?* Washington, DC: Island Press; Tucker, Hazel, and Colin Michael Hall. 2004. *Tourism and Postcolonialism: Contested Discourses, Identities and Representations.* New York: Routledge; Wearing, Stephen, and John Neil. 1999. *Ecotourism: Impacts, Potentials and Possibilities.* New York: Elsevier.

ENDANGERED SPECIES

Controversies about endangered species center around the value of species and the cost of protecting and preserving them and their habitats. There are battlegrounds about whether a particular species is going extinct and whether a particular policy actually does protect a designated species. Natural resource extraction (logging, mining, grazing), land and road development into wildlife habitats, and increased recreational use are all battlegrounds for this controversy.

BACKGROUND

Species extinctions have occurred along with evolution. As plant and animal species evolve over time, some adaptations fail. As human population has increased, along with our hunting, farming, and foraging capacities, plant and animal species have begun to disappear faster. Pollution, climate change, and other significant environmental impacts can destroy species in sensitive niches in the food chain. In most cases species are endangered because of human impacts, but each case can present a battleground.

The evidence for human impact on species in the United States is often based on successful eradication programs for problem pests. Knowledge about species extinctions grew as environmentalists, hunters, researchers, and others observed extinctions and near-extinctions of several species, such as the buffalo and pigeon. Endangered species create a great concern for productive bioregions and ecosystem integrity. They can represent a significant part of the food web, and their loss can forever weaken other parts of that food web. Eventually, this great social concern for endangered species found its way into law, now one of the main battlegrounds for this controversy.

THE ENDANGERED SPECIES ACT: OVERVIEW

The Endangered Species Act (ESA) is the policy implementation of the concern for species preservation. In its application the ESA is very controversial, touching on flash points related to long-term leases of public lands, private property, and defensible science. Scientific controversies include ecosystem risk assessment, the concept of a species and how it has been interpreted for ESA application, and conflicts between species when individual species are identified for protection and others are not. A prominent controversy was the preservation of the habitat for the spotted owl in Oregon, which prevented logging. Approximately 60 logging mills subsequently closed. There are current discussions about whether the mitigation plan really did preserve the spotted owl or not. Endangered species designation can have a dramatic effect on natural resource extraction (logging, mining) by prohibiting or limiting it.

Before a plant or animal species can receive protection under the Endangered Species Act, it must first be placed on the federal list of endangered and threatened wildlife and plants. The listing program follows a strict legal process to determine whether to list a species, depending on the degree of threat it faces. According to the ESA, an *endangered* species is one that is in danger of extinction throughout all or a significant portion of its range. A *threatened* species is one that is likely to become endangered in the foreseeable future. The federal government maintains a list of plants and animals native to the United States that have potential to be added to the federal list of endangered species.

Congress passed the Endangered Species Preservation Act in 1966. It was a small but very important first step toward species preservation by the United States. This law allowed listing of only native animal species as endangered and provided limited means for the protection of species so listed. The Departments of the Interior, Agriculture, and Defense were to seek to protect listed species and to preserve the habitats of such species. Land acquisition for protection of endangered species was also authorized by law. The next law was the Endangered Species Conservation Act of 1969. It was passed to provide additional protection to species in danger of worldwide extinction. Importation of such species was prohibited, as was their subsequent sale within the United States. This act also called for an international ministerial meeting to adopt a convention on the conservation of endangered species. A 1973 conference in Washington led to the signing of the Convention on International Trade in Endangered Species of Wild Fauna and Flora (CITES). This is a very important species preservation agreement. It restricts international commerce in plant and animal species believed to be actually or potentially harmed by trade. Later that year in the United States, the Endangered Species Act of 1973 was passed. This law combined and considerably strengthened the provisions of earlier laws and greatly expanded the reach of the ESA. This also had the effect of intensifying the controversy and increasing the battlegrounds.

Its principal provisions follow:

- U.S. and foreign species lists were combined, with uniform provisions applied to both categories of endangered and threatened.
- Plants and all classes of invertebrates were eligible for protection, as they are under CITES.
- All federal agencies were required to undertake programs for the conservation of endangered and threatened species and were prohibited from authorizing, funding, or carrying out any action that would jeopardize a listed species or destroy or modify its critical habitat.
- Broad taking prohibitions were applied to all endangered animal species, which could also apply to threatened animals by special regulation.
- Matching federal funds became available for states with cooperative agreements.
- Authority was provided to acquire land for listed animals and for plants listed under CITES.
- U.S. implementation of CITES was provided.

Significant amendments have been enacted in 1978, 1982, and 1988, while the overall thrust of the 1973 act has remained basically the same. Principal amendments are listed here.

1978

- Provisions were added to section 7, allowing federal agencies to undertake an action that would jeopardize listed species if the action were exempted by a cabinet-level committee convened for this purpose.
- Critical habitat was required to be designated concurrently with the listing of a species, when prudent, and economic and other impacts of designation were required to be considered in deciding the boundaries.
- The Secretaries of the Interior and Agriculture (for the Forest Service) were directed to develop a program for conserving fish, wildlife, and plants, including listed species, and land acquisition authority was extended to such species.
- The definition of *species* with respect to *populations* was restricted to vertebrates; otherwise, any species, subspecies, or variety of plant, or species or subspecies of animal remained listable under the act.

1982

- Determinations of the status of species were required to be made solely on the basis of biological and trade information, without any consideration of possible economic or other effects.
- A final ruling on the status of a species was required to follow within one year of its proposal unless withdrawn for cause.
- Provision was made for designation of experimental populations of listed species that could be subject to different treatment under section 4, for critical habitat, and section 7.
- A prohibition was inserted against removing listed plants from land under federal jurisdiction and reducing them to possession.

1988

- Monitoring of candidate and recovered species was required, with adoption of emergency listing when there is evidence of significant risk.
- Several amendments dealt with recovery matters: (1) recovery plans will undergo public notice and review, and affected federal agencies must give consideration to those comments; (2) five years of monitoring of species that have recovered are required; and (3) biennial reports are required on the development and implementation of recovery plans and on the status of all species with plans.
- A new section requires a report of all reasonably identifiable expenditures on a species-by-species basis that were made to assist the recovery of endangered or threatened species by the states and the federal government.
- Protection for endangered plants was extended to include destruction on federal land and other taking when it violates state law.

As of 2006 many species in the United States are listed as threatened and endangered or proposed for listing as threatened or endangered. Listed are 398 species of animals and 599 species of plants. Fourteen species of animals and one species of plant are currently proposed for listing. Currently, 473 U.S. species have designated critical habitat. Critical habitats are specific geographic areas, whether occupied by a listed species or not, that are essential for its conservation and that have been formally designated by rule. Additionally, 138 species of animals are candidate species and 144 species of plants are candidate species for listing as endangered. Some species are also proposed for delisting, another battleground in this controversy.

Also, 557 habitat conservation plans have been approved. According to the law, a habitat conservation plan (HCP) is plan that outlines ways of maintaining, enhancing, and protecting a given habitat type needed to protect species; it usually includes measures to minimize impacts and may include provisions for permanently protecting land, restoring habitat, and relocating plants or animals to another area.

Currently, 1,043 species have approved recovery plans. A recovery plan is a document drafted by a knowledgeable individual or group that serves as a guide for activities to be undertaken by federal, state, or private entities in helping to recover and conserve endangered or threatened species. Recovery priority is also determined in these plans. There can be differences of opinion as to how high a priority certain species should have in a recovery plan. A rank ranges from a high of 1 to a low of 18 and these set the priorities assigned to listed species and recovery tasks. The assignment of rank is based on degree of threat, recovery potential, taxonomic distinctiveness, and presence of an actual or imminent conflict between the species and development activities.

The regulations for protection of endangered species generally require protection of species habitat. As our population grows and development expands into natural areas, the protection of wildlife habitat becomes more important and more difficult. Preservation of riparian (water) migratory pathways, private conservation efforts, and applied scientific research all hold promise for species preservation. However, the need for wildlife habitat preservation will still impair the ability of some property owners use their land as they wish. The controversies around species preservation are likely to be around for a long time.

POTENTIAL FOR FUTURE CONTROVERSY

As human habitation grows into more formerly wild areas, more species are likely to become extinct. On September 12, 2007, the World Conservation Union issued a report called their *red list*. In 2006 there were 16,118 species on the list; in 2007 the number of species threatened with extinction worldwide increased to 16,306, according to the report. The organization itself states that the list is incomplete because the state of knowledge about all the species is inadequate. Their own assessment is underfunded and relies heavily on the volunteer efforts of scientists. They are concerned because they maintain that some species could be extinct before we know of them. For example, coral reefs are rapidly dying in many parts of the world and, with them, many of the species that make up a coral reef in a given location. It is difficult for the organization to assess species endangerment in nations at war. Some apes may be eaten in the bush-meat trade of central Africa, where warring groups have persisted in conflict. The report indicates that apes may be among the most endangered groups, known to date.

There are strong world conservation efforts but also strong political controversies. As more information on human environmental impacts on marine environments develops, so too will lists of endangered species. There is controversy whenever a species is added to or taken off the endangered species list.

See also Ecosystem Risk Assessment; Logging; Mining of Natural Resources

Web Resources

Endangered and Extinct Species Lists. Available at eelink.net/EndSpp.old.bak/ES.lists.html. Accessed January 20, 2008.

Endangered Species List for Navaho Nation. Available at www.natureserve.org/nhp/us/navajo/esl.htm. Accessed January 20, 2008.

Office of Protected Resources. Species under the Endangered Species Act (ESA). Available at www.nmfs.noaa.gov/pr/species/esa/. Accessed January 20, 2008.

Further Reading: National Research Council, Committee on Scientific Issues in the Environment. 1995. *Science and the Endangered Species Act.* Washington, DC: National Academies Press; Noss, Reed F. 1997. *The Science of Conservation Planning: Habitat Conservation under the Endangered Species Act.* Washington, DC: Island Press; Shogren, Jason F., ed. 1998. *Private Property and the Endangered Species Act.* Austin: University of Texas Press; Westley, Frances R., and Philip S. Miller. 2003. *Experiments in Consilience: Integrating Social and Scientific Responses to Save Endangered Species.* Washington, DC: Island Press.

ENVIRONMENT AND WAR

Environmental effects of war are different from environmental warfare. Both have enormous, surprising, and controversial impacts on the environment. Land mines, wildlife recovery, and postwar environmental responsibility clash with postwar political realities.

WHAT IS ENVIRONMENTAL WARFARE?

The ability to control weather and climate for any purpose has been the subject of much study, speculation, and controversy. To do so for the purposes of war is called *environmental war.* War is generally an official act of a government attacking another nation that is the enemy. *War* can be a technical term that is avoided in many countries even if they are in fact at war with other nations. Environmental war is defined as an intentional attack on an ecosystem. This includes climate and weather, the ionosphere, the magnetosphere, the tectonic plate system, and/or the triggering of earthquakes. The 1977 United Nations treaty against modification of the environment for war forbids it. Environmental war weapons systems can include chemical weapons systems (climate and weather modification) and electromagnetic weapons systems (climate and weather modification; seismic warfare). Each large weapon system has the potential to be classified as an environmental war weapon, and avoiding this classification is a battleground. However, the real-life battleground of war seldom observes environmental impacts as anything more than collateral damage. To add to the controversy, some definitions of environmental war are broader. For example, the use of depleted uranium in the Yugoslavian war has been termed environmental war.

It is the nature of war to use whatever force is necessary to overpower the enemy. As technology improves and population increases, these forces can take the form of nuclear, environmental, and/or conventional war. Enormous international controversy exists with respect to environmental war. While it could give a military advantage in war, its global impacts could be long-term and irreversible. The argument against environmental war is that it is unjust because many unknown innocents could be killed who are not a party to the respective conflict. The possibility also exists that environmental war could backfire. That is, there could be an environmental act of war, like weather change, that could go awry and negatively affect the very nation who sent it as a weapon.

RADIOACTIVE CONTAMINATION: DEPLETED URANIUM

One environmental war weapon is radioactive, depleted uranium. Depleted uranium (DU) is a radioactive heavy metal. It is the waste left over when the isotope uranium-235 is extracted from naturally occurring uranium to fuel nuclear power stations and build nuclear bombs. There is a large amount of nuclear waste and a major controversy over where and how to store, ship, and eliminate it. It is very inexpensive and easy to obtain. DU is roughly 60 percent as radioactive as naturally occurring uranium. It has a long half-life of 4.5 billion years. DU is a very good material for bombs because of its density. DU shells can pierce several inches of armor-plated steel on tanks, humvees, and most other military vehicles. They are radioactive and have a long-term impact wherever they land. Their use is very controversial because of the long-term human and environmental impacts. Their use in this manner allows for some reuse of low-level radioactive waste, benefiting industries involved with nuclear power, defense industries, and military manufacturing.

The effects of DU on health are the topic of an ongoing controversy. The effects of DU are less aggressive because DU is less radioactive than natural uranium. Heavy-metal poisoning from the bomb itself is regarded as posing a more serious health risk than its radiation. When a DU bomb strikes armor or burns, it detonates, explodes, and releases clouds of uranium dust, which can be inhaled, ingested through swallowing, and absorbed through the skin and wounds. There are many heavy-metal particles swirling in the air. These heavy-metal particles are also absorbed but are difficult to eliminate from the body. Once internalized the DU dissolves and enters the bloodstream. What happens next is also a battleground. Some scientists maintain that most uranium is excreted from the body through the kidneys. They argue that the human body is very effective at eliminating ingested and inhaled natural uranium. Furthermore, the low radioactivity per unit of DU means that the amount of uranium needed for significant internal exposure is impossible to obtain. Other scientists argue that exposure from radioactivity can be cumulative and adverse to human health, that not all human kidneys are efficient at removing DU (especially at war), and that the radioactivity from DU is significant enough to affect vulnerable and wounded people. They argue that most of the DU will be eliminated shortly after exposure, but some remains in the body. This can increase the risk of cancer and all other illnesses over time. DU left in the body is a source of low-level radiation. Radiation can damage cell structure for years, they argue.

CONCERN ABOUT NUCLEAR FACILITIES AS TARGETS

Controversy about nuclear facilities as targets in war is rooted in concern about the human and ecological devastation that would follow the bombing of such a facility. It could essentially be the same as a nuclear bomb, which is why it is crucial to know whether a facility is used in the manufacture of nuclear weapons or for other purposes (power, research). If it is not used for the creation of nuclear weapons, then it is less likely to be a target.

ECOLOGICAL DAMAGE

Environmental war weapons cause ecological damage. A heavy DU bomb creates temperatures of around 3,000°C, destroys all flora and fauna, and turns the lower layers of soil into a useless area that can take thousands of years to regenerate. Ecological damage includes destruction of vegetation, disturbance of migration corridors of migratory birds, crippling of the reproductive cycles of fish and amphibians through water pollution due to bombing of industrial complexes along waterways, destruction of reserves of biodiversity. The top layers of land and watersheds usually suffer the most damage once the fires of the bombing stop.

Generally, only immediate measures to stop the downstream flow of pollution will prevent an ecological catastrophe. A war that targets chemical factories and oil installations and that deploys radioactive weapons in towns and cities will have enormous environmental effects. These impacts will follow natural vectors like watersheds, air sheds, and tides. In moving through these paths they will move through soil, air, and water. Their penetration into natural ecosystems is why many of the environmental effects are considered irreversible.

REFUGEES AND IMPACTED INTERNAL CITIZENS

The clearest evidence that there are environmental consequences of war comes from the impact of large numbers of refugees and returnees. Cities become more overcrowded than they previously were, leading to environmental impacts, often with inadequate water, sewage, and waste disposal facilities. Human habitats with half a million, a million, or more people concentrate on their surrounding land. These refugee camps are often not designed for large human settlement and face water, sewage, and disease problems on top of the food and shelter loss characteristic of war. The environmental impact of the camps is enormous, with very few living things left. In the case of South Vietnam, people fleeing from the countryside sought shelter in Saigon, whose population increased from 250,000 to 3 million. The 500,000 Rwandans fleeing to Tanzania in 1994 created the second largest urban center after Dar es Salaam. Refugee-receiving countries frequently herd refugees into very large camps from which refugees set out to supplement their needs if these are not met by relief agencies. Refugees also go somewhere within the affected country as internally displaced persons, in which case the effects on the environment would be dispersed instead of concentrated.

What happens in the abandoned areas from which people flee? If humans leave, wildlife should recover. That is the impression given in the case of Afghanistan, in Nicaragua, in Angola, in Mozambique, in Liberia, and, to a lesser degree, in Rwanda. There are records of nature preservation due to war. In Nicaragua, on the Atlantic side of the country, indigenous groups maintained much of the forest for cover. In the Koreas, the demilitarized zone has become a wildlife haven and sports species seen nowhere else in either country, causing concern among conservationists today about an eventual reunification of the country. In the Central African Republic, one forest area made unsafe by rebel poachers gradually emptied of people, allowing forest cover to increase. In Angola, minefields prevented human encroachment and allowed nature to recover.

In Liberia, the civil war forced so many people into the forests that gun-based hunting decreased. Shooting a gun would tell potential enemies your location. Wildlife recovered until peace made gun-based forest hunting safe again. An equally understudied topic is that of the biodiversity impact of returnees. Adverse wildlife effects are closely correlated with the return path of refugees. The reopened highways and transportation corridors serve as encroachment vectors along which wildlife disappears.

FIREPOWER AND MOBILITY

The red-necked ostrich in Niger is probably extinct due to the AK-47 and the jeep; guns, gas, and poaching are a deadly combination for wildlife. Populations of elephants and rhinoceroses in the Central African Republic have plummeted due to incursions of well-armed and well-traveled poachers from southern Sudan seeking to finance the war there. Small arms and light weapons make poaching for bush meat easy, but it takes access to the world market to convert diamonds to dollars and cash to finance internal war. So it is not only the jeep but airstrips built right into the ecosystem that greatly increase

---

ZOOS IN A WAR: HOW ANIMALS SUFFER

Animals perceive their environment in different ways. Many have senses much more attuned to the natural environment than humans. As a result, their reaction to an ongoing war is severe.

On May 30, 1999, a zookeeper in Belgrade noted the effects the bombing was having on the animals in the zoo. In his own words he describes a weapons assault at a zoo:

> The noise starts around half an hour before the bombs fall as the animals in the Belgrade Zoo pick up the sound of approaching planes and missiles. It's a strange and disturbing noise. The Zoo was hard hit by NATO's air strikes, particularly when the alliance attacked Belgrade's power system, and indirectly the water supply. 1,000 eggs of rare and endangered species incubating, some of them ready to hatch in a couple of days, were all destroyed. Many birds abandoned their nests, leaving eggs to grow cold. A snake aborted some 40 fetuses, apparently reacting to the heavy vibration shaking the ground as missiles hit targets nearby. The worst night at the Zoo was when NATO hit an army headquarters only 600 meters away, with a huge detonation. The next day some of the animals had killed their young. A female tiger killed two of her four three-day-old cubs, and the other two were so badly injured they died. She had been a terrific mother until then, raising several litters without any problems. Armed guards patrolling the zoo are not there to keep people from harming or stealing the animals but to shoot the animals if the zoo gets bombed and some of them escape. Many animals exhibit signs of trauma; not eating, shaking, and physical wasting.

NATO said that the environmental damage caused by the attack was collateral damage. This region contained a very important refinery, thus was considered a and strategic target. NATO bombed the chemical complex for 23 days, hitting it with at least 56 missiles or bombs.

environmental impacts from war. The role of oil, mining, and timber companies and of private military companies in Angola, Sierra Leone, and Papua New Guinea is enormous. In fact, Economists Allied for Arms Reduction in the United States this year has begun to consider whether an equivalent to the Sullivan principles is needed to help set standards of conduct for businesses operating in conflict zones. (The Sullivan principles were used to boycott those U.S. companies doing business in or with apartheid South Africa that did not adhere to the principles.)

Peace can be more environmentally destructive than war. In the case of the Persian Gulf War, ornithologists discovered how little damage the war did to bird life. They also discovered how much irreversible damage economic development of the Saudi coastline along the Persian Gulf had done through tourism, fishing, and the oil industry.

FRONTIERS OF SCIENTIFIC INVESTIGATION AND KNOWLEDGE

The lack of funded research into the environmental effects of war fuels this controversy. War reparations are made to repair infrastructure and needed community facilities. Some of the best research is on how effective human pesticides were in killing. The record on Vietnam is ample but is almost exclusively focused on the effects of one dioxin-containing herbicide, Agent Orange, on human populations. This is an important topic in its own right, of course, but the result of this emphasis is that little is known about the environmental effects of the bombing, burning, and bulldozing of Vietnam's forests in the 1960s. Likewise, in the Persian Gulf War, the majority of the international scientific effort was devoted toward study of the effects of the oil-fire-related smoke plume that might be transported by the jet stream around the world to affect the health and well-being of Europeans and Americans. This possibility captured the world's headlines and funding. Further, the oil released into the Persian Gulf's marine environment made funding available to study the environmental effects of the Gulf oil spill. Almost nothing was spent to study the effects of the war on the desert flora, fauna, and ecology of Kuwait, Iraq, Afghanistan, and Saudi Arabia. Critics have charged Western nations with environmental ethnocentrism. If the potential environmental effects of a war might affect us, as in the case of airborne, transboundary pollution, we fund the necessary studies but if the effect is likely to remain local, as in the case of desert soil that does not travel well, we do not fund studies. This is confirmed in the case of Kosovo, where the United Nations funded DU-related studies because nuclear issues are always high-priority in Europe but neglected to similarly fund continued study of other environmental aspects of the war, even though scientists requested it. Other studies focus on Western conservationists' interests, especially charismatic megafauna, not necessarily on important ecological effects of war. Microbes, contamination paths and size, preservable species and ecosystems, and sustainable food production would help reduce the scientific uncertainty that is driving this controversy.

MAXIMUM RADIATION EXPOSURE LIMITS AND THEIR LIMITED APPLICATION TO URANIUM AND DEPLETED URANIUM

The general public should not receive a dose of more than one millisievert (mSv) in a year. In special circumstances, an effective dose of up to five mSv in a single year is permitted provided that the average dose over five consecutive years does not exceed one mSv per year. An equivalent dose to the skin should not exceed 50 mSv in a year.

Occupational exposure should not exceed an effective dose of 20 mSv per year averaged over five consecutive years or an effective dose of 50 mSv in any single year. An equivalent dose to the extremities (hands and feet) or the skin should not surpass 500 mSv in a year.

In the case of uranium or DU intake, the radiation dose limits are only applied to inhaled insoluble uranium compounds. For all other exposure pathways and the soluble uranium compounds, chemical toxicity is the factor that limits exposure to the underlying radiation.

Guidance on Exposure Based on Chemical Toxicity of Uranium

MONITORING AND TREATMENT OF EXPOSED INDIVIDUALS: When an individual is suspected of being exposed to DU at a level significantly above the normal background level, urine analysis can provide useful information. Fecal measurement can also give useful information on DU intake. However, fecal excretion of natural uranium from the diet is considerable (very variable), and this needs to be taken into account. Following war, levels of DU contamination in food and drinking water might be high in affected areas for years.

POTENTIAL FOR FUTURE CONTROVERSIES

War itself is always controversial. The military industrial complex, as originally described by President Eisenhower, drives both research and major corporate economic development. Transportation, communication, and manufacturing industries all increase production in war. These are very powerful industries with powerful federal agencies protecting them in the name of national security. Environmental protection is a low priority in this context. Communities have no say, even as they witness dangerous environmental practices that put them at risk. The overall environmental impacts of war, inclusive of this industrial ramping up of production, are larger than the direct impacts of war alone. Technology has allowed our powerful forces of destruction to have long-term effects and pinpoint accuracy. Technology has also developed instruments and measures that scientists can use to study and monitor ecological effects. In this way, the best way to mitigate the impacts of war can be developed for future, hopefully more peaceful, generations. Until then, this controversy will continue.

See also Citizen Monitoring of Environmental Decisions; Cumulative Emissions, Impacts, and Risks; Environmental Impact Statements: United States; True Cost Pricing in Environmental Economics; Watershed Protection and Soil Conservation

Web Resources

Effect of War on Environment. Available at http://www.ppu.org.uk/learn/infodocs/st_envi ronment.html. Accessed January 20, 2008.

Sierra Club of Canada. The Environmental Consequences of War. Available at http://www. sierraclub.ca/national/postings/war-and-environment.html. Accessed January 20, 2008.

War and Environment. Available at http://www.mapcruzin.com/war-and-environment/. Accessed January 20, 2008.

World Health Organization. Depleted Uranium. Available at http://www.who.int/ionizing_ radiation/env/du/en/index.html. Accessed January 20, 2008.

Further Reading: Austin, Jay E., ed. 2000. *The Environmental Consequences of War: Legal, Economic, and Scientific Perspectives.* Cambridge: Cambridge University Press; Harf, James E., and B. Thomas Trout. 1986. *The Politics of Global Resources: Population, Food, Energy, and Environment.* Durham, NC: Duke University Press; Woodbridge, Roy M. 2004. *The Next World War: Tribes, Cities, Nations, and Ecological Decline.* Toronto: University of Toronto Press.

ENVIRONMENTAL AUDITS AND ENVIRONMENTAL AUDIT PRIVILEGES

Industries that perform environmental audits do so for a variety of reasons: by law, voluntarily, and as part of other audits. These audits can disclose whether a particular plant is in compliance with environmental laws, areas of cost savings in environmental compliance, and multifacility environmental compliance measures. Industries want audits kept secret, or privileged. Small and medium-sized businesses especially want this legislation because they want to level the playing field with large industry, which that can use its own lawyers and hide information within attorney-client privilege. This secrecy prevents communities, environmentalists, and others from knowing about the environmental audit and any information that would relate to local environmental impacts or risks. Industry is concerned about environmental lawsuits and, if not protected by some legal privilege, would not perform any type of environmental audit. Since most U.S. environmental information about industry is self-reported, an independent audit carries much more credibility than the usual industry and government reports. Access to accurate environmental information is the crux of this controversy.

More than 20 states have enacted environmental audit privilege legislation. It takes different forms and is usually controversial. Regular audits should become a normal business management tool that assists compliance with complex environmental regulations and avoids unnecessary waste. Such audits provide valuable information about potential environmental noncompliance, suggest methods for reducing or eliminating waste streams, inform shareholders and customer queries regarding off-site liability, and could be used to create a green corporate image.

Self-auditing programs generate evidence that could be used against a company in an enforcement action. Any noncompliance reported in such a document may create a paper trail available to both enforcement agencies and private plaintiffs. Consequently, although numerous businesses undertake self-audits, many do not want information suggesting environmental noncompliance to be circulated or written down. The fear that this information will be discovered by a private party or a governmental agency discourages self-auditing programs at various companies. To environmental policy makers this fear is problematic because it distorts environmental information. Many communities distrust this secret audit process, preferring clear and transparent transactions.

Large industry always has relied on the common law attorney-client privilege, the work-product doctrine, and, more recently, common law self-evaluation to argue that audit documentation is privileged. These legal arguments give privileged protection to large companies with environmental self-audit programs. The claim of lawyer-client privilege will start a discovery dispute that results in an in camera review by a judge, who will determine whether to allow the government to use the audit document against the regulated business. In contrast, a small business does not have the financial and strategic capacity to engage a lawyer for an expensive judicial fight for secrecy. The primary controversy between large and small industries here is who gets to privilege environmental information. This is not a controversy shared with communities or environmentalists.

NEW STATE LAWS ON ENVIRONMENTAL PRIVILEGE

More than 30 states have considered legislation involving environmental audit privilege, and 20 have enacted such laws, including Arkansas, Colorado, Idaho, Illinois, Indiana, Kansas, Kentucky, Michigan, Minnesota, Mississippi, New Hampshire, Ohio, Oregon, South Carolina, South Dakota, Texas, Utah, Virginia, and Wyoming. These state laws essentially codify discovery-dispute procedures that large businesses always have enjoyed. By eliminating the requirement of hiring an attorney, small companies can afford to engage in the same type of self-audit process that most large companies currently take for granted when this legislation is enacted.

The environmental privilege is limited by law and not universally applied or available. A common legislative intent is to encourage owners and operators to conduct voluntary environmental audits of their facilities by offering a limited privilege to audited information. Proponents point out that it is infeasible and unnecessary for state and federal regulators to police each and every business in the state, and therefore self-auditing should be encouraged. Generally, a number of conditions must be met. Some of the conditions that are necessary for the state law on environmental audit privilege to apply are

1. All noncompliance identified during the audit is corrected in a reasonable manner;
2. The privilege is *not* asserted for fraudulent purposes; and
3. Information in the audit is *not* otherwise required to be reported.

Some legislation also provides that a person or entity making a voluntary disclosure of an environmental violation is immune from any administrative, civil, and criminal penalties in return associated with that disclosure. As discussed further on, the compliance focus of environmental law allows for rapid reduction of penalties in return for quick compliance and disclosed and remedied harms.

HOW BROAD IS THE INDUSTRIAL PRIVILEGE?

Proponents of audit privilege legislation state that it does not compel secrecy, because no privilege exists unless there is prompt disclosure and correction of the violation. Furthermore, unless the information falls within the very narrow scope of privileged information, it is decidedly vulnerable.

STATE LAWS

State laws can vary from state to state. Some conditions and exceptions of state privilege and immunity laws include:

- The audit must be scheduled for a specific time and announced prior to being conducted along with the scope of the audit. (AK)
- The company makes available annual evaluations of their environmental performance. (AZ)
- In exchange for a reduction in civil and/or administrative penalties a company implements a pollution-prevention or environmental management system. (AZ).
- Privilege is not applicable to data, reports, or other information that must be collected, developed, maintained, or reported under federal or state law. (AR)
- Audit report is privileged (secret) unless a judge determines that information contained in the report represent a clear, present, and impending danger to public health or the environment in areas outside the facility property. (CO)
- An environmental audit report is privileged (secret) and is not admissible as evidence in any civil or administrative proceeding with certain exceptions, if the material shows evidence of noncompliance with applicable environmental laws and efforts to achieve compliance were not pursued by the facility as promptly as circumstances permit. (WY)
- The Colorado Environmental Audit Privilege and Immunity Law does not affect public access to any information currently available under the Colorado Open Records Act. This information would include, but is not limited to, permits, permit applications, monitoring data, and other compliance/inspection data maintained by the Colorado Department of Public Health and Environment.
- Additionally, the audit privilege does not affect the Colorado Department of Public Health and Environment's authority to enter any site, copy records, inspect, monitor, or otherwise investigate compliance or citizen complaints.

Community interest in this issue is high, and environmental organizations advocate against these laws. It is likely that state laws on this controversy will change rapidly.

POTENTIAL FOR FUTURE CONTROVERSY

Many states that favor privileging environmental information argue that environmental protection efforts require that businesses, municipalities, and public agencies take self-initiated actions to assess or audit their compliance with environmental laws and correct any violations found. By getting to know all the industries affecting the environment and protecting their information better, compliance with the intent of environmental laws results.

Communities and environmentalists respond that most if not all industrial emissions are self-reported, in a context of very weak enforcement. They argue that environmental information is a common good to be shared. Keeping it secret promotes a high degree of distrust and breeds controversy. Advocates of sustainability and environmentalists want full and complete disclosure of all environmental impacts. The battleground for this controversy is state and local legislatures and federal environmental agencies like the EPA. The relationship between states and the EPA on this issue is contentious and unfolding.

See also Citizen Monitoring of Environmental Decisions; Community Right-to-Know Laws; Different Standards of Enforcement of Environmental Law; Good Neighbor Agreements; Supplemental Environmental Projects

Web Resources

Enforcement Response Policy for Treatment of Information Obtained through Clean Air Act Section 507 Small Business Assistance Programs. Available at es.epa.gov/oeca/ccsmd/file11.html. Accessed January 20, 2008.

Incentives for Self-Policing: Discovery, Disclosure, Correction and Prevention of Violations (Audit Policy). Available at http://www.epa.gov/compliance. Accessed January 20, 2008.

Policy on Compliance Incentives for Small Businesses Memorandum—Subject: Reduced Penalties for Disclosures of Certain Clean Air Act Violations. Available at es.epa.gov/oeca/ore/caa-tit.pdf. Accessed January 20, 2008.

Protocol for Conducting Environmental Compliance Audits for Hazardous Waste Generators under the Resource Conservation and Recovery Act (PDF-726K), Available at es.epa.gov/oeca/ccsmd/epcrafinal.pdf. Accessed January 20, 2008.

Protocol for Conducting Environmental Compliance Audits under the Emergency Planning and Community Right-to-Know Act and CERCLA Section 103 (3/01). Available at es.epa.gov/oeca/ccsmd/epcrafinal.pdf. Accessed January 20, 2008.

Further Reading: Dietz, Thomas, and Paul C. Stern, eds. 2002. *New Tools for Environmental Protection: Education, Information, and Voluntary Measures.* Washington, DC: National Academies Press.

ENVIRONMENTAL IMPACT STATEMENTS: INTERNATIONAL

International environmental impact statements are controversial because they have much higher standards than U.S. environmental impact statements. More and more nations are requiring them as a condition of doing business in their country. They include cumulative impacts and ecosystem-based boundaries for environmental assessment.

Environmental impact assessment is as varied as the environment. Environmental definitions, values, beliefs, and scientific capacities also differ greatly around the world. Many countries have natural resources they can trade but are concerned about environmental impacts of getting the natural resources out. There is a worldwide concern that some countries could be ravaged by foreign natural resource exploitation. Many countries have environmental disputes on their borders. For example, mining companies in Canada dump pollution into water headed for the United States. The U.S. illegal waste trade pollutes the Tijuana River in Mexico. In both cases, as in most, one country allows acts of environmental pollution it would not allow otherwise. River and lake boundaries for countries do not coincide with natural systems of watersheds. Many international agreements and treaties deal with trade and the environment. Some specifically deal with environmental impact assessment.

CONVENTION ON ENVIRONMENTAL IMPACT ASSESSMENT IN A TRANSBOUNDARY CONTEXT 1991

The Environmental Impact Assessment Convention lists countries' obligations to assess the environmental impact of certain activities at an early stage of planning for that activity. It is much easier to solve many potential environmental problems early in the process. The convention reaffirms the general obligation of states to notify and consult each other on all major projects under consideration that are likely to have a significant adverse environmental impact across boundaries. This international convention entered into force on September 10, 1997.

Under this agreement *environmental impact assessment* means a national procedure for evaluating the likely impact of a proposed activity on the environment. *Impact* means any effect caused by a proposed activity on the environment including on human health and safety, flora, fauna, soil, air, water, climate, landscape, and historical monuments or other physical structures or the interaction among these factors. It also includes effects on cultural heritage or socio-economic conditions resulting from alterations to those factors. *Transboundary impact* means any impact, not exclusively of a global nature, within an area under the jurisdiction of a country caused by a proposed activity the physical origin of which is situated wholly or in part within the area under the jurisdiction of another country.

CONTENT OF THE INTERNATIONAL ENVIRONMENTAL IMPACT ASSESSMENT

Environmental impact assessments can vary greatly in content. It is a battleground to get the environmental risk assessment that is acceptable to community, industry, environmentalists, and scientists. Generally, the more scientifically comprehensive the assessment is the more time it takes. Scientific site-applied research is very time-consuming and labor intensive. If the environmental assessment includes cumulative risks and ecosystem risk assessment, it could be an ongoing process.

The international consensus in the treaties regarding the content of an environmental assessment represents the ideal content if resources and time were unlimited. Information to be included in the environmental impact assessment documentation would be:

1. A description of the proposed activity and its purpose;
2. A description, where appropriate, of reasonable alternatives (for example, location or technological) to the proposed activity and also the no-action alternative;
3. A description of the environment likely to be significantly affected by the proposed activity and its alternatives;
4. A description of the potential environmental impact of the proposed activity and its alternatives and an estimation of its significance;
5. A description of mitigation measures to keep adverse environmental impact to a minimum;
6. An explicit indication of predictive methods and underlying assumptions as well as the relevant environmental data used;
7. An identification of gaps in knowledge and uncertainties encountered in compiling the required information;
8. Where appropriate, an outline for monitoring and management programs and any plans for postproject analysis; and
9. A nontechnical summary including a visual presentation as appropriate (maps, graphs).

Unlike the U.S. and U.S. tribal environmental impact statement requirements, international agreements require postproject analysis. This is an extremely important phase for evaluation of the effectiveness of any mitigating measures taken to soften or prevent environmental impacts. Some of these impacts can be irreversible if the mitigating measures do not work. They can be expensive and may slow or stop the project and its operations, and therefore industry usually avoids postproject analysis if it can. Without postproject analysis it is difficult to know true environmental impacts, so environmentalists and sustainability proponents generally advocate for this. It is a battleground.

Countries at the request of any other country determine whether, and if so, to what extent, a postproject analysis will be done. Any postproject analysis undertaken includes the surveillance of the activity and the determination of any adverse transboundary impact. *Adverse* is a term of art, and its definition differs greatly from country to country.

WHAT DO INTERNATIONAL ENVIRONMENTAL IMPACT STATEMENTS LOOK AT?

International environmental impact statements look at those activities that have negative impacts on the environment. The impact depends on the vulnerability of the environment affected, the biodiversity of a place, and other unique factors. These impacts can be cumulative, individual, collective, or some combination. They can be hard to measure and may engender additional scientific controversy. Therefore, the pragmatic focus is on activities that are known to have large environmental impacts:

- Crude oil refineries (excluding undertakings manufacturing only lubricants from crude oil) and installations for the gasification and liquefaction of 500 metric tons or more of coal or bituminous shale per day
- Thermal power stations and other combustion installations with a heat output of 300 megawatts or more
- Nuclear power stations and other nuclear reactors, including the dismantling or decommissioning of such power stations or reactors (except research installations for the production and conversion of fissionable and fertile materials, whose maximum power does not exceed one kilowatt continuous thermal load)
- Installations for the reprocessing of irradiated nuclear fuel
- Installations designed:

  - For the production or enrichment of nuclear fuel;
  - For the processing of irradiated nuclear fuel or high-level radioactive waste;
  - For the final disposal of irradiated nuclear fuel;
  - Solely for the final disposal of radioactive waste; or
  - Solely for the storage (planned for more than 10 years) of irradiated nuclear fuels or radioactive waste in a different site than the production site.

- Major installations for the initial smelting of cast iron and steel and for the production of nonferrous metals
- Installations for the extraction of asbestos and for the processing and transformation of asbestos and products containing asbestos; for asbestos-cement products, with an annual production of more than 20,000 metric tons finished product; for friction material, with an annual production of more than 50 metric tons finished product; and for other asbestos utilization of more than 200 metric tons per year
- Integrated chemical installations
- Construction of motorways, express roads, and lines for long-distance railway traffic and of airports with a basic runway length of 2,100 meters or more
- Construction of a new road of four or more lanes, or realignment and/or widening of an existing road of two lanes or less so as to provide four or more lanes, where such new road, or realigned and/or widened section of road, would be 10 km or more in continuous length
- Large-diameter pipelines for the transport of oil, gas, or chemicals
- Trading ports and also inland waterways and ports for inland-waterway traffic that permit the passage of vessels of more than 1,350 metric tons
- Waste-disposal installations for the incineration, chemical treatment, or storage of toxic and dangerous wastes

- Waste-disposal installations for the incineration or chemical treatment of nonhazardous waste with a capacity exceeding 100 metric tons per day
- Large dams and reservoirs
- Groundwater abstraction activities or artificial groundwater recharge schemes where the annual volume of water to be abstracted or recharged amounts to 10 million cubic meters or more
- Pulp, paper, and board manufacturing of 200 air-dried metric tons or more per day
- Major quarries, mining, on-site extraction, and processing of metal ores or coal
- Offshore hydrocarbon production. Extraction of petroleum and natural gas for commercial purposes where the amount extracted exceeds 500 metric tons/day in the case of petroleum and 500,000 cubic meters/day in the case of gas
- Major storage facilities for petroleum, petrochemical, and chemical products
- Deforestation of large areas
- Works for the transfer of water resources between river basins where this transfer aims at preventing possible shortages of water and where the amount of water transferred exceeds 100 million cubic meters/year
- In all other cases, works for the transfer of water resources between river basins where the multiannual average flow of the basin of abstraction exceeds 2,000 million cubic meters/year and where the amount of water transferred exceeds 5 percent of this flow
- In both cases transfers of piped drinking water are excluded
- Waste-water treatment plants with a capacity exceeding 150,000 population equivalent
- Installations for the intensive rearing of poultry or pigs with more than:

  - 85,000 places for broilers;
  - 60,000 places for hens;
  - 3,000 places for production pigs (over 30 kg); or
  - 900 places for sows.

- Construction of overhead electrical power lines with a voltage of 220 kV or more and a length of more than 15 km
- Major installations for the harnessing of wind power for energy production (wind farms)

Environmental impacts of all these projects are large and often controversial. There can be battlegrounds when these projects begin, as they operate, and as they close down and need to be cleaned up.

POTENTIAL FOR FUTURE CONTROVERSY

Environmental impact statements per se may be controversial because of their methodology or legal and political context. International requirements for environmental impact assessment may be seen as a restraint on trade by industry. Many industries and multinational corporations strongly resist these requirements. Many governments and communities ask for information on environmental impacts. When mitigation measures are promised by project proponents, they require follow-through and compliance to effectively reduce environmental impacts. International environmental impact statements have a postproject analysis phase that others do not. This allows for mitigation of the negative environmental consequences of the mitigation measures themselves (not required in the United States), and for environmental accountability. Since communities, governments, environmentalists, and industries all are far apart in what they all would agree is a minimally accurate environmental assessment, it is likely this particular controversy will intensify. On an international level, until the U.S. environmental impact process catches up with international standards, controversy will ensue in trade, travel, and global environmental issues.

See also Climate Change; Cumulative Emissions, Impacts, and Risks; Environmental Impact Statements: Tribal; Global Warming; Sustainability

Web Resources

Canadian Environmental Assessment Agency. Available at www.ceaa-acee.gc.ca/. Accessed January 20, 2008.

European Commission on the Environment. Environmental Assessment. Available at ec. europa.eu/environment/eia/home.htm. Accessed January 20, 2008.

United Nations Economic Commission for Europe. Guidance on the Practical Application of the Espoo Convention. Available at www.unece.org/env/eia/guidance/intro.html. Accessed January 20, 2008.

Further Reading: DiMento, Joseph F. C. 2003. *The Global Environment and International Law.* Austin: University of Texas Press; Farrell, Alexander E., and Jill Jäger. 2006. *Assessments of Regional and Global Environmental Risks: Designing Processes for the Effective Use of Science in Decision Making.* Washington, DC: Resources for the Future; Glasson, John, Rick Therivel, and Andrew Chadwick. 1999. *Introduction to Environmental Impact Assessment.* UK: Routledge; Harris, Paul G. 2001. *The Environment, International Relations, and U.S. Foreign Policy.* Washington, DC: Georgetown University Press; Wathern, Peter. 1988. *Environmental Impact Assessment: Theory and Practice.* UK: Routledge.

ENVIRONMENTAL IMPACT STATEMENTS: TRIBAL

Tribes are given control over their water, land, and natural resources. Tribes can be treated as states and limit the rights of other water users. They can also control the terms of development in Indian country. Tribal environmental impact statements can be controversial because they can be seen as limiting the rights of non-Indians to reservation development opportunities and water.

BACKGROUND

Tribal status was conferred upon indigenous people who agreed to move to a reservation. Those that did not are generally called bands. These reservations were often placed in land no one wanted, far from home. Tribes are given a limited sovereignty as trustees of the United States. About 10 percent of the 500 or so tribes have casinos. The context of this controversy is one of horrific oppression of indigenous people. As tribes economically self-develop, nontribal businesses want access to the reservation. Some tribes have leased their land to governments for long-term leases. These areas have become major roads, and some of the leases are up soon.

Tribal land is often sacred in intangible and holistic ways. Water from the river can be used in sweat lodges, which could intensify the exposure to any pollutants in the water. Tribes that rely on fish are very concerned about water quality. Currently, some tribes have been approached by energy companies to develop wind power along the wind paths of their reservations. Whether tribes have a formal environmental impact statement or not, environmental awareness and concern are very high on the reservation and in the minds of the inhabitants. Today a significant number of tribal governments are regulating their resources and managing environmental programs.

TRIBES, CASINOS, AND THE ENVIRONMENT

Many states regulate the environmental impact of tribal casinos based on casino receipts. Agreements between states and tribes are called compacts. These compacts have provisions for contributions to the state; traffic and environmental mitigation to local governments; and environmental analysis and mitigation of those impacts. In general, tribes must prepare a tribal environmental impact report (TEIR) before starting a new project. A project is broadly defined as any activity occurring on Indian lands to serve the tribe's gaming activities that may cause a direct or indirect physical change in the off-reservation environment. In contrast, The National Environmental Policy Act (NEPA) requires significant impacts on the environment before an environmental assessment is considered. NEPA has large categorical exclusions (cities) and frequently litigated language, which is different than many TEIRs.

Failure by the tribe to prepare a TEIR could result in the state going to court and stopping the project. Tribes are required to meet and negotiate with local governments. This can sometimes present a battleground. They usually need to adopt an enforceable written agreement that addresses all off-reservation environmental impacts, public safety, gambling addiction, and other issues. If an agreement is not reached within a specified amount of time, binding arbitration becomes the next battleground. This generally requires the arbitrator to consider the most reasonable offer from the tribe and the local government. One of the battlegrounds of most big projects is the requirement to mitigate environmental impacts. These can be very expensive but are unknown until some environmental assessment is made. The arbitration process could also determine the adequacy of the TEIR prepared by the tribe, which can determine the cost of mitigation and project viability.

States already require environmental impact assessments when casinos want to expand. The federal government requires tribes to prove they are able to regulate water or air. Tribes have a high environmental awareness because of state and federal environmental requirements, with or without casinos.

TRIBES PREFER THEIR OWN ENVIRONMENTAL ASSESSMENTS

Many actions on reservations are subject to federal NEPA requirements, such as roads and airports. But NEPA is limited in several ways that do not account for issues of concern to tribes. NEPA is limited to procedural requirements of the U.S. government and does not impose substantive limits on environmental effects. The Endangered Species Act (ESA) requires habitat conservation plans in areas with endangered species. Many tribes feel that these plans do not adequately protect valued species.

POTENTIAL FOR FUTURE CONTROVERSY

As tribes develop their natural resources and pursue economic development, they develop environmental codes. A clash of cultures can occur around issues of environmental assessment, creating controversy. As more tribes become recognized as such (about 245 applications are currently pending at the U.S. Department of Interior) and as more tribes seek status as states to control water and air as a state would, more people will be affected. Businesses that want to develop tribal lands, other water rights holders off the reservation, the bands and tribes themselves, local and state governments that work with tribes in other economic and noneconomic ventures, and environmental organizations must now engage the environmental aspects of sovereignty for indigenous peoples.

WHEN CAN TRIBES CONTROL THEIR WATER RESOURCES?

Tribal Environmental Impact Statements

ELIGIBILITY CRITERIA FOR DETERMINING THAT INDIAN TRIBES CAN IMPLEMENT SAFE DRINKING WATER ACT (SDWA) AND CLEAN WATER ACT (CWA) PROGRAMS: The CWA and the SDWA both use the same general criteria that an Indian tribe must meet in order to be eligible to implement programs that a state can implement. These criteria follow:

- the tribe must be federally recognized by the Secretary of the Interior (recognition criterion);
- the tribe must carry out substantial governmental duties and powers over a federal Indian reservation (governmental body criterion);
- the functions to be exercised by the Indian tribe must be within the area of the tribal government's jurisdiction (e.g., under the CWA the tribe must have appropriate authority over the surface waters of its reservation) (jurisdictional criterion); and
- the Indian tribe must be reasonably expected to be capable, in the administrator's judgment, of carrying out the functions to be exercised in a manner consistent with the terms and purposes of the CWA and the SDWA and of all their applicable regulations (tribal capabilities criterion).

Under the simplified process for determining that a tribe is eligible to administer a CWA or SDWA program, the EPA intends to ensure compliance with the statutory requirements as an integral part of the process of reviewing program approval applications. Once a tribe has met the four criteria for any CWA or SDWA program, only information unique to another program would need to be provided when a tribe is requesting permission to implement that program.

RECOGNITION AND GOVERNMENTAL BODY CRITERIA: The determinations that a tribe has been recognized by the Secretary of the Interior and has a governing body are essentially the same under the SDWA, the CWA, and also the Clean Air Act (CAA). To establish that it is federally recognized, a tribe needs only to indicate that it appears on the list of federally recognized tribes that the Secretary of the Interior publishes periodically in the Federal Register. A tribe can establish that it meets the governmental duties and powers requirement with a narrative statement describing the form of the tribal government and the types of functions it performs and identifying the sources of the tribe's governmental authority. The information to meet these criteria needs to be provided only the first time that a tribe applies to assume a program under the provisions of any of these laws. That is to say, once a tribe has been approved to implement any program that a state can implement under any of those statutes, it will not need to demonstrate in subsequent applications that it is federally recognized and governs a reservation. It will, however, need to meet the other criteria discussed further on.

However, a determination that a tribe has inherent jurisdiction to regulate activities in one medium (e.g., water) might not conclusively establish its jurisdiction over activities in another medium (e.g., air). The EPA could approve the portion of a tribal application covering certain areas and withhold approval of the portion of an application where tribal authority has not been satisfactorily established. This is a potential battleground.

TRIBAL CAPABILITIES CRITERION: A tribe may have the capability to carry out only some programs under the CWA or the SDWA. In evaluating tribal capability, the EPA will consider the following:

- the tribe's previous management experience;
- existing environmental or public health programs administered by the tribe;
- the mechanisms in place for carrying out the executive, legislative, and judicial functions of the tribal government;
- the relationship between regulated entities and the administrative agency of the tribal government that will be the regulator; and
- the technical and administrative capabilities of the staff to administer and manger the program.

See also Cumulative Emissions, Impacts, and Risks; Water Pollution

Web Resources

Indigenous Environmental Network. Available at www.ienearth.org/. Accessed January 20, 2008.

Inter-Agency/Tribal Coordination Agreement. Available at www.vhb.com/pvd/eis/pdf/Inter-Agency-Tribal_Agreement_12212005.pdf. Accessed January 20, 2008.

Further Reading: Gedicks, Al. 1993. *The New Resource Wars: Native and Environmental Struggles against Multinational Corporations.* Boston: South End Press; Kreske, Diori L. 1996. *Environmental Impact Statements: A Practical Guide for Agencies, Citizens, and Consultants.* New York: John Wiley and Sons; Vine, Jr., Deloria Wilkins, and David Eugene. 2000. *Tribes, Treaties, and Constitutional Tribulations.* Austin: University of Texas Press.

ENVIRONMENTAL IMPACT STATEMENTS: UNITED STATES

Environmental impact statements (EISs) are powerful regulatory tools that force proponents of projects that have significant impacts on the environment to assess those impacts. While advisory only, they are used in many environmental controversies.

As knowledge about the environment has grown, so too has concern for the impacts of major projects and processes. Technology and project scale can greatly increase the impact of large-scale development on the environment. Environmental impact statements are advisory in practice but nonetheless required. It is a process fraught with controversies at most of the major stages. The environmental impact process under the National Environmental Policy Act (NEPA) is one that every major project or process with a significant impact on the environment must undergo. Generally, anyone contesting the process must go through the internal agency process first. Some states and tribes have their own environmental impact statement rules and laws.

Flash points for controversies under the EIS requirements are lack of notice, lack of inclusion, and inadequate stakeholder representation. The actual area of impact, called the study area, can shift during the process. Many of the processes of the EIS are time driven, and there is often inadequate time to assess ecosystem or cumulative impacts. The underlying environmental decision, as in the spotted owl controversy, can lend fuel to the EIS process. In the case of the spotted owl, the fact that logging the habitat of this endangered species was a significant environmental impact triggered the requirement for a full EIS. Environmental impact assessment also brings in controversies of risk assessment generally. However, an increase in environmental impact assessment at all levels is inevitable. Assessment is necessary to measure impacts of new projects and to establish baselines with which to measure changes in the environment. Citizen monitoring is also prodding more environmental assessment. Another common frustration with the federal EIS process is that it is advisory only. The decision maker is free to choose more environmentally harmful alternatives. All these controversies are likely to continue as the range of environmental assessments continue to expand into ecosystem and cumulative approaches.

NEPA was signed into law on January 1, 1970. The act establishes national environmental policy and goals for the protection, maintenance, and enhancement of the environment, and it provides a process for implementing these goals within the federal agencies. The act also establishes the Council on Environmental Quality (CEQ). This act is a foundational environmental law. The complete text of the law is available for review at NEPAnet.

NEPA REQUIREMENTS

Title I of NEPA contains a Declaration of National Environmental Policy that requires the federal government to use all practicable means to create and maintain conditions under which humans and nature can exist in productive harmony. Section 102 requires federal agencies to incorporate environmental considerations in their planning and decision making through a systematic interdisciplinary approach. Specifically, all federal agencies are to prepare detailed statements assessing the environmental impact of and alternatives to major federal actions significantly affecting the environment. These statements are commonly referred to as environmental impact statements (EISs). Section 102 also requires federal agencies to lend appropriate support to initiatives and programs designed to anticipate and prevent a decline in the quality of humans' world environment. In 1978, the CEQ promulgated regulations under NEPA that are binding on all federal agencies. The regulations address the procedural provisions of NEPA and the administration of the NEPA processes, including preparation of EISs.

The NEPA process is an evaluation of the environmental effects of a federal undertaking including its alternatives. Any type of federal involvement, such as funding or permitting, can trigger NEPA regulations. There are three levels of analysis depending on whether or not an undertaking could significantly affect the environment. These three levels include categorical exclusion determination; preparation of an environmental assessment/finding of no significant impact (EA/FONSI); and preparation of an EIS.

At the first level, an undertaking may be categorically excluded from a detailed environmental analysis if it meets certain criteria that a federal agency has previously determined as indicating no significant environmental impact. A number of agencies have developed lists of actions that are normally categorically excluded from environmental evaluation under their NEPA regulations. The U.S. Army Corp of Engineers, the U.S. Environmental Protection Agency, and the Department of the Interior have lists of categorical exclusions. This is an area of policy controversy. One aspect of these lists is that the cumulative effects of their exclusion are not considered. Another is that some of the categories that were once thought to be insignificant may not be now. Projects having insignificant environmental impacts are not required to perform an EIS.

At the second level of analysis, a federal agency prepares a written environmental assessment (EA) to determine whether or not a federal undertaking would significantly affect the environment. Generally, an EA includes brief discussions of the following: the need for the proposal; alternatives (when there is an unresolved conflict concerning alternative uses of available resources); the environmental impacts of the proposed action and alternatives; and a listing of agencies and persons consulted. It may or may not describe the actual study area. There is no actual requirement for notice to the community. Some communities are environmentally assessed without their knowledge. If the agency finds no significant impact on the environment, then it issues a finding of no significant impact (FONSI). The FONSI may address measures that an agency will take to reduce (mitigate) potentially significant impacts. This is the first notice many communities receive about any evaluation of the impacts. Many communities feel that there are significant environmental issues and, had they known about the EA process, could have directed the agency to them.

If the EA determines that the environmental consequences of a proposed federal undertaking may be significant, an EIS is prepared. Significant environmental impacts can be threats to an endangered species, historic sites, or culturally significant areas. An EIS is a more detailed evaluation of the proposed action and alternatives. The public, other federal agencies, and outside parties may provide input into the preparation of an EIS and then comment on the draft EIS when it is completed.

Interested parties are allowed to submit draft alternatives. The agency calls this *scoping*. Scoping is when the agency selects the interested parties who can submit an alternative proposal. This is a controversial stage of the process. There may be groups who wanted to participate but were not selected. Often there are communities who did not know about the internal agency scoping decisions. Once interested parties are selected, the alternative selection begins. A controversy can occur about which alternatives are examined. One alternative that is always examined is the *no action* alternative. The alternatives are compared and contrasted. There may be public hearings and some scientific studies. The agency then produces a draft EIS. This document can be a trove of information because it includes all the alternatives considered. The agency administrator then selects one alternative, and it is published in the final environmental impact statement of the EIS, along with its justification. The final EIS can only be 150 pages in length. The decision maker does not have to prioritize environmental protection over economic considerations. The EIS process seldom stops the decision or project, but it can slow it down and focus the public's attention on the environmental controversy.

If a federal agency anticipates that an undertaking may significantly impact the environment, or if a project is environmentally controversial, a federal agency may choose to prepare an EIS without having to first prepare an EA. After a final EIS is prepared and at the time of its decision, a federal agency will prepare a public record of its decision addressing how the findings of the EIS, including consideration of alternatives, were incorporated into the agency's decision-making process. An EIS should include discussions of the purpose of and need for the action; alternatives; the affected environment; the environmental consequences of the proposed action; lists of preparers, agencies, organizations, and persons to whom the statement is sent; an index; and an appendix (if any).

FEDERAL AGENCY ROLES

The role of a federal agency in the NEPA process depends on the agency's expertise and relationship to the proposed undertaking. The agency carrying out the federal action is responsible for complying with the requirements of NEPA. In some cases, more than one federal agency may be involved in an undertaking. In this situation, a lead agency is designated to supervise preparation of the environmental analysis. Federal agencies, together with state, tribal, or local agencies, may act as joint lead agencies. A federal, state, tribal, or local agency having special expertise with respect to an environmental issue or jurisdiction by law may be a cooperating agency in the NEPA process. A cooperating agency has the responsibility to assist the lead agency by participating in the NEPA process at the earliest possible time; by participating in the scoping process; in developing information and preparing environmental analyses including portions of the EIS concerning which the cooperating agency has special expertise; and in making available staff support at the lead agency's request to enhance the lead agency's interdisciplinary capabilities. While there are cooperating federal agencies, there is some controversy about intergovernmental relationships with states and municipalities. Where there is a large federal government land presence, as in the western United States, some communities are excluded from important EIS processes.

EPA'S ROLE

The Environmental Protection Agency, like other federal agencies, prepares and reviews NEPA documents. However, the EPA has a unique responsibility in the NEPA review process. Under section 309 of the Clean Air Act, EPA is required to review and publicly comment on the environmental impacts of major federal actions including actions that are the subject of EISs. If the EPA determines that the action is environmentally unsatisfactory, it is required by section 309 to refer the matter to the CEQ. Also the EPA carries out the operational duties associated with the administrative aspects of the EIS filing process. The Office of Federal Activities in the EPA has been designated the official recipient of all EISs prepared by federal agencies.

THE PUBLIC'S ROLE

The public has an important role in the NEPA process, particularly during scoping, in providing input on what issues should be addressed in an EIS, and in commenting on the findings in an agency's NEPA documents. The public can participate in the NEPA process by attending NEPA-related hearings or public meetings and by submitting comments directly to the lead agency. The lead agency must take into consideration all comments received from the public and other parties on NEPA documents during the comment period.

Public participation can be time-consuming and costly for many stakeholders but especially community members. Receiving actual notice of when they can get involved in a particular EIS is generally a point of contention. Some communities consider the EIS decision already made and their participation a formality. Some EISs use complicated scientific analyses to measure different impacts, and these can be difficult to explain to some citizens. If a particular project is controversial, the agency can find that a significant impact itself, thus triggering the EIS requirement. Demand for community involvement can be part of a particular controversy. There is no public participation in the list of actions categorically excluded from the EIS requirements.

POTENTIAL FOR FUTURE CONTROVERSY

Environmental impact assessment is now an integral part of many environmental decisions. The process forces an assessment and includes the public and interested parties. It can also include human health risk assessments and ecological risk assessments, which can create controversies of their own.

EIS processes are necessary for the development and refinement of environmental policy at all levels. For sustainability purposes these assessments allow us to understand the environment around us. More communities and environmentalists demand them with the expectation of involvement and the hope that they are environmentally meaningful. To the extent these groups become more dissatisfied with both process and product, more controversy will develop.

See also Citizen Monitoring of Environmental Decisions; Cumulative Emissions, Impacts, and Risks; Ecosystem Risk Assessment; Human Health Risk Assessment; Public Participation/Involvement in Environmental Decisions

Web Resources

U.S. Environmental Protection Agency. National Environmental Policy Act home. Available at www.epa.gov/compliance/nepa/index.html. Accessed January 20, 2008.

U.S. Environmental Protection Agency. Submitting Environmental Impact Statement. Available at www.epa.gov/compliance/nepa/submiteis/index.html. Accessed January 20, 2008.

Further Reading: Eccleston, Charles H. 2000. *Environmental Impact Statements*. New York: John Wiley and Sons; Kreske, Diori L. 1999. *Environmental Impact Statements: A Practical Guide for Agencies, Citizens, and Consultants*. New York: John Wiley and Sons; Morris, Peter, and Riki Therivel. 2001. *Methods of Environmental Impact Assessment*. London: Spon Press.

ENVIRONMENTAL JUSTICE

Environmental justice refers to the distribution of environmental benefits and burdens. It includes fair and equal access to all decision-making functions and activities. Race and income shape the historic and present distribution of many environmental benefits and burdens.

PROXIMITY OF COMMUNITIES OF COLOR TO POLLUTION

African Americans are almost four-fifths more likely than whites to live in neighborhoods where industrial pollution is suspected of posing the greatest health danger. In 19 states, African Americans are more than twice as likely as whites to live in neighborhoods with air pollution. Controversies about racism between whites and African Americans, between other nonwhite groups and African Americans, and within environmental organizations and the government are inflamed by the proximity of African American communities to dangerous industrial pollution.

The Associated Press (AP) analyzed the health risk posed by industrial air pollution using toxic chemical air releases reported by factories to calculate a health risk score for all communities in the United States. The scores are used to compare risks from long-term exposure to industrial air, water, and land pollution from one area to another. The scores are based on the amount of toxic pollution released by each factory, the path the pollution takes as it spreads through the air, the level of danger to humans posed by each different chemical released, and the number of males and females of different ages who live in the exposure paths. The AP study results confirm a long string of reports that show that race maps closely with the geography of pollution and unequal protection. These data do not include many other sources of pollution known to affect all urban residents. It also does not consider possible synergistic and cumulative effects.

BACKGROUND

Historically, African American and other people-of-color communities have borne a disproportionate burden of pollution from incinerators, smelters, sewage treatment plants, chemical industries, and a host of other polluting facilities. Environmental racism has rendered millions of blacks invisible to government regulations and enforcement.

The risk scores also do not include emissions and risks from other types of air pollution, like trucks and cars. The AP research indicates residents in neighborhoods with the highest pollution scores also tend to be poorer, less educated, and more often unemployed. However, numerous other studies show blacks and other people of color concentrated in nonattainment areas that failed to meet EPA ground level ozone standards. This is pollution mainly from cars, trucks, and buses. It is substantial and affects African Americans and Hispanics more than others.

In 1992, 57 percent of whites, 65 percent of African Americans, and 80 percent of Latinos lived in one of the 437 counties that failed to meet at least one of the EPA ambient air quality standards. A 2000 study by the American Lung Association found children of color to be disproportionately represented in areas with high ozone levels.

Hispanics and Asians

According to the AP report in 12 states Hispanics are more than twice as likely as non-Hispanics to live in the neighborhoods with the highest risk scores. There are seven states where Asians are more than twice as likely as whites to live in the most polluted areas. In terms of air quality, other studies have shown that Hispanic neighborhoods are disproportionately affected by air pollution, particularly in the southwestern United States.

Income

Many hold that environmental proximity is a function of income. This assumes a free and flowing housing market without institutional barriers like racism. Higher-income neighborhoods have more political, legal, and economic power to resist industrial sites. The average income in the highest-risk neighborhoods was $18,806 when the census last measured it (2000), more than $3,000 less than the nationwide average. One of every six people in the high-risk areas lived in poverty, compared with one of eight in lower-risk areas. Unemployment was almost 20 percent higher than the national average in the neighborhoods with the highest risk scores.

Proximity to Pollution Increases Long-Term Exposure to Risk

Short-term exposure to common air pollution worsens existing lung and heart disease and is linked to diseases like asthma, bronchitis, and cancer. Long-term exposure increases these risks. Many potentially synergistic chemical reactions in waste in cities are unknown, and so are their potential or actual bioaccumulative risks to humans. The question is who bears the risk of risks not regulated by the government? Until recently, the costs of public health have been separate from the costs of production for industrial capitalism. As health costs mount, the stakeholders who pay for them are protesting.

Current EPA Response

More than 80 research studies during the 1980s and 1990s found that African Americans and poor people were far more likely than whites to live near hazardous waste disposal sites, polluting power plants, or industrial parks. Other studies of the distribution of the benefits and burdens of EPA environmental decisions also found a clear demarcation along race lines. The disparities were blamed on many factors, including racism in housing and land markets, and a lack of economic and political power to influence land-use decisions in neighborhoods. The studies brought charges of racism. Legally, one must prove the intent to be racist, not just the fact that a given situation is racist. It is very difficult to prove the intent of a city or town when they pass a racially or economically exclusionary zoning ordinance. They are very difficult legal issues to litigate, but litigation still happens. President Clinton responded in 1993 by issuing an environmental justice executive order (EO 12898) requiring federal agencies to ensure that people of color and low-income people are not disproportionately exposed to more pollution. Recent reports suggest little has changed.

The EPA does not intervene in local land-use decisions. The federal government has preemptive power over state and local government to take property it needs. The state governments tend to know about local land-use decisions in relation to environmental agencies. The weak intergovernmental relations between these branches of government allow this controversy to continue to simmer. There are often battles between state environmental agencies and the EPA over the requirements of EO 12898. State environmental agencies are resistant to incorporating environmental justice issues but accommodate regulated industries with one-on-one consultation and permit facilitation.

FIRST U.S. SENATE HEARING ON ENVIRONMENTAL JUSTICE: THE BRINK OF A NEW POLICY?

On July 25, 2007, Congresswoman Hilda L. Solis (CA-32) testified before the U.S. Senate Environment and Public Works Committee's Subcommittee on Superfund and Environmental Health about the state of environmental justice in the United States. Many are anticipating a round of environmental justice hearings in the House of Representatives in the fall of 2007. Here are highlights of Congresswoman Solis's testimony:

Environmental justice is about ensuring that the most vulnerable in our society have clean air, clean water, safe homes, and good health. My legislation directed the California EPA to ensure environmental policies protect the health of minority and low-income communities. . . . minority and underserved communities have been forced to live in close proximity to industrial zones, power plants, and toxic waste sites. More than 5.5 million Latinos and 68 percent of all African Americans live within the range where health impacts from coal powered plants are the most severe. More than 70 percent of African Americans and Latinos live in counties that violate federal air pollution standards.

In the district I represent, the water basin is contaminated, 17 gravel pits leave neighborhoods covered with gravel dust, there are three superfund sites and nearby is one of the largest landfills in the nation. Forty-three enforcement actions were taken against 39 facilities in LA County between October 2005 and May 2007. Ninety-two percent of people living within a three-mile radius of these facilities are minority and 51 percent live below the poverty level.

Environmental justice communities have been under attack as a result of the policies of the Bush Administration. Consider the following:

- Since 2004, the Bush Administration has requested at least a 25 percent cut in the environmental justice budget.
- Despite "reaffirming its commitment to environmental justice" in a November 2005 memo, the administration proposed significant weakening changes to the Toxic Release Inventory Program in early 2006.
- Despite promises that the EPA's "environmental justice considerations are accurately described to the public when proposed and final regulations are published after January 2007," a rule on locomotive emissions released this April failed to mention environmental justice even one time.

The rhetoric from the administration on environmental justice is an empty promise which leaves the health of vulnerable communities across our nation hanging in the balance. We must do better for the health of our communities.

We must also address the decision of the Supreme Court in *Alexander v. Sandoval*. As a result of this case, persons would have to prove discriminatory intent rather than disparate impact. While this may be a difficult issue to resolve, I believe it is one we must address. Finally, we must ensure that environmental justice communities are protected in the drafting of any global warming legislation.

Minority and low-income communities across this country are vulnerable to health impacts resulting from environmental conditions which have been largely ignored by this administration. Absent a real commitment to environmental justice, the health and welfare of these communities will continue to suffer.

Racial Disparities

The ways to measure race are themselves very controversial. The U.S. census undercounts urban residents of color frequently, and mayors file lawsuits every 10 years. Significant disparities in health and the actual quality of aspects of the urban environment exist at every level, an indicator of institutionalized racism.

- African Americans represent 12.7 percent of the U.S. population; they account for 26 percent of all asthma deaths.
- African Americans were hospitalized for asthma at more than three times the rate of whites (32.9 per 10,000 vs. 10.3 per 10,000) in 2001.
- The asthma prevalence rate in African Americans was almost 38 percent higher than in whites in 2002.
- African American females have the highest prevalence rates (105 per 1,000) of any group.
- African Americans are more likely to develop and die of cancer than persons of any other racial and ethnic group. During 1992–1999, the average annual incidence rate per 100,000 for all cancer sites was 526.6 for African Americans, 480.4 for whites, 348.6 for Asian/Pacific Islanders, 329.6 in Hispanics, and 244.6 in American Indians/Alaska Natives.
- African Americans are more likely to die of cancer than any other racial or ethnic group in the United States. The average annual death rate from 1997 to 2001 for all cancers combined was 253 per 100,000 for blacks, 200 for whites, 137 for Hispanic Americans, 135 for American Indians/Alaska Natives, and 122 for Asians/Pacific Islanders.
- Cancer kills more African American children than white children. Cancer is surpassed only by accidents and homicides as the number-one killer of African American children.
- While cancer mortality rates for all races combined declined 2.4 percent each year between 1990 and 1995, the decline for African American children (0.5%) was significantly less than the decline for white children (3.0%).

- African American men have the highest rates of prostate, lung, colon, oral cavity, and stomach cancer.
- African American men are more than 140 percent more likely to die from cancer than white men.
- More white women are stricken with breast cancer than black women, yet black women are 28 percent more likely to die from the disease than white women.
- The overall cancer cure rate, as measured by survival for over five years following the diagnosis, is currently 50 percent for whites but only 35 percent for blacks.
- Cancers among African Americans are more frequently diagnosed after the cancer has metastasized and spread to regional or distant sites.
- Minorities with cancer often suffer more pain due to undermedication. Nearly 62 percent of patients at institutions serving predominantly African American patients were not prescribed adequate analgesics.
- Many low-income, minority communities are located in close proximity to chemical and industrial settings where toxic waste is generated. These include chemical waste disposal sites, fossil-fuel power plants, municipal incinerators, and solid waste landfills.
- African Americans and other socioeconomically disadvantaged populations are more likely to live in the most hazardous environments and to work in the most hazardous occupations.
- Inner-city black neighborhoods are overburdened with pollution from diesel buses. In a 2002 EPA report, researchers concluded that long-term (i.e., chronic) inhalation exposure to diesel engine exhaust (DE) is likely to pose a lung cancer hazard to humans, as well as damage the lung in other ways, depending on exposure.
- There is a strong relationship between environmental exposure and lung cancer among African Americans, which accounts for the largest number of cancer deaths among both men (30%) and women (21%).
- People living in the most polluted metropolitan areas have a 12 percent increased risk of dying from lung cancer compared to people living in the least polluted areas.
- Smoking does not explain why lung cancer is responsible for the most cancer deaths among African Americans. While many black men identify themselves as current smokers, they typically have smoked less and started smoking later in life than white men.
- Rates are higher in urban areas because of increased air pollution and increased particulate matter in the air.
- Minority workers are at a higher health risk from occupational exposure to environmental contaminants.
- African American men are twice as likely to have increased cancer incidence from occupational exposure as white men.

Many feel that belated government efforts to control polluting industries have generally been neutralized by well-organized and well-financed opposition. Industry is challenged in lengthy court battles, during which time industry still has the right to maintain production and exposure of people to suspect materials. Since the environmental regulations themselves and laws apply on a per industrial plant basis, and it is hard to prove any one plant at any one time did directly cause the harm alleged, the process and controversy continue. Communities have also become organized around this issue and have been developing environmental information and data.

PROXIMITY OF COMMUNITIES OF COLOR TO POLLUTION

Environmental Justice Locator

Scorecard.org provides maps at the national, state, county, and census-tract levels that illustrate estimated cancer risks from outdoor hazardous air pollution and the location of three types of pollution-generating facilities: manufacturing firms reporting to the Toxics Release Inventory, facilities emitting criteria air pollutants, and Superfund sites. You can see whether your home, workplace, or school is located in an area where estimated cancer risks are higher, comparable to, or lower than in other communities. You can also see how many polluting facilities are located in your area of interest. Charts associated with the maps provide demographic information about an area, including the percentage of people of color, percentage of families living in poverty, and percentage of homeownership. You can also use Scorecard's mapper to access environmental data at the most local level (i.e., for each individual census tract in the United States).

Distribution of Environmental Burdens

Scorecard uses easy-to-understand bar charts to illustrate which demographic group bears the burden of different pollution problems. Four problems are evaluated: releases of toxic chemicals, cancer risks from hazardous air pollutants, Superfund sites, and facilities emitting criteria air pollutants. Scorecard analyzes the distribution of these problems using seven demographic categories: race/ethnicity, income, poverty, childhood poverty, education, Homeownership, and job classification. For example, Scorecard calculates whether whites or people of color live in areas with greater toxic chemical releases, and then graphically portrays the extent of the disparity, indicating which group is worse off. Further information about any environmental problems in an area can be found in Scorecard reports listed in the links section.

Locator for Unequal Impacts

For any burden or combination of burdens that you select, or any group you select, this locator will show you every county where that group of people experiences a higher impact than the rest of the population in the same county.

Distribution of Risks by Race, Ethnicity, and Income

Is race or income the driving factor accounting for disparate environmental burdens in your state? Scorecard examines the distribution of estimated cancer risks associated with outdoor hazardous air pollution to illustrate patterns of inequity by race/ethnicity and income. Scorecard calculates a population-weighted estimate of the average lifetime cancer risks imposed on each racial/income group by hazardous air pollutants. The Y-axis shows the estimated cancer risk per million persons, and the X-axis displays nine annual household income categories ranging from less than $5,000 to more than $100,000. Each line in the graph represents one of five racial/ethnic groups: whites, African Americans, Native Americans, Asian/Pacific Islanders, and Latinos. Gaps between the lines indicate potential racial/ethnic disparities in cancer risk burdens. Slopes in the lines indicate potential differences in cancer risk across income categories.

Environmental Hazards

Scorecard provides several measures of environmental hazards that can be used to compare states or counties within a state, including average cancer risks from hazardous air pollutants, the number of facilities per square mile that emit criteria air pollutants, the number of Superfund sites per square mile, and the number of Toxic Release Inventory facilities per square mile. Environmental hazard indicators for counties and states can be compared to demographic profiles in order to assess which communities bear the largest burden from pollution sources.

POTENTIAL FOR FUTURE CONTROVERSY

Racism in U.S. society is not news but a fact. Slavery is racist and the United States had African slaves that built the foundations of the country. These facts reach far into many present-day environmental dynamics that are as repulsive as slavery and racism seem to present day populations. And in the environmental area, just like history, the most pernicious racism is reserved for African Americans. After the Civil War three waves of African American people migrated north to the cities, seeking freedom and economic opportunity just as all other immigrants and migrants before and since. When urban industrialization expanded, it polluted the city. Many other people of color and migrants were able to melt into U.S. society. In the areas of housing, employment, health, education, and transportation this has not been the case with African Americans. Instead of moving out of the city, many African Americans stay because of foreclosed opportunities. Industry also stays in these neighborhoods. This controversy is the broken lock to a Pandora's box of unavoidable and necessary controversy. All discussions of cumulative effects, sustainability, and U.S. urban environmentalism must know about the true environmental past of every place. There are many reasons for this, the least of which is to know where to clean up first. The next set of policy controversies involves the prevention of industrial growth in areas that may be irreparably damaged.

Underneath this controversy is another set of issues. The primary reason for most environmental policy is to protect the environment and the public. In most

U.S. cities it is now fairly easy to establish which communities bore the brunt of cumulative and synergistic risks. These communities are now shown to have a disproportionate, adverse reaction to environmental stressors, expressing itself in a number of physical ways such as childhood asthma. New environmental policies such as sustainability and the precautionary principle will require information about past environmental conditions, but the question of reparative public health intervention for proximate communities is left dangling. This is also known as the canary-in-the-coal-mine phenomenon.

Currently, the National Environmental Justice Advisory Committee is meeting with the EPA to recommend ways to limit or mitigate harms to local communities from increased emissions of particulate matter and nitrogen oxide due to increased trade and movement of goods and related transportation infrastructure growth. Some feel that this will focus attention on commercial marine and locomotive engines and their emissions, a current battleground between environmentalists who want much stricter standards and industry that resists regulation. Ports, railroad depots, airports, and truck depots all create pockets of emissions, and many suspect these disproportionately affect low-income people and people of color. Concern over the impacts of the movement of goods has increased due to recent and projected increases in foreign trade. The assumption is that this increase will require substantial transportation expansion from coasts and ports to inland destinations, likely affecting many environmental justice communities that are already disproportionately affected by past and present pollution. It may be a sign of progress in some areas that the canaries in the coal mine are actively resisting all activities that increase their pollution exposure. It promises to be a large environmental justice battleground in the near future, especially as scientists begin to explore the ecological restoration of coastal waters and rivers. Environmental information will be highly scrutinized, there will be scientific debate about risk and causality, and government regulators will eventually enforce much stricter emission standards at multimodal transportation hubs.

ROOTS OF ENVIRONMENTAL INJUSTICE: BLACK FARMERS AND LAND LOSS

African American farmers have suffered the loss of large amounts of land in the Southeastern United States. White farmers are offered loans to help them stay in business, comply with soil and watershed conservation measures, and avoid foreclosure. Black farmers are not offered these loans and do not get them when they apply. Institutionalized racism by the federal agencies in charge is one cause. There are other controversial reasons. After recent federal litigation and a failed class action settlement, black farmers are moving this controversy into Congress.

Background

Farmers have significant environmental impacts. Farming is a complex business with many cash flow problems for small farmers competing against agribusiness. Efforts to preserve the small family farm are widespread and included federal and state legislation that created agencies. According to the U.S. Department of Agriculture 14 percent of working farms were owned by African Americans in 1920, but today, only 1 percent of farms are. Overall the small, family-owned farm has declined. However, it has occurred three times faster for African American farmers. The USDA Commission on Small Farms admitted that "[t]he history of discrimination by the U.S. Department of Agriculture . . . is well documented," finding that "indifference and blatant discrimination experienced by minority farmers in their interactions with USDA programs and staff . . . has been a contributing factor to the dramatic decline of Black farmers over the last several decades."

Land Loss Facts

The loss of agricultural land is considered a destructive dynamic by sustainability proponents. The further away crops are from the point of consumption, the more energy is consumed, in transportation, preservation of food, and retail sale. As reported by the Land Loss Fund Web site (http://members.aol.com/til lery/llf.html):

In 1920, 1 in every 7 farmers was black.
In 1982, 1 in every 67 farmers was black.
In 1910, black farmers owned 15.6 million acres of farmland nationally.
In 1982, black farmers owned 3.1 million acres of farmland nationally.
In 1950, black farmers in North Carolina owned 1/2 million acres.
In 1982, black farmers in North Carolina owned only 40,000 acres.
In 1920, there were 926,000 farms operated by blacks in the United States.
In 1982, the total number of black farms had dropped to 33,000 and is
  steadily declining.

What caused the loss of black-owned farmland?

- Farmers Home Administration and other lending institutions' failure to provide farmers the proper and adequate assistance they were supposed to give.
- Failure on the part of black landowners to make wills that would secure hired land for future generations. When many heirs share one piece of land, but no one knows which portion belongs to them because a will was not made out, the land is easily lost to the tax office or the real estate speculator.
- Failure of heirs to pay taxes on family-owned land.
- Discriminatory practices by both public and private lending institutions toward African American landowners. In 1982, African Americans received only 1 percent of all farmownership loans, only 2.5 percent of all farm operating loans, and only 1 percent of all soil and water conservation loans. Despite some new regulations by the Farmers Home Administration (FmHA) meant to offset historically discriminatory lending practices, black farmers are still not getting adequate funding.

- Inadequate technical, marketing, and research assistance from the U.S. Department of Agriculture. Less than 5 percent of the research funds spent by land grant colleges and extension programs are directed toward the problems of limited-resource farmers.
- Lack of access to land. Existing African American farmers cannot get adequate or timely funding to expand their operations, and young blacks are unable to secure funding to purchase new farms.
- Lack of knowledge of legal rights as landowners. Most rural African American farmers do not have access to essential legal assistance and thus fall prey to land speculators and unscrupulous lawyers.

Rough Litigation

Legal resources were at a minimum for African American farmers. Lawyers willing to sue the federal government in a protracted and controversial case are very expensive. As a result, lawyers from large law firms in the Washington, D.C., area took on a large number of cases on a pro bono basis to assist the farmers' originally. All parties, including the pro bono attorneys, expected a streamlined mediation process, not a lengthy trial. The trial was as rough as they come. It was characterized by excessive motions practice, USDA interlocutory appeals that repeatedly interrupted the cases, numerous evidentiary objections, delays, and aggressive litigation tactics. Such tactics included agreeing to postponements, then seeking dismissal for failure to prosecute, or seeking recusal of the arbitrator when faced with sanctions or adverse rulings. The pro bono work was more time-consuming and complicated than expected.

Denials and Delays Prevail

Of the nearly 100,000 farmers who came forward with racial discrimination complaints, 90 percent were denied any recovery from the settlement. As a result, instead of the $2.3 billion, USDA provided $650 million in direct payments to farmers. The farmers who had the best chance at achieving justice were the 230 who opted out of the settlement.

Of the 73,747 African American farmers who sought late entry into the settlement process, 65,947 filed their applications on time. The overwhelming majority of the farmers who did apply on time, some 63,816 farmers, were ultimately denied entry into the settlement. Their claims were never heard on their merits, and they will never again have a chance to seek relief for their discrimination complaints.

Overall, USDA spent $12 million by 2002 for assistance from the Department of Justice in disputing individual farmers' claims. These numbers are extraordinarily high for a settlement that was intended to provide a virtually automatic payment to farmers through an abbreviated procedure. Instead, it appears that African American farmers were treated as adversaries rather than as partners in a cooperative settlement.

Congressional Intervention into This Controversy

Right now, those trying to compensate for and prevent land loss for black farmers are moving the controversy into congressional forums. They are asking the following:

1. Congress should order USDA to provide full compensation to the nearly 9,000 farmers who were denied relief after being accepted into the settlement class;
2. Congress should order USDA to reevaluate the merits of the nearly 64,000 farmers' claims that were shut out due to lack of notice of the settlement.
3. Congress should direct the USDA to institute accountability measures to monitor and enforce civil rights standards throughout the agency, requiring that in the future the USDA shall exert best efforts to ensure compliance with all applicable statutes and regulations prohibiting discrimination; and
4. Congress should ensure the full implementation of outreach and financial assistance programs that support minority farmers.

The history of discrimination that led to the lawsuit tells the tale of deeply entrenched institutionalized racism. The discrimination that led to the suit still persists in many forms, including even the administration of a civil rights settlement. Instead of a fair facilitation of the settlement, the victimization continues with delay tactics and aggressive litigation strategies. A settlement is a cooperative process, not a small-scale litigation battle. Ultimately, the farmers have not fared substantially better than they would have at a civil trial. Eighty-six percent of the farmers with discrimination complaints have been unsuccessful and walked away from the settlement with no money and no ability to redress their grievances in a court of law. It is a continuation of the disenfranchisement of the African American farmer at the hands of the USDA.

POTENTIAL FOR FUTURE CONTROVERSY

Small farmers, the group of farmers to which most African American farmers belong, are the backbone of our sustainable agricultural future. By contributing a heightened awareness of the needs of the land, utilizing sustainable practices such as multicropping, and supporting the growth and wealth of their local communities, small farmers provide an invaluable resource to the agricultural system. Government-subsidized loans and grants are designed to support the small farmer and provide vital resources to this important segment of the farming industry. In order for this system to operate effectively, it must operate equitably. To discriminate against small farmers, and to further marginalize particular small farmers with racially discriminatory practices in the administration of financial assistance, contravenes the spirit and purpose of these USDA programs. To the extent that federal agencies alter congressional intentions they act illegally, outside the scope of powers delegated to them. Actions of this nature can prevent the advancement of environmental policy.

See also Children and Cancer; Cumulative Emissions, Impacts, and Risks; Different Standards of Enforcement of Environmental Law; Ecosystem Risk Assessment; Farmworkers and Environmental Justice; Indigenous People and the Environment; Sustainability; Toxics Release Inventory

Web Resources

Environmental Working Group. Available at http://www.ewg.org/featured/172. Accessed January 20, 2008.

Is It Race or Income? Available at http://www-personal.umich.edu/~bbryant/enujustice.htm. Accessed January 20, 2008.

Land Loss Fund. Available at http://members.aol.com/tillery/llf.html. Accessed January 20, 2008.

National Black Farmers Association. Available at http://www.blackfarmers.org/. Accessed January 20, 2008.

National Cancer Institute. Cancer Mortality Maps and Graphs. Available at www3.cancer.gov/atlasplus/charts.html. Accessed January 20, 2008.

Further Reading: Bittker, Boris I. 2003. *Case for Black Reparations.* Boston: Beacon Press; Bullard, Robert. 2000. *Dumping in Dixie: Race, Class, and Environmental Quality.* New York: Westview Press; Bullard, Robert. 1993. *Confronting Environmental Racism: Voices from the Grassroots.* MA: South End Press; Ficara, John Francis, and Juan Williams. 2006. *Black Farmers in America.* KY: University Press of Kentucky; Frazier, John W. 2003. *Race and Place: Equity Issues in Urban America.* New York: Westview Press; Gerrard, Michael B., ed. 1999. *The Law of Environmental Justice: Theories and Procedures to Address Disproportionate Risks.* Chicago: American Bar Association; Rechtschaffen, Clifford, and Eileen Guana. 2002. *Environmental Justice: Law, Policy, and Regulation.* Durham, NC: Carolina Academic Press.

ENVIRONMENTAL MEDIATION AND ALTERNATIVE DISPUTE RESOLUTION

The primary controversy with environmental mediation and dispute resolution is whether it is effective. Environmentalists, industry, and increasingly communities can go to court for an attempt at a formal resolution.

BACKGROUND

Environmental mediation and alternative dispute resolution describes two problem-solving processes borrowed from other areas of controversy. Most of the areas borrowed from are legal areas like family law, employment law, and labor law. The Clean Water Act and Clean Air Act both give environmentalists special access to federal courts to help with the enforcement of the law. Other environmental laws also allow for citizen suits. Resolutions of legal controversies about environmental laws have therefore tended to be resolved in federal and some state courts. If the lawsuit somehow fails, then there is no motivation for the winning party to mediate and engage in informal dispute resolution practices. If parties to an environmental controversy do engage in these processes, and then go to court, the incentive for mediation and dispute resolution also dissipates. Environmental dispute resolution includes most nonlegal approaches that allow parties to meet face to face. The goal is to resolve the environmental issue of acts at hand between the parties to each person's satisfaction. It is not facilitation, which is more focused on multistakeholder deliberation on environmental policy issues.

Environmental mediation and alternative dispute resolution are not used frequently in the United States, but they are being used more by state environmental agencies. However, expensive and inaccessible judicial processes for environmental legal disputes may make alternative dispute resolution seem attractive for some types of environmental disputes. These types of services are generally not available in most communities. Communities and environmentalists prefer the finality and confidentiality of judicial processes. As citizen monitoring of environmental decisions increases, and new environmental policy developments like good neighbor agreements and community-based environmental planning evolve into implementable policy, communities are more engaged in environmental decisions. These decisions involve major environmental controversies that are settled in a court of law.

Another general aspect of alternative dispute resolution processes generally is that they lack the formality of judicial proceedings. Rules of evidence and general procedural due process may not apply. Records of the proceedings are not kept as a judicial transcript. It may be difficult to appeal an alternative dispute process, but that may not stop losing parties from seeking judicial intervention anyway. The amount of deference courts give to environmental dispute resolutions is unclear, itself a looming judicial controversy. Some groups do not trust the informality and lack of finality in alternative dispute resolution. Many U.S. minority groups do not trust this process for these reasons. Environmental justice has mobilized many communities around race and class issues in the distribution of environmental benefits and burdens. Environmental justice communities are uneasy about outside approaches to their environmental issues that may cause them to lose rights or political power.

FEDERAL AGENCY PERSPECTIVE

Environmental decision makers in government continue to face intrinsic controversies of balancing competing public interests and federal agency responsibilities when striving to accomplish national environmental protection and management goals. This is a fundamental governance challenge. This challenge can manifest itself through the following.

- Protracted and costly environmental litigation;
- Unnecessarily lengthy project- and resource-planning processes;
- Costly delays in implementing needed environmental protection measures;
- Forgone public and private investments when decisions are not timely or are appealed;

- Lower-quality outcomes and lost opportunities when environmental plans and decisions are not informed by all available information and perspectives; and
- Deep-seated antagonism and hostility repeatedly reinforced between stakeholders by unattended conflicts.

DO ENVIRONMENTAL MEDIATION AND ALTERNATIVE DISPUTE RESOLUTION WORK IN ENVIRONMENTAL DECISION MAKING?

The battleground of environmental controversy is very contentious. Many environmental controversies reflect deep-rooted issues locally and nationally. Industries contend that they need to comply with the law, period. There is very little analysis or empirical research on environmental mediation efforts in the environmental management literature. Therefore, there are many unanswered questions. What is the cost versus benefit of environmental mediation when ecological impacts and public health costs are tabulated? Is it inclusive of urban areas or focused on wealthy suburban communities and insular negotiated environmental-rule making processes? How are public notice and participation handled, if at all? How does this approach handle emerging concepts like sustainability, collaborative environmental decision making, environmental justice, and cumulative impacts? Collaborative environmental decision making and community-based environmental decision making do include these modern concepts and may displace the U.S. initiative of environmental mediation and dispute resolution.

THE ROLE OF MEDIATORS AND FACILITATORS

Many advocates for environmental alternative dispute resolution feel that a mediator or a facilitator can be significantly helpful in resolving complex, multiparty issues. They can help analyze of case studies of hazardous waste siting, site cleanup, and regulatory negotiation, assuming they have the substantive expertise. The appropriate roles for the environmental mediator or the environmental facilitator are dynamic and often situation specific.

A mediator or facilitator can be useful in identifying interested parties and determining whether they are willing to negotiate in good faith. Multistakeholder environmental decisions and controversies are complex, and if a stakeholder is there simply to slow down the process, then it is impossible to negotiate with them with. Mediators should lay out the costs and benefits of mediation for the parties. This is especially important in environmental controversies because of the availability of lawsuits for some environmental issues. A major role of the environmental mediator is to deal with the inequality of information among parties, the differences in environmental background of the stakeholders, and developing issues and areas of agreement and disagreement. Environmental justice is particularly sensitive to neighborhood inclusion and community capacity building. Industry is very sensitive to any one sector or site being held liable for accumulated pollution, whereas communities and environmentalists increasingly demand that these be accounted for in modern dispute resolution.

IS ENVIRONMENTAL ALTERNATIVE DISPUTE RESOLUTION CONFIDENTIAL?

Industry is a very large stakeholder in environmental dispute policy development. When industry lawyers are discussing case settlement, one very important provision is confidentiality. Settlements are not court decisions or judgments, but rather consent decrees that admit no wrongdoing and often have nondisclosure components. Attention must be given to the impact of mediation on the confidentiality of settlement discussions. Historically, rules relating to the confidentiality of settlement discussions have attempted to balance the competing policy interests of promoting settlements and ensuring public access to the judicial system. However, this tension is now being heightened by two facts. First, most court-based mediations receive more confidentiality protection than traditional settlement discussions. Second, the settlement of environmental cases often impacts individuals or groups who are not involved in the underlying litigation. When litigation involving matters such as oil spills or Superfund site clean-ups is settled through mediation, there is a concern that these impacted nonparties may not have sufficient access to information needed for accountability and decision making.

The increased use of court-based mediation has enhanced the amount of confidentiality protection afforded settlement discussions in many environmental cases. While this fact may cause some initial concern, it must be viewed in light of the mediation process itself. Because mediation promotes the exchange of information and allows parties to view the dispute in broader terms than in traditional litigation, it offers some unique advantages to the parties impacted by, but outside of, the mediation. These advantages include:

1. allowing parties outside of the litigation to directly participate in settlement discussions;
2. increasing the likelihood of disclosure of information and scientific data produced in the mediation; and
3. increasing the likelihood that settlements represent the interests of parties outside of the litigation.

These advantages are partly due to the participation of a neutral third party in the settlement process who not only ensures procedural fairness but who may also have an affirmative obligation to disclose certain information to parties outside of the mediation.

POTENTIAL FOR FUTURE CONTROVERSY

Environmental mediation and dispute resolution is a process everyone would like to work, not necessarily because of the process but because there are many ongoing unresolved environmental controversies. Environmental justice, community-based environmental planning, and collaborative environmental decision making all deal with unresolved environmental controversies in more holistic and community-inclusive ways. And courts still deal with issues of environmental law with greater formality, finality, and certainty. Environmental mediation and alternative dispute resolution will be refined through government use but remain controversial at the community and industry levels.

See also Collaboration in Environmental Decision Making; Community-Based Environmental Planning; Litigation of Environmental Disputes; Public Participation/Involvement in Environmental Decisions

Web Resources

Ecodirections Environmental Mediation Booklet. Available at www.ecodirections.com/pdf/Mediation_Booklet_Sept2003.pdf. Accessed January 20, 2008.
Texas Commission on Environmental Quality. Alternative Dispute Resolution for Disputed Environmental Permit Applications. Available at www.tceq.state.tx.us/comm_exec/dispute_res/. Accessed January 20, 2008.

Further Reading: Blackburn, J. Walton, and Willa Marie Bruce. 1995. *Mediating Environmental Conflicts: Theory and Practice.* Westport, CT: Quorum/Greenwood; Levy, Paul F., Lawrence E. Susskind, and Jennifer Thomas-Larmer. 2000. *Negotiating Environmental Agreements: How to Avoid Escalating Confrontation, Needless Costs.* Washington, DC: Island Press; Macnaughton, Ann L. 2002. *Environmental Dispute Resolution: An Anthology of Practical Solutions.* Chicago: American Bar Association; O'Leary, Rosemary, ed. 2003. *The Promise and Performance of Environmental Conflict Resolution.* Washington, DC: Resources for the Future.

ENVIRONMENTAL REGULATION AND HOUSING AFFORDABILITY

Environmental regulation of housing generally requires changes in construction methods and materials. Configuration of the housing unit, location of the property, minimum required lot size for housing construction, sensitive environmental areas, sprawl, and endangered species also sometimes arise as environmental controversies around housing.

Purchasing a house is the single biggest purchase for most people. It is often the time when young families seek to establish roots in a community. Housing can have a big environmental impact. Communities have sought ways to mitigate environmental impacts of housing; from a builder's perspective most are controversial. Many environmentalists also consider them controversial in that they do not go far enough to reduce environmental impacts. Many people are involved with the sale and purchase of a house and real property. Lawyers, real estate agents, banks, mortgage corporations, title insurance corporations, local land-use authorities, and others all have vested and mainly financial interests in this transaction.

Environmental regulations implemented at the environmental level are how most good regulations take effect. Decisions are made about the best way to proceed environmentally and passed on from the federal government to the state government for possible implementation. Eventually, some degree of environmental sensitivity and literacy reaches the local elected levels of government, who pass regulations about land use and building codes. These are the very codes of law that many sustainability advocates say need major revision. Often these regulations have the most effect on those creating the most impact on the environment, such as builders. New regulations, longer waits, and more permits and enforcement all increase the cost of doing business. Builders say they just pass that cost along. Banks, insurance companies, mortgage lenders, developers, and real estate brokers all say the same thing. Others point out that these stakeholders have large profit margins and can more readily absorb the cost of doing business in an environmentally sensitive or sustainable way. Profit margins are often an important piece of information in this battleground but are not usually shared outside of industry circles.

Current demographic trends in the United States are creating pressures on communities to provide adequate services for a growing population. Roads, sewers, drinking water, schools, hospitals, and police and fire stations are all services that greatly impact the environment. Many of these basic public works projects are now classified as environmental. Homeownership rates are the highest since measured in the 1940s, an indication of strong market demand. Homeownership includes almost 68 percent of U.S. households. Housing prices have continued to increase, although there is wide variation from region to region of the United States. There is also a wide variation in homeownership rate by age, race, gender, and income. Builders and developers respond to this demand by constructing housing units in suburban fringe and rural areas. Environmental regulations are barriers to development and impediments to affordability in the view of some stakeholders. Required cleanup responsibilities are also costs of land acquisition in urban areas. Many banks and mortgage lenders now require some certificate of due diligence in environmental risk evaluation. This is a battleground discussed in the entries on brownfields and Superfund ecological risk assessment. In practice right now it affects primarily industrial and commercial properties. As a battleground here, it is very likely to extend to all property. Most environmentalists and proponents of sustainability want to begin to clean up the environment from any pollution they can. The context of housing affordability is part of the battleground of whether environmental regulations decrease affordable housing.

HOMEOWNERSHIP RATES, PRICES, AND AFFORDABILITY

The median cost of a new home in the United States went up 32 percent from 1995 to 2005. Many factors go into the home price increases. Some houses are bigger, occupy bigger lots, or have modern technological features. Profit increases in building supplies and distribution, labor cost increases, and financing shifts in the primary and secondary mortgage markets are also factors. Government land-use regulations, designed by local elected officials to protect the public's health, safety, and welfare, may require permits and conditions that can increase time and cost involved in building a residential unit. When these land-use requirements, such as minimum lot size, become environmental requirements is a vague line. However, builders, developers, and their financiers attribute much of this increase to environmental regulations. Most housing researchers conclude that higher home prices are most likely the result of building bigger homes with more expensive modern amenities.

RADON CONTROVERSIES IN HOUSING

Radon is generally a naturally occurring mineral isotope. Radon gases trapped in homes and occupants' exposure to them are a controversy and public health concern. After years of debate and research most scientific researchers agree that radon causes cancer. It is a gas that can accumulate in enclosed areas in the ground. Many of the early studies were on miners who could be exposed to it. In 1984, a nuclear power plant employee in Pennsylvania set off the dosimeter (radiation detector) at the plant. His home was found to have radon concentrations well beyond those permitted at work. A database on indoor radon concentrations showed that the problem was widespread in residences.

Many public health professionals feel the need to improve the understanding of the risks posed to the general population by indoor radon is paramount.

Radon hazards in the home can cause physical illness, compromise growth and development, and lower school performance in children. Hazardous radon levels occur at all economic levels. However, some health hazards such as lead poisoning, asthma, and fatal injuries occur at disproportionately high rates in poor-quality homes of children in low-income families. Research is necessary to understand how hazards affect children's health and to develop interventions that can ameliorate or eliminate them.

Some argue that there is no direct connection between increased environmental regulation and house price increases. The homeownership rate in the United States is at a high of 66.8 percent, despite all environmental regulations. This is an important finding—markets adjust to environmental regulations. Much of this controversy takes place in the battleground of the marketplace in that the question often becomes who in the supply side will absorb the cost of environmental construction processes and housing products.

POTENTIAL FOR FUTURE CONTROVERSY

Citizen monitoring of environmental decisions, the Toxics Release Inventory, concern about cumulative impacts, and a societal push for sustainability all make knowledge about local environmental impacts widespread. One of the first industries noticed by most communities is new residential construction. Builders are at the point of interface with the environment and are required to partially implement some environmental policy. The goal of these environmental regulations is to protect the public welfare and the environment. It is probably true that environmental regulations will increase the relative cost of some housing. Builders and developers must comply with federal laws that include the Clean Water Act, Federal Water Pollution Control Act, Marine Mammal Protection Act, and the Endangered Species Act. They have long assumed that these environmental regu-

---

MOLD AND HOUSING

Mold contamination is an emerging controversial issue in housing. Mold in both old and new houses has been difficult to eradicate and is sometimes dangerous and unhealthy. When tenants have mold, it is generally the landlord's or property owner's responsibility to eradicate it. Some public housing tenants have complained about mold and then been retaliated against by their housing authority. Some new house buyers have had to destroy their houses due to mold infestation. Mold is an environmental condition that creates controversy when it remains untreated.

Molds are simple organisms that are found everywhere, indoors and outdoors. The potential health effects of indoor mold are a growing concern. Mold can cause or worsen certain illnesses.

Some molds are worse than others. *Stachybotrys* is a greenish-black mold and is especially dangerous to vulnerable populations. Inside your home you can control mold growth by:

- Keeping humidity levels between 40 percent and 60 percent
- Promptly fixing leaky roofs, windows, and pipes
- Thoroughly cleaning and drying after flooding
- Ventilating shower, laundry, and cooking areas

Mold growth can be removed from hard surfaces with commercial products, soap and water, or a bleach solution of no more than 1 cup of bleach in 1 gallon of water. Mold growth, which often looks like spots, can be many different colors and can smell musty. No matter what type of mold is present, it should be removed. The best practice is to remove the mold and work to prevent future growth.

Source: EPA Guide to Mold. Available at www.epa.gov/iaq/molds/moldguide.html. Accessed January 21, 2008.

---

lations which seek to improve air and water quality and protect biodiversity, have suppressed development and driven the costs of new homes beyond what buyers can afford. However, it could also be the case that these regulations increase the energy efficiency of the housing, thereby decreasing its cost over time. The role of technology is rapidly expanding in this battleground. It may be entirely possible that low-impact environmentally sensitive housing will be less expensive. Housing and building markets are captive to their suppliers. The U.S. housing material supply is characterized by a lack of diversity in its inventory, historically relying on wood and wood products. Other countries pioneer more economical and environmentally sensitive building materials, even using recycled materials. The U.S. inventory is rapidly catching up to world trends in the building supply inventory. However, as long as environmental protection remains a highly held public value and as long as population and housing demand increases, this will remain an environmental controversy.

See also Citizen Monitoring of Environmental Decisions; Community-Based Environmental Planning; Cumulative Emissions, Impacts, and Risks; Sprawl; Sustainability

Web Resources

Calculating the Cost of Environmental Regulation. Available at http://ideas.repec.org/p/rff/dpaper/dp-03-06. Accessed January 21, 2008.

Environmental Valuation and Cost Benefit News: Real Estate. Available at envirovaluation.org/index.php?cat=76. Accessed January 21, 2008.

Further Reading: Freeman, A. Myrick, III. 2003. *The Measurements of Environmental and Resource Values: Theory and Methods.* Washington, DC: Resources for the Future; Greenberg, Michael R. 1999. *Restoring America's Neighborhoods: How Local People Make a Difference.* Piscataway, NJ: Rutgers University Press; Greenberg, Michael R., and Dona Schneider. 1996. *Environmentally Devastated Neighborhoods: Perceptions, Policies, and Realities.* Piscataway, NJ: Rutgers University Press; Kelly, Eric Damian. 2004. *Managing Community Growth.* 2nd ed. Westport, CT: Praeger/Greenwood; Thompson, J. William, and Kim Sorvig. 2000. *Sustainable Landscape Construction: A Guide to Green Building Outdoors.* Washington, DC: Island Press; Watkins, Craig. 2002. *Greenfields, Brownfields and Housing Development.* London: Blackwell Publishing.

ENVIRONMENTAL VULNERABILITY OF URBAN AREAS

As underscored by the impact of Hurricane Katrina on New Orleans, urban areas are vulnerable to natural disasters. Cities are where most of the pollution is located and where most people live. Natural disasters like floods, hurricanes, earthquakes, and tidal surges can exacerbate already-polluted conditions. Climate change may deluge coastal cities as ocean levels rise. Controversies ensue about liability for cleanup, poor environmental planning, evacuation, and lack of monitoring.

GLOBAL URBANIZATION

Urbanization now dominates the human habitat. The environmental impacts of urbanization are powerful. Some nations, under the sponsorship of the United Nations, are pursuing policies of urban sustainability. Water, air, and land quality can erode to a point affecting the entire environment, including the public health.

According to population statistics, as of 2005 most of the people in the United States and in the world lived in cities. Cities are places where economic opportunity presents itself to many stakeholders. They are crossroads of transportation systems and engines of commerce, and have large environmental impacts. Many are located along the coasts or inland along water courses. Many cities began as ports, or developed at the point where a river ceased to be navigable. Global warming and climate change could affect the environmental vulnerability of these cities by increasing their exposure to changes in weather and water levels.

URBANIZATION DEFINED

Urban areas are characterized by large concentrations of people. Urban areas tend to be a unit of local governance and service delivery. Fire, police, sanitation, education, and public works are municipal services organized around the city. Urbanization involves an increase in the number of people in urban areas, as people move from rural, suburban, or squatting communities. Generally urbanization is associated with industrialization and economic development. Cities are seen as offering a greater range of municipal services, economic opportunity, and cultural diversity.

Urbanization and its potential environmental impacts are increasing in both developed and developing countries. Urban areas make up about 3 percent of the available landmass. Some statistics show that more than half of the world's 6.6 billion people live in urban areas. The proportion of the world's population living in urban areas is expected to reach 65 percent in 2030. It is likely that population counts underestimate actual numbers of the new urban humanity. In most cities in the world the city officially stops where the municipal water provision stops. Huge squatting populations congregate around the outskirts of megalopolitan area where water quality and quantity are a daily environmental battleground. Urbanization in Africa and Asia is expected to increase from 40 percent to 54 percent urban by 2025. Much of this urbanization will occur in areas that could suffer severe climatic changes under emerging global warming scenarios. Drought and famine could increase the environmental vulnerability of many people in many cities.

The full environmental implications of rapid urbanization for water supply and sanitation, especially the disposal of wastes that the cities produce, are unknown and increasingly unavoidable. Each area presents environmental battlegrounds of its own, but the underlying social dynamic is rapid urbanization. Some environmentalists now challenge the value of growth for the sake of growth. Most models of economic development are premised on growth of markets. Limitless growth could drain the Earth of valuable natural resources. This is one big concern of antisprawl proponents. The battleground for this controversy then becomes the ecological carrying capacity of a particular region.

POTENTIAL FOR FUTURE CONTROVERSY

The environmental vulnerability of cities, especially coastal ones, will likely continue to increase. A looming controversy is which cities to save under what conditions. Should government flood insurance be offered to people who rebuild in environmentally vulnerable areas? If a city is rebuilt, who should be allowed back in? There is a gnawing social concern over who is liable for vulnerable people in vulnerable urban environments who get caught in a natural disaster. Life-and-death choices face emergency responders when transportation is limited and communication intermittent under these conditions. These personal, policy, and professional choices are difficult and unresolved but must be faced. While some of the subsequent controversies will be about liability for the harm, much will be about how to adapt to specific changes in specific places.

MAJORA CARTER: SUSTAINABILITY IN THE SOUTH BRONX

For many outside observers, the image of the South Bronx of New York City is that of an urban holocaust. The media-inflamed and politically manipulated image of the South Bronx has it crime ridden, gang controlled, and without the rule of law. Its environment is a waste-land, its land, air, and water all toxic beyond human tolerance.

To Majora Carter this image presents a challenge. In 2001, she formed Sustainable South Bronx. This nonprofit organization is designed to achieve environmental justice by sustainable projects informed by the needs of the South Bronx. This is one of the many environmental accomplishments of Majora Carter. An early environmental battle developed when she learned that New York City was going to process 40 percent of its solid waste through and in her neighborhood. This means trucking in waste to waste transfer stations where it can sit for months. It means incinerating of some of it, adding to an already-bad air pollution problem. Exercising considerable community-organizing skills she helped to stop it. She has a vision for the land, air, and water in the South Bronx with a sustainable community. Parks, bike paths, greenroof development, community health and environmental education, and community food markets are all part of that vision. With Carter's focus, energy, and enthusiasm some of these are now viable projects.

A self-described lifelong resident of Hunts Point, Majora Carter graduated from the Bronx School of Science in 1984, earned her bachelor's degree from Wesleyan University in 1988, and earned her masters of fine arts (MFA) from New York University in 1997. Hunts Point has been at the leading edge of environmental and environmental justice issues for decades. A multicultural community with large historic and ongoing industrial emissions and municipal waste flows, it has also been the site of community research into cumulative risks and especially community empowerment. Hunter College, and now Columbia University, work with various groups in the South Bronx, in Hunts Point, and with Sustainable South Bronx.

Carter's leadership role in her community expressed itself in sustainable initiatives like the community composting project, air pollution studies called "greening for breathing," the South Bronx Green and Cool Roofs Demonstration Project, and others. She has earned numerous awards, including the John D. and Catherine T. MacArthur Foundation "Genius" award.

As the U.S. environmental movement matures it will include more cities and more women of color in leadership positions. Issues and advocacy postures could change. Sustainability and grassroots environmentally based community organization and mobilization may reshape some of the battlegrounds of environmentalism away from federal courts, legislation, and press conferences. When gifted young urban leaders like Majora Carter are given some support, their environmental work proves the necessity of their inclusion.

Source: Sustainable South Bronx. Available at ssbx.org/. Accessed January 21, 2008.

The ability to adapt to climatic changes will depend on the integration of local land-use law with state and federal environmental law, on the accuracy and depth of knowledge of the local ecosystem, and on developments and applications of science and technology. There are rapid changes in all three areas around the world. The United States, with the relatively recent formation of the Environmental Protection Agency, faces entrenched and controversial policy challenges in the first two areas. Nonetheless, human populations continue to swell and surge into cities. Global climatic conditions are changing and may further increase the environmental vulnerabilities of cities. Battlegrounds of emergency preparedness, timely evacuation, flood insurance, and community relocation policies will flare up again.

MEXICAN–U.S. BORDER CITIES

Maquiladora Communities, Environmental Vulnerability, and Health

Cities in countries with little or no environmental protection that border on countries with environmental protection can be especially vulnerable to pollution. If the unregulated towns are near water, roads, or rail lines, they become very attractive as large industrial sites. Such is the case with the Mexico and U.S. border towns.

The maquiladora industries, concentrated along the U.S.–Mexican border, are examples of business exploitation of the lack of environmental and occupational safety regulations. These industries have not incorporated the environmental protection or health promotion strategies required of U.S. industries north of the border. The continuing failure to build sufficient housing, water and sewage systems, or waste treatment facilities in these communities has created an environmental battleground.

See also Citizen Monitoring of Environmental Decisions; Climate Change; Environmental Justice; Evacuation Planning for Natural Disasters; Sprawl; Sustainability

Web Resources

Borden, Kevin Allen. "Comparing the Vulnerability of the Built Environment among U.S. Cities." Available at www.cas.sc.edu/geog/hrl/borden_ths.html. Accessed January 21, 2008.

Natural Resources Institute. Environment, Vulnerability and Policy. Available at www.nri.org/research/ds-evp.htm. Accessed January 21, 2008.

Further Reading: Pelling, Mark. 2003. *The Vulnerability of Cities: Natural Disasters and Social Resilience*. London: Earthscan; Riddell, Robert. 2004. *Sustainable Urban Planning*. Oxford, UK: Blackwell Publishing; Takashi, Inoguchi. 1999. *Cities and the Environment: New Approaches for Eco-societies*. New York: United Nations University Press; Walter, Bob, Lois Arkin, and Richard Crenshaw. 1992. *Sustainable Cities: Concepts and Strategies for Eco-city Development*. Los Angeles: Eco-Home Media.

EVACUATION PLANNING FOR NATURAL DISASTERS

Environmental controversies concerning all types of natural disasters follow their impact on people and places. Survivors generally demand more notice and monitoring of potential natural disasters. All communities want better evacuation planning. Forced evacuations are controversial, especially where there are no clear evacuation routes.

BACKGROUND: GETTING OUT!

Natural disasters are just that—disasters. Loss and suffering of loved ones, terror, anger, great fear, confusion, and pain can last for days, and precede one's own death. Natural disasters are large forces of nature like floods, hurricanes, earthquakes, fires, ice storms, avalanches, tsunamis, and volcanic eruptions. They pose great risk of catastrophic damage. They can often occur together. Hurricanes and floods often accompany each other. Earthquakes, tsunamis, and urban fires often accompany each other. Avalanches, landslides, and mudslides can follow earthquakes months afterwards. Wildlife, human life, and ecological order can be forever changed following significant natural disasters. With global warming and climate change some areas are predicted to have more severe weather of greater extremes. There is concern that the number of natural disasters will increase and that communities are not adequately prepared.

There is a fundamental controversy about responsibility and solutions to natural disaster planning and relief. Most agree that the government has some responsibility, and there are laws to that effect in most places in the United States. However, the government includes municipal, state, and federal response levels. Intergovernmental relations are very poor in the United States, especially in environmental areas. In the context of a disaster, communications quickly become tangled at best. In communities where a large industrial or nuclear facility may be at risk in a natural disaster, that facility has some responsibility to help plan for evacuation in case of catastrophe. Some communities have resisted the siting of a nuclear facility because of the failure of the site to provide emergency evacuation routes. The question of individual responsibility also arises, especially in the battleground of mandatory evacuations. In mandatory evacuations police and other emergency personnel will go door to door evicting people from their homes. The amount of time to prepare or evacuate will depend on the natural hazard. If the event is a weather condition that can be monitored, then there is some time to evacuate beforehand if there are sufficient routes. However, many disasters can happen with no notice, such as earthquakes. Early warning systems for tsunamis, hurricanes, floods, and some avalanches help reduce the risks to humans to the extent that enough notice is given, that individuals know and can respond appropriately, and that evacuation routes remain moving. Early warning systems are part of planning ahead. But they are insufficient alone because they rely on notice and human behavior and adequate evacuation routes. Planning ahead for natural disasters now focuses on evacuation routes. This is often much more difficult than it appears because most roads are not built for a mass exodus. Depending on the extent of the natural disaster, rail lines, waterways, and airports may be shut down. Most evacuation plans, if they exist at all, rely on the individual car. If you do not have access to a car and receive very little notice, then the options for evacuating from a natural disaster decrease.

EVACUATION: TRADITIONAL EMERGENCY PLANNING

Evacuations occur hundreds of times each year. Transportation accidents with hazardous chemicals and industrial accidents release toxic substances. This forces thousands of people to leave their homes. Fires and floods cause evacuations more frequently. In the western United States, fires cause hundreds to evacuate their homes every year. Almost every year, people along the Gulf and Atlantic coasts evacuate because of approaching hurricanes. The main battleground is evacuation. In the case of fire, an immediate evacuation to a predetermined area away from the fire may be necessary. In a hurricane, evacuation could involve the entire community and take place over a period of days.

Controversy can follow emergency evacuation plans. Not everyone can be evacuated at the same time. Routes can quickly become completely clogged. Not only does this stop evacuees, it also prevents efficient use of emergency vehicles. A developing solution for application to hurricanes is to open up all lanes in one direction on federal and state highways. Which routes evacuees and emergency vehicles should use is a question that can prioritize certain neighborhoods over others.

ENVIRONMENTAL JUSTICE AND EVACUATIONS

Timeline of an Unnatural Disaster: Hurricane Katrina

Environmental justice refers to the unequal distribution of environmental benefits by the government based on race or income. The failure of federal, state, and local government to evacuate or provide relief for five to seven days for the primarily African American residents of New Orleans, in the face of the threat of two massive hurricanes, is considered a glaring example of environmental injustice. A basic chronology gives one view of the controversial and disputed series of events around Hurricane Katrina.

Friday, August 26

Louisiana Governor Kathleen Blanco declared a state of emergency in the state. Gulf Coast states including Louisiana began requesting National Guard support and other federal assistance. As the United States is embroiled in a war with insurgents in Iraq and is employing the National Guard at a very high rate, many wonder if they will be available for their usual missions of firefighting and aiding communities in times of national emergency. Katrina confirmed the fear that a lack of National Guard troops would make U.S. cities more vulnerable to natural disasters.

Saturday, August 27

At the request of Louisiana Governor Kathleen Blanco, President Bush declared a federal state of emergency in Louisiana, specifically authorizing the Federal Emergency Management Agency (FEMA) to coordinate all disaster relief and to identify, mobilize, and provide, at its discretion, equipment and resources necessary to alleviate the impacts of the emergency. Some employees of FEMA allege that they began a series of increasingly desperate e-mails and telephone calls to director Michael Brown, a close friend of President Bush. Mr. Brown chose to ignore these e-mails and discussed getting a new suit as he went to dinner instead of activating emergency plans.

Sunday, August 28

New Orleans Mayor Ray Nagin ordered the first-ever mandatory evacuation of the city. President Bush, Department of Homeland Security (DHS) Secretary Michael Chertoff, and (FEMA) Director Brown were briefed about the danger of levee failure by the national hurricane director.

Monday, August 29

Hurricane Katrina made landfall as a category 4 storm. FEMA Director Brown requested that DHS send 1,000 FEMA employees into the area within two days. The 17th Street Canal levee was breached.

When President Bush flew over the area in a helicopter a few days afterwards, citizens fired guns at him from the ground. What few evacuation plans existed were designed to get only private vehicles out. Low-income people and people of color in the New Orleans area relied heavily on public transportation when they had to travel anywhere. Many people in nursing homes and hospitals, and in many historically African American neighborhoods, either simply did not travel or walked. Post-Katrina mortality statistics bear this out. Elderly African American women died disproportionately. An interesting blip in the statistics indicates the difficulty with mandatory evacuations. The second biggest death rate was for elderly white men, primarily from a historically exclusive white community.

For a complete description of the state of emergency planning, the problems with evacuation, the environmental justice implications, and the policy ramifications, see *An Unnatural Disaster: The Aftermath of Hurricane Katrina* by member scholars of the Center for Progressive Reform, http://www.progressivereform.org/Unnatural_Disaster_512.pdf.

To develop an evacuation policy and procedure it is necessary to:

- Determine the conditions under which an evacuation would be necessary.
- Establish a clear chain of command.
- Identify personnel with the authority to order an evacuation.
- Designate evacuation wardens to assist others in an evacuation and to account for personnel.
- Establish specific evacuation procedures. Establish a system for accounting for personnel.

- Consider employees' transportation needs for community-wide evacuations.
- Establish procedures for assisting persons with disabilities and those who do not speak English.
- Post evacuation procedures.
- Designate personnel to continue or shut down critical operations while an evacuation is under way. They must be capable of recognizing when to abandon the operation and evacuate themselves.
- Coordinate plans with the local emergency management office.

EVACUATION ROUTES AND EXITS

Designate primary and secondary evacuation routes and exits. Have them clearly marked and well lit. Post signs. Install emergency lighting in case a power outage occurs during an evacuation.

Ensure that evacuation routes and emergency exits are:

- Wide enough to accommodate the number of evacuating personnel
- Clear and unobstructed at all times
- Unlikely to expose evacuating personnel to additional hazards
- Evaluated by someone not in your organization
- Consider how you would access important personal information about employees (home phone, next-of-kin, medical) in an emergency

Given the overall lack of emergency preparedness in U.S. cities, it is likely that they will be battlegrounds for improved preparation, evacuation routes, and emergency responses.

POTENTIAL FOR FUTURE CONTROVERSY

While no one opposes evacuation planning for natural disasters, very few come forward with the resources necessary to plan and implement realistic evacuation plans. Federal agencies and states are rapidly developing models for evacuation. It is not yet, however, a policy that is effectively implemented. It is clear that more research needs to be undertaken to investigate specifically the effects of a mass evacuation on current transport networks. With inadequate local knowledge of transportation and unimplemented policies, it is likely this controversy will continue from one natural disaster to the next. Some island nations, such as Barbados, that experience frequent hurricanes and wave surges have sophisticated evacuation plans that include a smooth transition from evacuation point to safety point. Family services, social services, religious services, and municipal services (fire, police, and sanitation) continue. Hurricanes trigger other disasters such as fires and floods. These can be mitigated with thorough transition planning.

In the United States, FEMA lost its separate agency status and merged with Homeland Security under the Bush administration. In 2005 Hurricane Katrina devastated New Orleans and the surrounding area, leaving many to die slow deaths as federal assistance never came, even though federal resources were nearby. It was five days before meaningful federal help arrived, and more than 1,000 people were known dead at the time. During the hurricane the oil refineries along the Mississippi River and in Texas, which are allowed to use sulfuric acid to clean the refineries, released everything, as did all the large number of industries there. The Mississippi River there has long been known as "cancer alley" because of the pollution and is now a toxic stew spread as far as Hurricane Katrina could carry it. African Americans from New Orleans were referred to as *refugees* by FEMA and callously shunted from place to place. Many sought refuge in the New Orleans Superdome. Crowded and with no sanitation, some people died there and the bodies were not moved for days. Many communities and churches across the United States voluntarily came to the aid of the refugees. The bitterness of this controversy remains. The Bush administration has subsequently worked to relax already weakly enforced environmental cleanup laws. One of the largest private insurers of flood risk in the United States, State Farm Insurance Corporation, announced it was pulling out of Mississippi altogether because the risks are too high.

The incidence of natural disasters has increased over the past 30 years. Coupled with an increase in the populations located in the path of these natural disasters, the imminent danger posed by naturally occurring phenomena has also grown. Given the potential dangers, most communities want appropriate emergency plans in place that minimize the negative environmental impacts from these disasters. Effective emergency planning and management should successfully combine the skills and knowledge of law enforcement agencies and transport planners as well as the knowledge and skills of emergency planning professionals. It has not yet done so in the United States.

EVACUATION PLANNING: HURRICANE KATRINA AND THE EVACUATION OF NEW ORLEANS

FEMA has provided nearly $6 billion in assistance directly to Hurricane Katrina victims for housing and other needs through the Individuals and Households Assistance Program. The more than $6 billion provided to victims of Hurricane Katrina is the most ever provided by FEMA for any single natural disaster. In all, nearly 950,000 applicants were determined to be eligible for assistance under the Individuals and Households Assistance Program.

- Housing—$4.2 billion: Housing assistance covers temporary housing, repair, replacement, and permanent housing construction.

  - FEMA has provided nearly $4.2 billion to nearly 950,000 applicants for housing assistance following Hurricane Katrina.
  - FEMA has paid out $1.3 billion to nearly 550,000 applicants in Louisiana and Mississippi under the DHS transitional housing program for homes that were inaccessible to inspectors due to persistent flooding.

- Other Needs Assistance—$1.9 billion: This money is used to cover medical, mental health, transportation, and other expenses related to disasters.

  - Assistance to nearly 420,000 households provided nearly $ 1.9 billion total for "Other Needs Assistance."

- Hotel Motel Program—85,000 households and $650 million

  - FEMA has paid more than $650 million for hotel/motel rooms to date; at its peak in October 2005, FEMA provided hotel/motel rooms for 85,000 households in need of short-term shelter.

- Housing Inspections and Repair—1.3 million inspections

- Since Hurricane Katrina 1.3 million housing inspections have been completed in Alabama, Louisiana, and Mississippi.

Travel Trailers and Mobile Homes—101,174 households (currently occupied) *Section revised on 8/25/06

- There are 101,174 travel trailers and mobile homes serving as temporary housing for Hurricane Katrina victims, outnumbering any housing mission in FEMA's history. The following shows the number of units currently occupied as of August 17, 2006:

  - Louisiana Total—64,150
    - Mobile Homes—3,169
    - Travel Trailers—60,981
  - Mississippi Total—36,127
    - Mobile Homes—4,709
    - Travel Trailers—31,418
  - Alabama Total—897
    - Mobile Homes—0
    - Travel Trailers—897

Of those households occupying travel trailers and mobile homes, there are:

- 15,000 households in group/commercial sites:

  - 9,344 in Louisiana
  - 5,507 in Mississippi
  - 149 in Alabama

- 83,962 households in private sites:

  - 52,594 in Louisiana
  - 30,620 in Mississippi
  - 748 in Alabama

- 2,212 households in industry sites located in Louisiana.

Travel Trailers and Mobile Homes: 121,922 households (Cumulative Leases)

- 121,922 travel trailers and mobile homes have served as temporary housing for Hurricane Katrina victims, outnumbering any housing mission in FEMA's history. The following shows the cumulative number of units used as of August 25, 2006.

  - Louisiana Total—71,134
    - Mobile Homes—3,514
    - Travel Trailers—67,620
  - Mississippi Total—48,274
    - Mobile Homes—6,300
    - Travel Trailers—41,974
  - Alabama Total—2,514
    - Mobile Homes—0
    - Travel Trailers—2,514

Of those households occupying travel trailers and mobile homes, there were:

- 19,074 households in group/commercial sites:

  - 10,976 in Louisiana
  - 7,242 in Mississippi
  - 856 in Alabama

- 99,959 households in private sites:

  - 57,269 in Louisiana
  - 41,032 in Mississippi
  - 1,658 in Alabama

- 2,889 households in Industry Sites located in Louisiana.

Cruise Ships—Housing over 7,000 households

- For the initial six months after Hurricane Katrina, FEMA used cruise ships to house evacuees, workers from the City of New Orleans and St. Bernard Parish, and first responders and their families, totaling more than 7,000 workers and families.

Public Assistance Projects—$4.8 billion

- More than $4.8 billion has been "mission assigned" for public infrastructure projects, such as debris removal and restoration of roads, bridges, and public utilities. This total nearly doubles the combined total of $2.6 billion allocated for public assistance projects from the 2004 hurricanes across 15 states, Puerto Rico, and the U.S. Virgin Islands.

Debris Cleanup—99 million cubic yards

- Since Hurricane Katrina, more than 99 million cubic yards of debris have been removed in Alabama, Mississippi, and Louisiana, costing out $3.7 billion to date.

Crisis Counseling—$126 million

- After Hurricane Katrina, all 50 states, Puerto Rico, and the District of Columbia were eligible to apply for CCP grants to serve victims in the disaster area and displaced evacuees in other locations. Currently, more than $126 million in federal crisis counseling support has been approved thus far. This funding allows states to have the flexibility to develop service programs including outreach, counseling, support groups, and public education most appropriate for the hurricane evacuees within their state.

Evacuation Reimbursement—$735 million to evacuee host states

- FEMA public assistance reimbursed more than $735 million to 45 states and the District of Columbia for sheltering and emergency protective measures taken during the evacuation of the Gulf Coast and for ongoing sheltering initiatives directly following Hurricane Katrina. This is in addition to funds obligated to Louisiana, Mississippi, and Alabama totaling nearly $1.75 billion for emergency sheltering operations.

Expedited Assistance—$1.6 billion

- Through expedited assistance, FEMA's accelerated method of disbursing disaster assistance, FEMA provided more than $1.6 billion to 803,470 individuals and households to help the evacuees meet immediate emergency needs, such as housing, food, and clothing.

Disaster Unemployment Assistance—$410 million

- FEMA has obligated more than $410 million to support expenditures disaster unemployment assistance to Hurricane Katrina victims.

See also Environmental Vulnerability of Urban Areas; Floods; Global Warming; Hurricanes

Web Resources

Transportation Research Board. Framework for Modeling Emergency Evacuation. Available at trb.org/news/blurb_detail.asp?id=5023. Accessed January 21, 2008.

U.S. Department of Transportation. Emergency Preparedness and Individuals with Disabilities. Available at www.dotcr.ost.dot.gov/asp/emergencyprep.asp. Accessed January 21, 2008.

Further Reading: Balog, John N. 2004. *Public Transportation Security: A Guide for Decision Makers.* Washington, DC: Transportation Research Board; Bullard, Robert D., ed. 2007. *Growing Smarter: Achieving Livable Communities, Environmental Justice, and Regional Equity.* Cambridge, MA: MIT Press; Daniels, Ronald J., Donald F. Kettl, and Howard Kunreuther, eds. 2006. *On Risk and Disaster: Lessons from Hurricane Katrina.* Philadelphia: University of Pennsylvania Press; Nicolet-Monnier, Michel, and Adrian V. Gheorghe. 1996. *Quantitative Risk Assessment of Hazardous Materials Transport Systems: Rail, Road, Pipelines and Shipping.* New York: Springer.

F

FARMWORKERS AND ENVIRONMENTAL JUSTICE

About 4.2 million seasonal and migrant farmworkers and their dependents work across most of the United States, although this is probably an undercounted population. This population is largely minority (90% Hispanic), medically underserved, and at risk for a variety of environmental health problems. Controversies ensue around illegal immigration and the cost of providing services to illegal immigrants. Environmental controversies involve the use of pesticides, labor camp conditions, and agricultural industrial practices.

Mexicans come to the United States to work on large farms. Most farmwork is seasonal and temporary. Farmworkers are among the poorest of all workers. Nationally, the median annual income of a single worker is below $7,500 and, for families, between $10,000 and $14,000. Federal and state laws requiring overtime pay do not apply to agricultural workers. Part of the controversy around farmworker treatment is that many of them are not citizens but availing themselves of U.S. benefits.

Although farmworkers are essential to the production of food in the United States, they have little power to control their work conditions. Farmworkers often make little more than minimum wage, seldom receive any employment benefits, and in many areas are not organized. Most farmworkers are immigrants to the United States. The national farmworker population has become increasingly Latino and Mexican during the past decade. In 1998, 81 percent of all migrant and seasonal farmworkers in the United States were foreign-born, and 95 percent of those were born in Mexico. Although some areas of the United States (e.g., California, Florida) have routinely employed large numbers of Latino seasonal and migrant farmworkers, other areas have recently experienced a dramatic increase in these workers as family labor gives way to hired labor. In North Carolina, which ranks fifth in the size of its farmworker population, most farmworkers years ago were African American. Today only 10 percent are African American; most U.S. farmworkers are Latino.

CESAR CHAVEZ

Cesar Chavez was a courageous and powerful labor organizer for Mexican and Mexican American farmworkers. He was born in 1927 on a small farm outside of Yuma, Arizona. Later his family moved to California and worked as farmworkers. In 1942 he graduated from eighth grade and became a full-time farmworker to help support his family. He served in the U.S. navy from 1944 to 1946. In 1962 he founded the United Farmworkers with Dolores Huerta. He was also assisted by Dr. Marion Moses, a pesticide researcher and author.

Cesar Chavez spoke out against injustices to all people but focused his activism and nonviolent approach on the conditions of farmworkers. His brave activism, inspired by Mahatma Gandhi and Martin Luther King, included farmworker strikes, targeted boycotts and picketing, media-grabbing fasts to bring attention to the condition of farmworkers, and a 340-mile protest march from Delano to the state capitol of Sacramento. Along the way he empowered thousands of oppressed farmworkers, who previously had little or no hope. He raised state and national public awareness of the labor conditions under which food was grown, harvested, and marketed. He got results in terms of forcing growers to contract with farmworker unions and state laws to help protect farmworkers. He improved the quality of life of oppressed people by improving working and living conditions for them.

Cesar Chavez died on April 23, 1993. His family and friends created the Cesar E. Chavez Foundation and a model educational curriculum around nonviolent activism.

For more information see the Cesar E. Chavez Foundation, www.cesarechavezfounda tion.org and United Farmworkers, www.ufw.org.

PESTICIDES

Farmworkers, their families, and often their communities have large pesticide exposures and injuries. As pesticides are altered to have less environmental and public health impact on the consumer of the food, they become more acutely hazardous to the farmworker. They do this by releasing their chemicals on the plant in more intense dosages that do not last as long. Because farmworkers often live near the farm, they share water sources. As pesticide and other by-products of industrial agricultural processes run off the land and accumulate over time in the water, farmworker communities suffer disproportionate health risks due to their proximity to the ecosystems shared with the agricultural site.

Contemporary U.S. agriculture uses large amounts of pesticides. Agricultural pesticides include those chemicals intended to kill insects, plants, fungi, rodents, and other organisms that interfere with the production, storage, and distribution of agricultural produce. Most agricultural pesticides now being used can have detrimental effects on human health. The nature of farmwork exposes everyone who works on a farm to pesticides, farm owners and managers as well as farmworkers. However, farmworker pesticide exposure must be considered separately because of the extensive manual labor that most farmworkers perform and because farmworkers have limited power to influence workplace safety. The health effects of pesticide exposure can be immediate and include rashes, headaches, nausea and vomiting, disorientation, shock, respiratory failure, coma, and, in severe cases, death. Pesticide exposure can also have long-term effects on health in the form of cancer and neurologic and reproductive problems.

FARMWORKERS' PERCEPTIONS OF RISK: SPEAKING FOR THEMSELVES

Farmwork is risky business. Large agricultural business organizations use seasonal and migrant laborers to produce food. Pesticides, fertilizers, emissions from industrial equipment, and poor health care make it especially risky for farmworkers. There are important questions about the cultural and educational appropriateness of the farmworker regulations and the materials developed to implement them. In general these materials are prescriptive, telling farmworkers how to behave, but they fail to tell how such behaviors will reduce risk. A variety of theories have been developed to provide frameworks for understanding and predicting change in health behaviors.

The health belief model (HBM) is particularly useful for the study of farmworker pesticide safety behavior because of its simplicity and parsimony. The relationship of perceived susceptibility to taking a health action is modified by perceived severity of the outcome, the perceived benefits of a health behavior to modify the risk of the outcome, and the perceived barriers to taking action. Beyond these, cues to action and self-efficacy can also modify the relationship of perceived susceptibility to action. Self-efficacy is confidence in one's ability to reduce risk through behavior change. Most of the work to date on farmworkers has addressed the constructs of perceived risk and perceived control/self-efficacy without linking it to knowledge.

Effective communication with farmworkers is necessary for their own protection, as well as protection of the food supply. Linking knowledge to perceived self-efficacy to reduce environmental risk is a very important perception that can differ greatly by race, gender, income, and education. These differences in risk perception can be the basis for controversies about public safety and emerging environmental controversies around sustainability.

The U.S. agricultural industry has always relied on marginalized workers to reduce labor costs. African American slaves in the Southeast built much of the infrastructure and some of the universities there (the University of Virginia, for example). Irish immigrants built much of the infrastructure of New York City and Boston. Chinese workers on the West Coast built the railroads. Mexican immigration began in the early twentieth century. It exploded around World War II when Congress passed the Labor Importation Program, known as the *bracero program.* This program brought about 4.8 million Mexican workers to the United States. It was specifically designed to bring in cheap labor for U.S. agribusiness interests. It was ended in 1964 as an official government program. However, many farmworkers and many farms came to rely on each other.

The use and abuse of migrant and temporary labor is an issue of global proportions. Costa Rica is the country with the most illegal immigrants per capita. It is a prosperous country directly south of war-torn Nicaragua. As war and natural disasters affect whole regions, the search for food and work can emerge. Certain industrial sectors, such as technology and information systems, can use migrant and temporary labor less expensively in other countries. The concern is that these new workers do not have the same environmental protections or living conditions of U.S. workers.

OREGON: AN EXAMPLE

Each state approaches farmworker controversies in its own way. There are complex and dynamic relationships with federal and state agencies. Oregon's elected leaders have proposed various guest-worker statutes over the years, and the practices toward farmworkers have been scrutinized. Although each state is different, some common themes in this controversy are exemplified by the experience in Oregon.

Basic worker standards, initiated as federal reforms in the 1930s, excluded farmworkers. Agriculture argued that the industry needed to be protected from strikes and higher labor costs because of the importance of agriculture to the national economy and the need for food security. Others maintain racism against African American workers in the former confederate South was a strong factor in the exclusion of farmworkers from these laws. They argue that migrant laborers, now predominantly Mexican, are treated as African Americans were, in law, in the past.

Many African Americans sharecropped farmland to live after the Civil War and emancipation from slavery. Others worked as farmhands. The primary farmworkers were African Americans in the eyes of the 1930s' Congress. They were excluded from many of the sweeping reforms of the 1930s, and farmworker protection was one of them. Hispanic farmworkers do not identify as African American. Most Hispanics consider themselves white, when asked to choose between African American or white. They are counted as a community of color in the United States, although there are substantial differences between farmworker communities and many other diverse Hispanic cultures and communities. This can be the unspoken source of ethnic political tension between Latino and African American groups. African Americans may consider Latinos as white, meaning that African Americans are once again lower in status than the latest wave of immigrants.

CURRENT FARMWORKER STATUS

Farmworker designation is a work status defined by the U.S. government. The legal definition is "any person who works for pay in the production and harvesting of agricultural commodities, including crops, animals, and horticulture specialties." It is a broad definition that is refined by the following rules about other work-status definitions. These are current operating definitions for the U.S. government and most states:

- Permanent workers have year-round jobs in agriculture for 150 days or more a year.
- A migrant farmworker moves from his home location to one or more work locations. He is absent from his permanent residence for months at a time.
- A seasonal worker works for part of the year in agriculture within driving distance of his home. These can be very long distances.

FARMWORKERS AND ENVIRONMENTAL JUSTICE

EPA INNOVATION: An internal Region 8 workgroup composed of members from Region 8 Technical Enforcement Program, Pesticides, Water, Geographic Information Systems, the Environmental Justice Program and the EPA laboratory developed a database of migrant farmworker camps to establish a structured resource for future use by EPA programs, state and federal agencies, and health organizations in a quest to improve the quality of life within migrant farmworker communities. The database contains information about camp sites, drinking water sources, size of the camps, and current addresses. The current database emerged from a list of fields the internal workgroup compiled as a guide for project research and is, consequently, the result of an extensive information search. Among the individuals and organizations contacted about farmworker camp information were community health centers, various state and federal departments, Farmworker Health, Farmworker Justice Fund, Clean Water Fund, university programs, local governments, and a number of U.S. EPA programs. Each contact was given a brief explanation of the project and then surveyed on the list of fields developed by the workgroup. Information from this database is currently being used by the Department of Labor and the Region 8 Pesticides Worker Protection Program.

The environmental justice program has partnered with several state and federal agencies, other outside groups, and other EPA programs to begin collection of water samples from various camps throughout Colorado. The Region 8 lab trained Department of Labor (DOL) inspectors in water sampling techniques and chain of custody procedures. The DOL collected samples at several camps and shipped the results back to the lab. Results of those samples showed that one site was high in total coliform and another site was high in *Escherichia coli (E. coli)*. Nitrates were elevated at several of the sites tested, but were not over the maximum contaminant level. Through various funding opportunities, two summer interns were also trained in sample collection and were able to accompany the DOL inspectors and take samples at additional camps. Analysis of the samples taken by the two summer interns is still in process.

By accurate monitoring of the actual environmental conditions in the labor camps and fields, the EPA can better enforce safety conditions that benefit everyone.

Average life expectancy for a migrant farmworker is 49 years as opposed to the average U.S. life expectancy of 73. Infant mortality is at least 25 percent higher than the national U.S. average. Farmworkers' dental health is very poor. From their work farmworkers frequently have back and muscle problems as they grow older. Bilingual staff, community outreach, family health education, and free and accessible transportation to health services are all important in serving farmworkers. However, these services are controversial when provided to farmworkers. The farm owners and agribusiness managers do not like these services to interfere with work and are known to retaliate by evicting a whole family.

POTENTIAL FOR FUTURE CONTROVERSY

Given the reliance on migrant and seasonal labor for food production, the lack of effective implementation of environmental or worker protection law, and the increased knowledge of pesticides and their cumulative risks for farmworker populations this controversy will continue. It is enmeshed with another debate on immigration policy in the United States. Strict enforcement of immigration rules against employers and illegal immigrants would have big impacts in food production, nurseries, and other necessary work areas. It could also force the United States into greater reliance on foreign food producers, increasing cost and environmental impacts. The immigration debate does form the larger battleground for this controversy with economic and environmental undercurrents.

Another crinkle in this controversy is that many environmentalists are concerned that farmworkers are being exploited to hide environmentally harmful practices. Farmworkers are reluctant to report illegal environmental practices for fear of retaliation against them and their family. Industrial agricultural practices are large in scope but often outside of public purview. Environmental justice proponents are concerned that farmworkers are carrying a disproportionate share of environmental burdens. As right-to-know laws move into the agricultural area, farmworkers could have more protection from retaliatory practices.

See also Cumulative Emissions, Impacts, and Risks; Pesticides; Toxic Waste and Race

Web Resources

Ecology Center. Cesar Chavez. Available at www.ecologycenter.org/chavez/environmental ism.html. Accessed January 21, 2008.

Marentes, Carlos. "Farm Workers Fight against Environmental Racism and Neo-liberalism." *Synthesis/Regeneration* 33. Available at www.greens.org/s-r/33/33–06.html. Accessed January 21, 2008.

Further Reading: Bullard, Robert D., ed. 1993. *Confronting Environmental Racism: Voices from the Grassroots.* Boston: South End Press; Hahamovitch, Cindy. 1997. *The Fruits of Their Labor: Atlantic Coast Farmworkers and the Making of Migrant Poverty, 1870–1945.* Chapel Hill: University of North Carolina Press; McWilliams, Carey. 2000. *Factories in the Field: The Story of Migratory Farm Labor in California.* Berkeley: University of California Press; Parks, Lisa Sun-Hee, and David Naguib Pellow. 2002. *The Silicon Valley of Dreams: Environmental Injustice, Immigrant Workers, and the*

High-tech. New York: New York University Press; Thompson, Charles Dillard, and Melinda Wiggins. 2002. *The Human Cost of Food: Farmworkers' Lives, Labor and Advocacy.* Austin: University of Texas Press; Weber, Devra. 1996. *Dark Sweat, White Gold: California Farm Workers.* Berkeley: University of California Press.

FEDERAL ENVIRONMENTAL LAND USE

The United States is the owner of vast tracts of land. Many of the landholdings are for environmental or conservation purposes. Access to public lands also plays an increasing role in the exploration and development of hydrocarbon resources and other uses such as recreation. Federal environmental land-use planning is extensive and will dictate which lands will be available for leasing and what kind of restrictions will be placed on use of those lands. Public lands are also leased to ranchers, loggers, and miners for natural resource use. These uses are considered harmful to the environment. Many tenants on public lands have sense of entitlement due to generations in a given profession, such as ranching or logging, in one place. Many environmentalists press for laws that force the federal government to use these lands sustainably with little or no environmental impact.

PUBLIC LANDHOLDINGS OF THE U.S. GOVERNMENT

The federal government owns about 29 percent of the land in the United States. There is a maze of federal agencies, and some landholdings are not public lands. Four federal agencies administer most of the 657 million acres of federal land. The National Park Service (NPS), the Fish and Wildlife Service (FWS), and the Bureau of Land Management (BLM) in the Department of the Interior, and the Forest Service (FS) in the Department of Agriculture, make most of the environmental and land-use decisions. The majority of the federal lands (92%) are in 12 western states. The federal government owns more than half of the land in those states.

CHALLENGING ASSUMPTIONS: FEDERAL GOVERNMENT AND TAKING OF PRIVATE PROPERTY

Land-use regulation is the United States is done mainly by local government with state oversight. Although the federal government has eminent domain power to control land it is not exercised at the local level. This view is wrong, according to Bruce Babbitt, former Secretary of the Interior and author. The federal government has long played a powerful role in local land-use decisions.

Development has scarred U.S. landscapes and destroyed ecosystems, Babbitt argues. Much of the environmental damage has been done through the environmental land-use decisions of the federal government. He says we should gather the federal government's resources to mitigate the problems that are the legacy of these decisions. Babbitt argues that federal policies have dispersed too many people into landscapes too sensitive to tolerate heavy use. People are allowed to develop on federal lands in places that are subject to natural hazards such as fires and avalanches. They are allowed to mine, log, and graze livestock on federal lands in wildlife areas. The most damage is done by the construction of

SCIENTISTS RETALIATED AGAINST FOR RANGE MANAGEMENT REPORT

A major part of federal land management is ranching. Ranches are leased to private corporations and individuals, and these leases are enforced by the landlord—the U.S. federal government. This federally owned land dominates the environment in many Western states. These leases cover vast tracts of land, and are often controlled by long term leases held for generations in some communities. Reports from government scientists on the environmental impact of ranching on federal lands are controversial because they stand up to powerful vested special-interest ranching industries. These government reports may be the only reliable information available about the environmental condition of our federal lands because ranchers treat this land as private property. Ranch management scientists work for the federal government and the ranchers to increase the productivity of the range for ranchers. Traditionally they have not focused on ecological preservation or accurate and cumulative environmental impacts. When the environment is rapidly degrading, the productivity of the range for ranchers goes down. These government reports are very powerful documents because they put limits on the abuse of natural resources in federal lands. Some government biologists say the George W. Bush administration is interfering with the science of range management to promote its pro-ranching agenda.

In one internal report released to the media, scientists reported that "the cumulative effects...will be significant and adverse for wildlife and biological diversity in the long-term." "The numbers of species listed or proposed for listing under the Endangered Species Act will continue to increase in the future under this alternative." According to the government range scientists, that language was removed from the scientific analysis that accompanied the new grazing regulations.

There are many environmental controversies around government scientists about government censure and misrepresentation. In many environmental controversies scientists follow principles and ethics, often to their own personal detriment in terms of careers and professional black listing. Sometimes government scientists can find protection under a law called whistleblower protection. These are rules and laws designed to protect the independent professional judgment of scientists and other professionals such as lawyers, engineers, and accountants employed by the U.S. government. They coincide with the professional ethics of a given profession because exercising independent professional judgments is the essence of being a professional. The two previous scientists made their statements after they retired. Other scientists in government are laterally transferred. The effectiveness of the whistleblower protection laws is controversial. They only apply to government professionals. Scientists' ability of scientists to exercise independent professional judgment is a very important battleground, especially in the context of federal environmental land-use planning. The manipulation of environmental facts for profit compounds the intensity of this battleground.

roads needed by the loggers, oil drillers, miners, and ranchers. Where the roads go, development often follows. Abuse and lack of federal enforcement of these uses creates, and continues to create, severe threats to wildlife and wilderness habitat. As examples, Babbitt cites the Everglades, farmlands where streams and grasses have been ravaged by farm policies, and watersheds degraded by abuse and pollution.

"The notion that land use is a local matter has come to dominate the political rhetoric of our age," and it is just wrong, Babbitt writes. He recommends congressional action to assess current federal landholdings and create a system of "permanent national-interest lands." As secretary of the Department of the Interior, Babbitt had some success in this effort, through the protection of lands designated as national monuments.

For 80 years, the federal government has regulated the ranchers that graze across millions of acres of federal land in the West. It has been a controversy between preserving a rural way of life and responding to a growing environmental concern that values watersheds and biodiversity.

The recent federal land regulations make it easier for ranchers to use federal lands without concern for environmental impacts. The new rules give ranchers more time, up to five years, to reduce the size of their herds if the cattle are damaging the environment. Proving the herds are damaging the environment is a difficult proposition because the ranchers do not let people on the federal land, treating it as if it were their own private property. Without being able to access the environment it is impossible to discover abuses and ecological damage until it is severe. The George W. Bush administration gave shared ownership in the water rights and some structures on federal land to private ranchers. The regulations also decrease the opportunities for public involvement in deciding grazing issues on federal lands.

Critics note that the mining, logging, and federal grazing program cost the United States because they are leased at below-market rates. It is a huge battleground to determine the cost to the United States of the subsidized industries operating on federal lands. Part of the cost is the mitigation and cleanup of their environmental impacts. It is a cost that the responsible industries will not voluntarily bear, that the small western communities cannot afford, and that western states allow. Over the last 80 years or so the cumulative environmental effects of large- and small-scale resource exploitation operations on federal lands may begin to undermine their mission of conservation and environmental protection.

POTENTIAL FOR FUTURE CONFLICT

The battleground for these controversies is those states dominated by federal landowners. These agencies represent an employment base in many communities with an ebbing natural resource base, such as timber-or-mining dependent communities. These industries still want to continue logging and mining. Concessions in federal land areas such as national parks exert a strong voice in these policy debates. On a national level, environmentalists and proponents of sustainability want the federal landholdings to lead the way in environmental policy development. They press for ecosystem risk assessment, environmental impact statements, and limited environmental impacts. Some demand wilderness areas and then press for the reintroduction of wolves and grizzly bears. Many other diverse federal land users want more access to federal lands. They include snowmobilers, hang gliders, mountain and rock climbers, hunters, and surfers. All these groups have different environmental priorities that will be debated in future federal land policies.

See also Cumulative Emissions, Impacts, and Risks; Endangered Species; National Parks and Concessions; "Takings" of Private Property under the U.S. Constitution

Web Resources

Bureau of Land Management, Department of the Interior. Public Land Statistics 1998. Available at www.blm.gov/natacq/pls98/. Accessed January 21, 2008.

Public Lands Foundation. Available at www.publicland.org/. Accessed March 2, 2008.

Public Lands Information Center. Available at www.publiclands.org/home.php?SID=. Accessed January 21, 2008.

The Wilderness Society. Available at www.wilderness.org. Accessed January 21, 2008.

Further Reading: Babbitt, Bruce. 2005. *Cities in the Wilderness: A New Vision of Land Use in America.* Washington, DC: Island Press; Davis, Charles, ed. 2001. *Western Public Lands and Environmental Politics.* Oxford: Westview Press; Nelson, Robert H. 1995. *Public Lands and Private Rights: The Failure of Scientific Management.* MD: Rowman and Littlefield.

FIRE

Fire as a natural disaster often accompanies other disasters, such as hurricanes, floods, and tsunamis. Human causes of fire, intentional or unintentional, are also controversial. Humans are seven times more likely to start a wildlands fire than lightning. The environmental consequences of large fires can be far-reaching. Controversy erupts about the appropriate intervention by humans, about salvaging the burned logs, and about fire as a technique for deforestation in rain forests.

Fires are very powerful forces of nature. The context for fire to be a disaster is its impact on humans and the environment. It can be a natural and expected part of an ecosystem. In some ecosystems it is essential for regeneration of certain species of plants. Some trees actually require a fire to procreate. Some pinecones need the fire to burn off a resinous coating, and other tree seeds in the ground need a fire to germinate. Some communities use fires to clear fields and forests and gain direct economic benefits from doing so. Uncontrolled, large-scale fires can have large environmental and economic impacts.

Many of the environmental controversies around fire as a natural disaster involve wildlands. Fires in towns and cities are handled by local fire departments.

In rural areas the only local fire protection may be a volunteer fire organization. The first wildland fire prevention program was established in the United States in 1885 in the Adirondack State Park. That first policy focused on stopping all fires. For a century this has been the dominant fire management strategy. Sometimes lightning-caused fires were not suppressed if the fuel load was high and it was an unpopulated area.

SMOKEY BEAR

Smokey Bear is the longest-running advertising campaign in U.S. history. It began in 1944 to prevent forest fires. Part of the motivation for its creation was fear of enemy attack. In 1942 a Japanese submarine fired into an oil field outside of Los Padres National Forest in southern California. Some feared that such an attack could destroy a valuable natural resource, wood, when it was necessary for national defense, being used in ships, guns, and military transport packing. Although Bambi was the first antifire forest animal, Smokey Bear was the final choice. In a cooperative effort between the War Advertising Council and the U.S. Forest Service, Smokey Bear was created on August 2, 1944. It is officially Smokey Bear, not Smokey the Bear.

Smokey Bear is a protected name under federal law. Misuse of the name can result in a prison term. The United States Code, Title 18, Part I, Chapter 33, s 711 states in part:

> Whoever, except as authorized under rules and regulations issued by the Secretary of Agriculture after consultation with the Association of State Foresters and Advertising Council, knowingly and for profit manufactures, reproduces, or uses the character "Smokey Bear" originated by the Forest Service, U.S. Department of Agriculture, in cooperation with the Association of State Foresters and Advertising Council for use in public information concerning the prevention of forest fires, or any facsimile thereof, or the name "Smokey Bear" shall be fined under this title or imprisoned not more than six months, or both.

As one of the most recognizable images in the United States, Smokey Bear has been a success in conveying his message that "Only you can prevent wildfires!" Although Smokey Bear was born in an era when the primary fire policy was suppression, his message was an early one of prevention and personal responsibility. As recreational use of wilderness and other natural areas has increased since then the message still holds true, especially in the prevention of wildfires.

Many people contributed to Smokey Bear's creation. The first to paint a forest-fire-prevention bear after Bambi was Albert Staehle. Then the Forest Service used their own artist, Rudy Wendelin, to produce and disseminate the Smokey Bear symbol. Mr. Wendelin emerged from retirement to paint the 40th anniversary Smokey Bear commemorative postage stamp. One of the last artists to work on Smokey Bear's image was Charles Kuderna.

Smokey Bear has a dedicated following of fans. There is a Smokey museum, many commemorative patches, books, tapes, and memorabilia.

For further information, visit Smokey's Vault, www.smokeybear.com/vault/default.asp. See also Ruthven Tremain, *The Animals' Who's Who* (Ann Arbor: University of Michigan Press, 2007).

U.S. WILDFIRES

There are many federal agencies that have some jurisdiction over wildfires. The main ones are the Bureau of Land Management, Bureau of Indian Affairs, the National Park Service, the U.S. Fish and Wildlife Service, and the Forest Service. States also have their own agencies with wildfire responsibilities. A minor battleground can develop when agencies contest jurisdiction over wildfires. For the most part firefighting agencies work hard to cooperate and allocate resources where they are needed. Fire suppression costs are expensive. Moving water, chemicals, people, and equipment around a wildfire requires split-second decision making and adequate resources.

The number of acres consumed by wildfires has steadily risen, although this may in part be due to prescribed fires. According to the National Interagency Fire Center, there were 103,387 fires in 1960 and they consumed 4,478,188 acres. In 1980 there were 234,370 fires that consumed 5,260,825 acres. In 1990 the number of fires decreased to 122,763 but they consumed 5,452,874 acres. This trend continues. In 2005 there were 66,552 fires that consumed 8,686,753 acres.

WELCOME FOREST FIRES?

Forest fires are not welcome in all ecosystems. In humid, tropical forests, fires play a minimal role in the natural process of the ecosystem. Dry tropical forests rely on them for species regeneration. Forest fires become controversial when they threaten wilderness, timber, mining, grazing, and residential development interests. The very human forces that are threatened by fire can also be the cause of it. Motorized equipment, dry conditions, and ignitable fuels can all unintentionally cause fires. Some locations in the United States periodically become so dry that fires become a constant threat. A single careless cigarette or campfire can start an inferno. When the grass on gravel and dirt roads is allowed to grow and then dies it creates a fire threat because the heat of engines and generators can ignite them. This is part of the battleground of permitting all-terrain vehicles and other motorized vehicles in wildlands. They increase fire risk by the vehicles themselves and their presence in fire-prone environments.

Fire management presents many challenges. In the United States, most efforts have been to suppress fires. In some places this has increased the amount of fuel for fires, creating superfires. Some U.S. national parks are especially prone to this fire risk because old fire suppression policies actually increased the fuel load for decades. Fire suppression in now less automatically applied, and a decision is made about whether to let a fire burn or to control the fire to burn out in a certain manner.

Fire has been used as a land management tool for centuries. It is very inexpensive, easy to apply, and works quickly and thoroughly.

OREGON FIELD BURNING

Oregon is the grass seed capital of the world, growing large fields of grass to produce the seeds for sale worldwide. They specialize in the grass grown for golf course greens. As logging has declined, other industries such as agribusiness have grown in Oregon. An agricultural practice of burning the fields to clear the way for new growth was practiced there by indigenous people before European settlement. Today, the Oregon Department of Agriculture Smoke Management program determines the safe days to burn the fields. They keep track of smoke impact hours and enforce burn permits for open-field burns. They also permit stack burning but exert less control over that practice. Many people object to the field burning, and as population in the impacted communities has increased, so too has the profile of this battleground.

In Oregon three types of field burning are permitted. All acreage proposed to be burned must be registered with the Oregon Department of Agriculture. In regular open fields burning perennial or annual grass and some cereal grain residue is permitted if the weather conditions are right. Some strains of grass seed cannot be profitably raised without burning. Industry clams that this thermal sanitation is necessary for at least three identified species, and this is another type of field burning. The last type of open-field burning is on steep terrain. Grass is grown there, but the land is too steep for anything but open-field burning.

According to the Oregon Department of Agriculture, 114,297 acres were registered for open-field burning in 2006. 2006 was the tenth year of operation of the smoke management program. Of this, 96,962 acres were classified as regular burns, 16,294 acres as identified-species burns, and 1,041 acres as steep-terrain burns. State law only permits a maximum of 40,000 acres of regular burns a year so permits were allocated. The other categories were allocated 100 percent, so there is a temptation to maximize profits by moving burns into other categories. Some have questioned the strength of the enforcement efforts. They received 1,182 complaints in 2006 from 24 burn days. They made five enforcement contacts that resulted in warning letters and no penalty assessments.

In the 1980s and 1990s field burning was so prevalent that the smoke caused car accidents on the interstate highways nearby. Although field burning has been continuously restricted since then, it is still permitted. Oregon has among the worst air quality in the United States, after New York and California. These other two states have huge populations, many more vehicle emissions, and a history of industrialization. By comparison Oregon is sparsely populated, with fewer roads per capita and much less industry. Part of the problem is that the smoke management program must rely on complex weather conditions to get the smoke up and out from communities. Mistakes are common, which means a community is filled up with additional pollution from field burning. Some communities and environmentalists claim that the grass seed agribusiness industrial complex is making a profit off their health and environment, and that no field burning should be permitted. Grass growers also use inefficient burn techniques, often incorporating propane burners. This increases smoke production. Oregon has been a battleground for several major environmental controversies, such as logging and protection of the endangered spotted owl. Agricultural communities and agribusinesses resist environmental regulation of traditional methods such as field burning. They point out that field burning is pesticide free and may be ecologically sustainable with appropriate practices and advancements in agritechnology.

HOW THE ENVIRONMENT INTERACTS WITH FIRE

Essentially, fire is a chemical reaction called combustion. It requires fuel, oxygen, and heat to the ignition point of the fuel. There are basically three types of wildfires. They are categorized as surface, crown, or ground fires. The intensity of the fire and the type of fuels consumed determine its type. Ground fires are the most intense, leaving nothing but dirt behind. They are large and suck in so much oxygen so quickly they create winds. They are rare but occur more frequently with high fuels loads. They can also cause a second category of fire called crown fires. These occur in the tops of trees. The last category of wildfire is a surface fire. These are the most common and cause the least damage to mature trees. The burn at lower heat and rates and tend to consume minimal fuel.

Controlled fires are intended to deplete the fuel sources before they accumulate. The amount of fuel in a wildlands area is called the *fuel load*. This is measured in terms of weight of fuel, or vegetation. Vegetation can make a large difference in the fuel load of a given wildland. The more surface area the plant offers, the lower its ignition temperature. Leaves, forest floor debris, grasses, and resinous trees all increase the fuel load. If secondary timber growth is not thinned then it dramatically increases the fuel load. Many factors can affect a given fuel load. It is used to determine whether a preemptive burn is necessary, or whether to let a particular fire burn. This computation has recently created a battleground because timber companies argue that it is better to log an area that is scheduled for a preemptive burn. Environmentalists are very concerned about using preemptive burn rules to circumvent controversial logging.

Another factor in wildfires is the moisture content of the vegetation and air. Dead plants have about 30 percent moisture at most. Live plants can hold three times their weight in water. Lower moisture in the fuel load causes the ignition temperature to decrease. Drought can greatly increase the risk of wildfires. With climate change certain areas could see more droughts, and wildfires could increase. An area that typically gets only small surface fires from lightning strikes will risk experiencing a more dangerous ground fire during a drought.

Moisture in the air can have a dramatic effect on wildfire formation and activity. Dry, strong winds decrease the moisture in the vegetation. One example of these is the Santa Ana winds of California. These hot, dry winds blow west out of the deserts and to the Pacific Ocean, which is opposite of the prevailing wind patterns. When they do this wildfires increase in number and duration. Under these circumstances these fires can create their own weather systems, creating pyrocumulus clouds.

AFTER THE FIRE: ENVIRONMENTAL IMPACTS

Depending on the type of fire and the topography, fire can increase the risk of other natural disasters such as landslides and mudslides. Without vegetation, especially on steep mountainous slopes, soil quickly erodes. It goes into water flows and can affect the aquatic environment. If the fire was an intense ground fire it can actually burn the soil, leaving it unable to even absorb water. Intense ground fires are difficult to extinguish. They can follow tree roots down and smolder for days, reigniting if fuel and conditions are right.

The smoke from a fire can have substantial environmental impacts. Scientists are still trying to determine the content of the smoke. While wildfires cannot be regulated, controlled burns can. EPA researchers have found many carcinogens in this smoke. As a fire burns it emits large amounts of particulate matter, or soot, which can affect the lungs. Fires also produce gases. Nitrogen oxides, ozone, volatile organic compounds, heavy metals, and carbon dioxide are all emitted as gases and may create risks. There is increased concern about carbon monoxide and mercury emissions.

Atmospheric mercury is absorbed by vegetation. This mercury can come from many sources, and it is battleground as to how much of it comes from industrial emissions. When the vegetation burns as in a wildfire this mercury is converted to a gas and sent back into the atmosphere. It again falls as rain. The concern is that this mercury will fall as rain on the watershed and form methyl mercury. Methyl mercury is a deadly neurotoxin. Mercury requires interaction with certain microbes in the water to form methyl mercury. Carbon monoxide is a deadly gas. Big, intense ground fires emit large amounts of carbon monoxide. This gas will last in the atmosphere for weeks.

BATTLEGROUNDS: SALVAGE LOGGING

Salvage logging refers to salvaging timber after a fire. Logging itself is a controversial issue, especially on public lands. Harvest systems include tractor, cable, and helicopter logging. There are some emerging policy guidelines, but state and federal regulatory agencies present differing information. Some scientists contend that salvaging timber can eliminate or reduce future fire intensity by decreasing fuel load. Others disagree. Environmentalists claim that any impacts from salvage logging are not justified because of the impacts already created by the wildfire. Some claim that the remaining timber is necessary for ecosystem recovery. Others distrust the timber companies to log only suitable trees. Loggers claim it is necessary to quickly get the remaining timber out because delay causes a loss in the quantity, utility, and economic value of the dead trees. They claim that the length of time needed to do an environmental impact statement, often required for timber leases on public lands, decreases their profitability.

As this battleground develops new policies are being developed on a site-specific basis. Government agencies look to fees and leases for salvage logging as a source of revenue for other projects, and this can create additional battlegrounds. Some of the terms and conditions that shape this battleground are:

- What levels of snags and coarse woody debris should be kept in place. Snags are very hazardous for loggers but provide shelter for animals.
- The size of the area. Salvage logging occurs in patches of fire-killed trees between 3 and 10 acres. Within each of these patches, a minimum of two to three acres is excluded from salvage.
- Further environmental impact analysis prior to implementation of any restoration process.
- They could include limited road improvements necessary to conduct salvage logging.
- Only trees that are considered dead would be salvaged. A fire-killed tree is defined as "a tree with no apparent sign of green foliage."

The restoration projects could include fish habitat improvement, roads, culverts, and plantings. Environmentalists try to protect severely burned areas, fragile soils, riparian areas, steep slopes, or sites where accelerated erosion is possible.

Salvage of fire-killed roadside trees that are or could be a hazard to drivers is usually encouraged. Trees felled within riparian areas or needed for wildlife habitat are excluded from salvage. Other areas excluded from logging can be negotiated.

Salvage logging opens up old wounds in the logging controversies of the United States. Questions about how terms such as the previous list could actually be enforced are raised.

SLASH-AND-BURN FOREST PRACTICES IN THE RAIN FOREST

Rain forests present a different battleground for fire. Rain forests are thought to play an important role in combating global warming because they use carbon dioxide and convert it to oxygen. Slash-and-burn techniques of land clearance have been used in many of these areas for centuries by indigenous people. Recently, the scale of slash-and-burn techniques has increased dramatically as plantation-style crop rotation and ranching grow and require more land. A battleground of traditional land rights for indigenous people can arise in these contexts.

POTENTIAL FOR FUTURE CONTROVERSY

Human population around wilderness and forested areas increased about 10 percent in the 1990s in the United States alone. Preventing the development of fuel loads in wooded areas is less possible with private property, which tends to increase as human populations increase. Although it is in the property owners' interest to reduce fire risk, especially if they are far from fire services, some do not do so. Fire does not recognize political or property boundaries. Private property owners can be large timber, mining, and ranching corporations. They can also be subdivisions and mountain log cabins.

See also Acid Rain; Drought; Rain Forests

Web Resources

Firewise Communities. Risks in Fire Prone Areas of the United States. Available at www.Firewise.org. Accessed January 21, 2008.

The Global Fire Monitoring Center. Available at www.gfmc.org. Accessed January 21, 2008.

U.S. Government Accounting Office Report on the Biscuit Fire Salvage Logging Operations. Available at www.gao.gov/htext/d06967.html. Accessed January 21, 2008.

Further Reading: Allison-Bunnell, Stephen, and Stephen F. Arno. 2002. *Flames in Our Forest: Disaster or Renewal?* Washington, DC: Island Press; Biswell, Harold. 1999. *Prescribed Burning in California Wildlands Vegetation Management.* Berkeley: University of California Press; MacLean, John N. 2003. *Fire and Ashes: On the Front Lines of American Wildfire.* New York: Henry Holt Publishers; Pyne, Stephen J. 2004. *Tending Fire: Coping with America's Wildland Fires.* Washington, DC: Island Press; True, Alianor. 2001. *Wildfire: A Reader.* Washington, DC: Island Press.

FLOODS

Rapidly rising water rushing through communities is a dreaded natural disaster. Water control systems such levees and dams evolve to mitigate floods. When these systems do not work and human environmental impacts mix with large water flows, controversy can ensue.

HUMAN CAUSES OF FLOODING

In many ecosystems seasonal flooding is a natural and anticipated event. In some tropical regions the forests remain flooded for months. Springtime flooding is common in many mountainous regions. Floods are also associated with other powerful natural disasters, such as hurricanes, tsunamis, and landslides. As unplanned or poorly planned development increases the amount of impervious surface areas (i.e., trading fields for roofs, roads, and parking lots) in our watersheds, the quantity and speed of stormwater runoff increase. As large parking lots associated with big-box retail development and shopping malls replace pervious surfaces with impervious surfaces, stormwater can mix with automobile emissions and wastes, and controversies around responsibility for water quality can develop.

Climate change could also change the frequency and severity of inland flooding, particularly near the mouths of rivers. Some climate change models suggest that some regions of the United States may have more rainfall that would increase river and lake levels. Combined with rising sea levels vulnerability to inland flooding could drastically increase. Moreover, increased flooding could occur even in areas that do not become wetter: According to the U.S. Environmental Protection Agency (http://yosemite.epa.gov) some of the impacts of climate change could be more flooding. The U.S. marine infrastructure—the docks, piers, levees, canals, seawalls, sewer and water pipes, waste treatment plants, and port buildings—would be vulnerable. Because the costs of any infrastructure changes would be enormous, most places resist making these changes. That is one reason why the forecast of global warming, and rising sea levels, and more inland flooding is a battleground. The EPA's climate change forecast vis-à-vis flooding is as follows.

1. Earlier snowmelt could worsen spring flooding while diminishing summer water availability;
2. Some climate models suggest wetter winters and drier summers;
3. The need to ensure summer/drought water supplies could lead water managers to keep reservoir levels higher, thereby limiting the capacity for additional water retention during unexpected wet spells;
4. Warm areas generally have a more intense hydrologic cycle and thus more rain in a severe storm; and
5. Many areas may receive more intense rainfall.

The most flood-prone communities of the United States are partly protected by levees and reservoirs. However, the large continuing loss of wetlands increases the threat of floods particularly in these communities. The typical flood-control system is designed to prevent floods with at least a 1 percent chance of occurring in any given year. However, these computations of risk of future floods do not take into consideration rising water levels due to climate change. Not only do water levels rise, but the power of tidal surges and storms increase. These weak flood-control systems are not designed for a long-term, gradual rise in water or an increase in powerful storms.

The cost of flood-protection infrastructure is very expensive. In the Netherlands, where much of the land is reclaimed from flooded sea plains, $14 billion a year is spent on just the pumps alone. Levees, levee maintenance, wetlands and wetlands preservation, dredging, and water monitoring are generally too expensive for small and medium-sized communities. The environmental impacts of flood-protection infrastructure can be significant. The wetlands and riparian areas that rely on the flood stage could be eliminated, which would dramatically affect species of plants and aquatic life dependent on these areas, such as salmon. Most communities are not prepared for a flood and have taken very few measures to prevent one. Most communities that have done anything for flood protection rely on land-use regulations to prevent flood damages. Almost 1,000 communities along rivers and lakes are part of the National Flood Insurance Program. This program is designed to preclude construction in most 100-year noncoastal floodplains. Again, however, these computations are made without incorporation of climatic change models, a small but continuing battleground. Changing climate will shift floodplain boundaries, so real estate and economic development may be taking place today in areas that will be in floodplains. Private property owners, developers, land speculators, and related finance industries all like to see building and growth. However, the risks are so high that private insurers will not accept them. When hurricanes hit, insurance companies may pay the insured for either wind or flood damage. When the damage is so devastating that it is difficult to separate wind and flood damage, bitter controversies ensue. The battleground for this controversy right now is the courts. Some Katrina (2005) victims have successfully sued their insurance companies for failure to pay for the damage. The insurance companies will not pay for flood damage if it is in a floodplain. High winds mark a hurricane but can also cause floods. Insurance and other industries seek relief from high judgments in state legislatures by supporting legislation that prevents juries from awarding high awards or awards for punitive damages.

FLOOD PROTECTION: ARE WETLANDS WORTH IT?

Wetlands are extremely important in most ecosystems. Water and soil meet to form the breeding grounds for plants, bacteria, and animals. Wetlands moderate water movement through ecosystems. They trap and slowly release surface water, rain, snowmelt, groundwater, and flood waters. Vegetation in riparian areas slows the speed of flood waters and distributes them more gradually over the floodplain. This combined water-storage and braking action moderates water flow, which can lower flood heights and flow rates and reduce erosion. The holding capacity of wetlands helps prevent waterlogging of crops. Another battleground in flood protection emerges with contrasting policy approaches. Can wetlands provide the level of flood control otherwise provided by dredge operations and levees? The problem is that these wetlands are owned by someone as private property. It can be a citizen, corporation, state, or arm of the federal government. Protecting their wetland functions often means no real estate development, agricultural use, economic development, or industrial development. The riparian wetlands along the Mississippi River once stored at least 60 days of floodwater, acting similar to a huge sponge. Now they store only 12 days' worth because most of the wetlands have been filled or drained to develop their profit potential.

THE FEDERAL FLOOD INSURANCE CONTROVERSY

Historically, flooding has been the most common natural disaster in the United States, costing more in property damages and human life than any other natural disaster. Congress created the National Flood Insurance Program (NFIP) in 1968. The object was to reduce future flood losses through flood-hazard identification, floodplain management, and insurance protection. NFIP coverage is available to all owners and occupants of insurable property in a participating community. Not all communities that experience floods or could experience floods with rising water levels are participating communities.

Property damage caused by flooding is explicitly excluded under most homeowner insurance policies sold in the private sector. Property insurance companies insist that flood insurance is not commercially profitable. As a general rule, property insurance markets will provide coverage when insurers are confident that they can identify the risk and set insurance rates that cover expected losses.

WHY NOT USE POROUS SURFACES? PRINGLE CREEK COMMUNITY

According to the National Weather Service, Oregon experienced record-breaking rainfall throughout the month of November 2006. The Portland metro area received 11.61 inches while Salem alone received more than 15 inches. Yet while many regional streets and sidewalks flooded as a result of clogged storm drains, Pringle Creek Community, a 32-acre sustainable living community located in the Willamette Valley, had no flooding due to the success of its state-of-the-art porous pavement or green street system.

With 7,000 feet of green streets and 2,000 feet of green alleyways, Pringle Creek is the nation's first full-scale porous pavement project. The use of porous pavement within Pringle Creek retains at least 90 percent of the rainwater and returns it to an aquifer. These environmentally-conscious landscaping initiatives are a leading model of eco-conscious construction and design.

Porous pavement retains stormwater runoff and replenishes local watershed systems. Pringle Creek's porous pavement system is used to capture rainwater. Rain seeps from the pavement into nearby gardens of plants selected specifically for their water-retention qualities in this climate. The resulting rain gardens allow stormwater runoff to infiltrate into the underlying soil, creating a natural water-retention structure for the entire community. The water-retention structure uses retention ponds, bioswales, and soil berms throughout the community. The streets are narrower than conventional roads and use fewer materials to create roads, curbs, and gutters.

"We are excited about the success of our green street system," said Don Myers, president, Sustainable Development, Inc. "The use of these sustainable materials has allowed our project team the ability to design and build an elegant solution to a complex problem. The use of porous pavement emphasizes our 'wholistic' approach to maintaining ecological and aesthetic benefits to create healthy homes and healthy lifestyles for all residents of Pringle Creek."

Repetitive Loss Problem: Controversial Solutions

According to the Federal Emergency Management Agency, a small number of properties account for a large proportion of paid flood claims. A total of 112,540 properties nationwide have sustained repetitive losses, but only 50,644 of these properties had insurance, as of September 30, 2004. Of these 50,644 repetitive loss properties (RLPs), 11,706 are considered severe repetitive loss properties (SRLPs) that were placed in FEMA's Target Group Special Facility. In total, there were 4,498,324 flood insurance policies, so RLPs are 1 percent of the total policies nationwide. Yet, according to FEMA, this 1 percent accounts for an annual average of 30 percent of amounts paid in claims. RLPs exist in all 50 states. The top 10 states accounted for 78 percent of all repetitive loss claims; and the top 25 states account for 96 percent of all repetitive loss claims. The majority of existing flood-prone structures are residences grandfathered into the National Flood Insurance Program when the program was created. These properties have been flooded and repaired multiple times.

Insurers generally lack the ability to spread flood risk sufficiently to safeguard their assets against catastrophic flood losses.

The first step in assessing a community's flood hazards is identifying and mapping the special flood-hazard areas. Flood maps provide the basis for establishing floodplain management ordinances (i.e., building standards), setting insurance rates, and identifying properties whose owners are required to purchase flood insurance.

ACCURACY OF FLOOD MAPS: GOOD ENOUGH FOR THE PRIVATE SECTOR?

An important policy issue for state and local officials, insurers, mortgage lenders, and property owners is that many flood maps have not been updated with detailed topography or more accurate methodologies and do not reflect real estate growth. They also do not account for building codes that mitigate flood damage, climate changes that affect water levels, or drainage capacities. Growth tends to increase runoff and alter drainage patterns on floodplains and, thus, increase flood-hazard risk.

FLOOD MAP MODERNIZATION

Flood map modernization is the locus of much scrutiny and some controversy. However, so little is known about the general environmental conditions of U.S. cities that until more is known this controversy will simmer. Accurate floodplain maps would show the true extent of development and real estate land use, including loss of wetlands and increase in impervious surfaces. An analysis of the local water systems could provide evidence to environmentalists who want to constrain growth, especially if they want to stop a particular project with large environmental impacts.

POTENTIAL FOR FUTURE CONTROVERSY

Flooding is part of many natural disasters, as well as being exacerbated by land development practices. The damage caused by floods is extensive and not always predictable or preventable. Controversies occur as a result of allocating liability for damage and for cleanup. Poor local land-use control and enforcement, inadequate local environmental knowledge, and insurance disputes mark the battleground for this controversy. Rising sea levels and inland water rises will also increase the intensity of flooding controversies.

See also Climate Change; Sprawl

Web Resources

Link Claimed between Disease, Floods, and Global Warming. Available at www.sepp.org/Archive/controv/controversies/linkclaim.html. Accessed January 21, 2008.

Ottowa Riverkeeper. Canadian Flood Plain Approaches. Available at ottawariverkeeper.ca/news/currents_of_controversy/. Accessed January 21, 2008.

Further Reading: Heathcote, Isabel. 1998. *Integrated Watershed Management: Principles and Practice.* NJ: John Wiley and Sons; Marsalek, J. 2000. *Flood Issues in Contemporary Water Management.* New York: Springer; Newson, Malcolm David. 1997. *Land, Water, and Development: Sustainable Management of River Basins Systems.* UK: Routledge; William, J. Mitsch, and James G. Gosselink. 2000. *Wetlands.* NJ: John Wiley and Sons.

GENETICALLY MODIFIED FOOD

Plants that are used for food and medicine can be genetically modified and patented.

WHAT ARE GENETICALLY MODIFIED (GM) FOODS?

Genetically modified food results from a special set of technologies that alter the genetic makeup of living organisms such as animals, plants, or bacteria. Biotechnology is a more general term and refers to using living organisms or their components, such as enzymes, to make products that include wine, cheese, beer, and yogurt. Combining genes from different organisms is known as recombinant DNA technology, and the resulting organism is called genetically modified, genetically engineered, or transgenic. Genetically modified products include medicines and vaccines, foods and food ingredients, feeds, and fibers.

Locating genes for important traits, such as those conferring insect resistance or pesticide receptivity, is the most challenging part of research in the process. Massive amounts of research are under way to set up genome characterization and sequencing for hundreds of different organisms and traits. They are creating detailed DNA maps.

GENETICALLY MODIFIED FOOD

Food is grown all over the world, although not all food can grow anywhere. In some places that would otherwise be good for food production enhanced local biodiversity can decrease productivity. Insects, animals, weeds, and other environmental factors can make it impossible to grow crops of any kind. The appeal of GM crops in these places is very high and is often the context of both subsistence and economic development. In 2003, about 167 million acres cultivated by seven million farmers in 18 countries were planted with GM crops. The main GM crop choices were herbicide- and insecticide-resistant soybeans, corn, cotton, and canola. Another crop grown commercially or field-tested is a sweet potato resistant to a common virus. Similar to the potato blight that swept Ireland and was never eliminated, the sweet potato virus could destroy most of the African harvest. GM rice with increased iron and vitamins may reduce chronic malnutrition in Asian countries. GM field testing of a variety of plants able to survive weather extremes continues.

In 2003, six countries grew 99 percent of the global transgenic crops. They were the United States (63%), Argentina (21%), Canada (6%), Brazil (4%), China (4%), and South Africa (1%). The growing of GM food is expected to increase in developing countries. The scale of GM projects could increase quickly. Some environmentalists and public health experts fear that too much growth too quickly could overlook destructive environmental and human health effects.

Global biotechnology crop acreage grew to 222 million acres in 2004. In 1996, when the first biotech crops were commercially grown, 7 million acres of biotech crops were grown worldwide. In 2004, a total of 222 million acres of biotech crops were planted in 21 countries by 8.5 million farmers. Of the 8.5 million farmers, 90 percent are resource-poor farmers in developing countries. Developing countries account for more than one-third of the global biotech crop acreage. Of the 21 countries growing biotech crops, five are in the European Union.

Food security is very important in most parts of the world. In some countries, food security can lead to mass population migrations and armed conflict. Genetically modified organisms (GMOs) also pose some risks, both known and unknown. Controversies surrounding GM foods and crops commonly focus on human and environmental safety, labeling and consumer choice, intellectual property rights, ethics, food security, poverty reduction, and environmental conservation. Each one of these areas is a battleground in this controversy.

GM PRODUCTS: BENEFITS AND CONTROVERSIES

The researchers and manufacturers of GM products claim there are many benefits depending on the specific product. Here is a brief listing of them.

- Crops
  - Enhanced taste and quality
  - Reduced maturation time
  - Increased nutrients, yields, and stress tolerance
  - Improved resistance to disease, pests, and herbicides
  - New products and growing techniques

- Animals

  - Increased resistance, productivity, hardiness, and feed efficiency
  - Better yields of meat, eggs, and milk
  - Improved animal health and diagnostic methods

- Environment

  - Friendly bioherbicides and bioinsecticides
  - Conservation of soil, water, and energy
  - Bioprocessing for forestry products
  - Better natural waste management
  - More efficient processing

- Community

  - Increased food security for growing populations

CONTROVERSIES

Others contest some of the claims made by GM product manufacturers. These are each a battleground and are briefly listed here.

- Safety

  - Potential human health impact: allergens, transfer of antibiotic resistance markers, unknown effects
  - Potential environmental impact: unintended transfer of transgenes through cross-pollination, unknown effects on other organisms (e.g., soil microbes), and loss of flora and fauna biodiversity

- Access and intellectual property

  - Domination of world food production by a few companies
  - Increasing dependence on industrialized nations by developing countries
  - Biopiracy: foreign exploitation of natural resources

- Ethics

  - Violation of natural organisms' intrinsic values
  - Tampering with nature by mixing genes among species
  - Objections to consuming animal genes in plants and vice versa
  - Stress for animals

- Labeling

  - Not mandatory in some countries (e.g., the United States)
  - Mixing GM with non-GM crops confounds labeling attempts

These controversies involve a trust in science to produce enough food safely. Our impacts on the environment are larger in scale then ever imagined. With profit as the primary goal of production, would the entire world have a safe food supply?

WHAT ARE THE CONTROVERSIES ABOUT GM FOODS?

A large controversy exists as to whether genetically modified foods should be labeled as such. Some contest industry claims that GM foods are the equivalent of ordinary foods not requiring labeling. They claim there are many dangers, hazards, and problems, all controversial. What is the actual record of GM foods? Has profit motivation decreased morality in food production? Is labeling enough?

> Recombinant DNA technology faces our society with problems unprecedented not only in the history of science, but of life on Earth. It places in human hands the capacity to redesign living organisms, the products of three billion years of evolution. Such intervention must not be confused with previous intrusions upon the natural order of living organisms: animal and plant breeding. . . . All the earlier procedures worked within single or closely related species. . . . Our morality up to now has been to go ahead without restriction to learn all that we can about nature. Restructuring nature was not part of the bargain. . . . this direction may be not only unwise, but dangerous. Potentially, it could breed new animal and plant diseases, new sources of cancer, novel epidemics. (Dr. George Wald, Nobel Laureate in Medicine, 1967 Higgins Professor of Biology, Harvard University)

About 25 percent of people in the United States have adverse reactions to foods. Eight percent of children and 2 percent of adults have food allergies. Genetically modified food manufacturers doing proprietary research may not want to find dangerous human health reactions after large financial investments. There is concern that they may know about them and compute the cost of death into their profit margins. Risk assessments become complicated and manipulated.

> Genetic Engineering is often justified as a human technology, one that feeds more people with better food. Nothing could be further from the truth. With very few exceptions, the whole point of genetic engineering is to increase sales of chemicals and bio-engineered products to dependent farmers. (David Ehrenfeld, Professor of Biology, Rutgers University)

POTENTIAL FOR FUTURE CONTROVERSY

Appropriate biotechnologies offer considerable potential for food security, but given the risks and uncertainties about the effect of GMOs on human health and the environment, environmentalists are concerned. Genetically modified organisms present a hotbed of controversy with scientists, ethicists, activists, farmers, and industry all engaged in lawsuits, legislation, and environmental policy development. Population increases, will drive demand for food, which will push this controversy even further. If climate change decreases land available for food production in areas of population growth, practical realities of survival will form the contours of this controversy.

See also Cumulative Emissions, Impacts, and Risks; Hormone Disrupters: Endocrine Disruptors; Human Health Risk Assessment

Web Resources

The Campaign. Available at www.thecampaign.org/. Accessed January 21, 2008.
Genetically Modified Food: News. Available at www.connectotel.com/gmfood/. Accessed January 21, 2008.
Human Genome Project Information. Genetically Modified Foods and Organisms. Available at www.ornl.gov/sci/techresources/Human_Genome/elsi/gmfood.shtml. Accessed January 21, 2008.

Further Readings: Lurquin, Paul F. 2002. *High Tech Harvest: Understanding Genetically Modified Food Plants.* Boulder, CO: Westview Press; McHughen, Alan. 2000. *Pandora's Picnic Basket.* Oxford: Oxford University Press; National Academy of Sciences. 2004. *Safety of Genetically Engineered Foods.* Washington, DC: National Academies Press; Toke, Dave. 2004. *The Politics of GM Food.* New York: Routledge.

GEOTHERMAL ENERGY SUPPLY

Geothermal energy supplies are themselves fairly noncontroversial given their current low usage. However, as citizens and environmentalists demand more clean power, alternative energy sources such as this will be used.

WHAT IS GEOTHERMAL ENERGY?

Our earth's interior provides heat energy from deep within its core. This heat is called *geothermal energy.* At Earth's center 4,000 miles deep, temperatures may reach over 9,000°F. The heat from the earth's core continuously flows outward and transfers to the surrounding layer of rock near the surface called the mantle. The mantle rock can melt and become magma. The magma rises slowly toward the earth's crust. Sometimes magma can reach the surface as lava. Usually the magma remains below earth's crust, heating nearby rock and water. Some of this hot geothermal water travels up through faults and cracks in the rocks and reaches the earth's surface as hot springs or geysers. Most of it stays deep underground. This natural collection of hot water is called the *geothermal reservoir.*

The size of the geothermal reservoir is unknown. Its role in the ecosystem is also unknown. This uncertainty makes environmentalists and others reach for more information.

HOW GEOTHERMAL ENERGY IS ATTAINED

Wells are drilled through the rock into the geothermal reservoirs to bring the hot water to the surface. Locating underground areas that contain these geothermal reservoirs takes a team of experts. The hot water and/or steam is used to generate electricity in geothermal power plants. In geothermal power plants, steam, heat, or hot water from geothermal reservoirs spins the turbine generators and produces electricity. Ideally, the used geothermal water is returned to the reservoir.

BENEFITS

Geothermal power is considered a renewable energy source, as opposed to petrochemical energy sources. Geothermal plants do not have to burn fuels, such as coal, to manufacture steam to turn the turbines. Because they do not have to burn coal or other fuels, they pollute less. Geothermal power plants are very efficient and designed to run 24 hours a day, all year. A well-designed geothermal power plant sits directly over its energy source and therefore is more reliable than other energy sources. Other energy sources can be interrupted by weather, natural disasters, or political controversies that can stop transportation of fuels. Geothermal energy is also relatively inexpensive to set up and operate. Environmentalists hope that geothermal projects can help developing countries economically develop without reliance on nonrenewable energy sources.

HOW MUCH ELECTRICITY IS GEOTHERMAL?

The use of geothermal energy for electricity has grown worldwide to about 7,000 megawatts in 21 countries. The United States produces 2,700 megawatts of electricity from geothermal energy. That is comparable to burning sixty million barrels of oil each year.

HOW CLEAN IS GEOTHERMAL ENERGY?

One battleground for all energy sources is the amount of pollution they generate. Coal, gas, and oil as energy sources all generate powerful pollutants. According to the Geothermal Energy Trade Association, emissions of nitrous oxide, hydrogen sulfide, sulfur dioxide, particulate matter, and carbon dioxide are extremely low, especially when compared to fossil fuel emissions. The geothermal plant produces nearly zero air emissions.

Lake County, California, has met all federal and state ambient air quality standards for 18 years. It is downwind of the geysers, which normally cause poor air quality because of natural sulfur emissions. Since geothermal power has tapped into the heat source for the geysers, air quality has improved, because hydrogen sulfide passes through an abatement system at the geothermal plant that reduces hydrogen sulfide emissions by 99.9 percent. In nature, it would be released into the air by the numerous geysers, hot springs, and fumaroles.

Geothermal power plants emit very low levels of one of the most significant gases known to induce global warming: carbon dioxide. According to the Energy Information Administration (EIA), carbon dioxide accounts for 83 percent of U.S. greenhouse gas emissions. Geothermal power plants emit only a small fraction of the carbon dioxide emitted by traditional power plants on a per-megawatt-hour basis and can help reduce the overall release of carbon dioxide into the atmosphere.

According to the U.S. Department of Energy, geothermal energy uses less land than other energy sources, both fossil fuel and renewable. No transportation of geothermal resources is necessary, because the resource is tapped directly at its source. In terms of land impacts, a geothermal facility uses 404 square meters of land per gigawatt hour, while a coal facility uses 3,632 square meters per gigawatt hour. There are different types of geothermal plants, with different emissions. Geothermal plants use five gallons of freshwater per megawatt hour, while binary air-cooled plants use no freshwater. This compares with 361 gallons per megawatt hour used by natural gas facilities.

POTENTIAL FOR FUTURE CONTROVERSY

This controversy, similar to many alternative energy controversies, is only starting to play out because of the extreme dependence of the United States on oil. Many view the current political and financial structure of utilities as exclusionary and oppressive of alternative energy entrepreneurs. If government acquisition of geothermal wells results in taking of private property, there could be those controversies as well.

Some sector of society, public or private, needs to assist in the development of alternative and renewable energy sources. Fundamental relationships between utilities, communities, and the states may change. These are all powerful stakeholders with long-term interests. Not all communities will have the option of geothermal energy. Of those that do, some may not want it. Some communities with deep geothermal reservoirs could seek to export the electricity for

---

GEOTHERMAL ENERGY AND HEALTH CARE SAVINGS?

Geothermal energy can reduce health impacts and health care costs. A recent analysis assesses the health impacts related to power plant emissions. Reducing power plant nitrogen emissions by one million tons and sulfur emissions by four million tons as of 2010 would mean:

- The number of related deaths would be reduced by 8,714, with an associated health care savings of almost $53 million.
- The number of related cases of chronic bronchitis would be reduced by 5,997, with an associated health care savings of almost $2 million.
- The number of related heart attacks would be reduced by 13,924, with an associated health care savings of almost $2 million.

profit. The same could be said of private property owners. There are many unanswered questions about energy policy generally. Environmentalists are uncomfortable with the lack of information about environmental impacts. The powerful stakeholders, lack of policy formation, and uncertainty about environmental impact information all contribute to an uneasy battleground. Geothermal energy production may be the first type of energy production that will challenge the status quo. The rising concern about sustainability will also renew many aspects of this battleground, especially local control of environmental decisions.

See also Mining of Natural Resources; Sustainability; "Takings" of Private Property under the U.S. Constitution

Web Resources

Geothermal Energy Association. Available at www.geo-energy.org/aboutGE/basics.asp. Accessed January 21, 2008.

U.S. Environmental Protection Agency. Geothermal Power. Available at www.eia.doe.gov/kids/energyfacts/sources/renewable/geothermal.html. Accessed January 21, 2008.

Further Reading: Chandrasekharam, D., and J. Bundschuh. 2002. *Geothermal Energy (Resources) for Developing Countries.* Oxford, UK: Taylor Francis; Elliott, David. 2003. *Energy, Society and Environment.* New York: Routledge; European Renewable Energy Council. 2004. *Renewable Energy in Europe: Building Markets and Capacity.* London: James and James/Earthscan; Smeloff, Ed, and Peter Asmus. 1997. *Reinventing Electric Utilities: Competition, Citizen Action, and Clean Power.* Washington, DC: Island Press; Sterrett, Frances Sterrett S. 1994. *Alternative Fuels and the Environment.* New York: CRC Press; Tester, Jefferson W. 2005. *Sustainable Energy: Choosing Among Options.* Boston: MIT Press.

GLOBAL WARMING

The topic of global warming is one that engenders intense controversy. The controversy has shifted from whether it exists to its extent and what to do about it.

The concept of global warming is controversial in its inception. The scale of human impact has never been so large and dangerous. The rising ocean levels will put many coastal communities and natural areas under water. Whole ecosystems and cultures are at risk. The extended reach of a range of natural disasters combined with ever-increasing human population increases the cost of disasters and creates more controversy. The cost of natural disasters can be large, and controversies ensue regarding emergency planning; federal, state, local, and private responsibility; and cleanup. Insurance companies, governmental agencies, energy industries, and impacted communities all have a large stake.

In the global context, the industrialized nations of the north, such as the United States, Germany, and Japan, are considered to have reaped the benefits of the type of industrial economic development that is partially the cause of global warming. The less-developed nations want the same economic benefits and want to use polluting energy sources to attain them. In the U.S. context, global warming will affect where we farm, fish, live, and work. Some of the Arctic nations have begun to work collaboratively on the issue because of the dramatic effect it has on them. As the ice caps melt, the salinity of the water decreases, which affects the entire aquatic ecosystem. The permafrost begins to melt. Some, including former vice president Al Gore, believe that there are large methane pockets in the Arctic tundra, and that once they melt they will release much larger amounts of methane than previously assumed in all climate change models. Methane is a powerful greenhouse gas. Methane traps over 21 times more heat per molecule than does carbon dioxide. The concern is that climate change could occur much more rapidly than predicted.

The Earth's surface temperature has risen by about 1°F in the past century. There is accelerated warming in the past two decades in the context of centuries. Most of the global warming over the last 50 years is attributable to human activities. Human activities have altered the chemical composition of the atmosphere through the buildup of greenhouse gases. Much is unknown about exactly how Earth's climate responds to them. There is no doubt this atmospheric buildup of carbon dioxide and other greenhouse gases is the result of human activities, although this is an early historic battleground.

There are other sources of methane emissions. Ungulates such as cattle and horses generate about 16 percent of global emissions of methane. Some environmentally leaning farmers and the University of Wales in Aberysywyth are experimenting with feeding their sheep and cattle garlic. Early reports are that methane emissions are down 50 percent. There are other natural sources of carbon dioxide emissions. An estimated 60 billion metric tons of carbon dioxide are released globally from the soil to the air, much of it facilitated by microbes, bacteria, and insects. A particular species of worm in the tropics is attracting current attention. *Pontoscolex corethrurus* is a small earthworm that is rapidly spreading through many equatorial forests. It seems to thrive in areas that have recently been deforested. Some researchers estimate that it can release carbon dioxide from the soil 20 to 30 percent faster than other worms. Humans do exert powerful control over the environment, and these can result in increases in non-industrial emissions from these activities, such as cattle grazing and deforestation of equatorial forests with the loss of trees and the increase in aggressive composting worms.

Scientific understanding of these and other factors remains incomplete. One battleground is the cooling effects of pollutant aerosols. The political world voiced its strong concern when, in 1987, many countries signed the Montreal Protocol on Substances That Deplete the Ozone Layer to limit the production and importation of a number of CFCs. The United States and other countries renewed their pledges to phase out ODSs by signing and ratifying the Copenhagen Amendments to the Montreal Protocol in 1992. Under these amendments, these countries committed to ending the production and importation of halons by 1994 and of CFCs by 1996. The Kyoto Treaty was not signed by the United States partially because of controversy around global warming.

Many controversies in science are related to uncertainty and the proof necessary to show causality. These controversies can become magnified in politics, especially when the interests of vested and powerful stakeholders are threatened. Early scientific controversies concerned whether global warming was actually occurring. These controversies were later echoed in political circles. In exploring issues of global warming, scientists gathered data from what sources were available about greenhouse gases in a way that allowed them to evaluate their impact. Scientists and engineers developed the concept of a global warming potential (GWP) to compare the ability of each greenhouse gas to trap heat in the atmosphere relative to another gas. The dynamic interaction of gases in the atmosphere is still an area of cutting-edge scientific research and controversy. The definition of a GWP for a particular greenhouse gas is the ratio of heat trapped by one unit mass of the greenhouse gas to that of one unit mass of carbon dioxide over a specified time period.

HISTORICAL RECORDS OF WEATHER AND GLOBAL WARMING

The historical record of weather and climate is based on analysis of air bubbles trapped in ice sheets. By analyzing the ancient air trapped in these bubbles trends in climate change can be observed over very long time periods. This record indicates that methane is more abundant in the Earth's atmosphere now than at any time during the past 400,000 years. Since 1750, average global atmospheric concentrations of methane have increased by 150 percent from approximately 700 to 1,745 parts per billion by volume (ppbv) in 1998. Over the past decade, although methane concentrations have continued to increase, the overall rate of methane growth has slowed. In the late 1970s, the growth rate was approximately 20 ppbv per year. In the 1980s, growth slowed to 9–13 ppbv per year. The period from 1990 to 1998 saw variable growth of between 0 and 13 ppbv per year. An unknown factor, and developing battleground, is the amount of methane in the polar ice caps. It is unknown how much methane in a currently inert form could be released as a gas into the atmosphere if the ice caps melt into water. If a large amount of polar methane gas is released faster than current models predict, then global warming could accelerate.

ATMOSPHERIC MEASUREMENT DATA: GOOD ENOUGH FOR GLOBAL WARMING MEASURES?

As international concern over global warming has mounted, pressure on the scientific community to come up with useful answers has increased. Governments have responded with an increase in monitoring and analysis of meteorological data. The U.S. Global Change Research Program (USGCRP) has identified as a priority research activity the development of global monitoring sites to measure atmospheric methane levels. The USGCRP provides access points to atmospheric measurement data related to methane. The National Oceanic and Atmospheric Administration's (NOAA) Climate Monitoring and

Diagnostics Laboratory (CMDL) Carbon Cycle Greenhouse Gases group also makes ongoing atmospheric measurements from land and sea surface sites and aircraft, as well as continuous measurements from baseline observatories and towers. Measurement records from international laboratories are integrated and extended to produce a globally consistent cooperative data product called GLOBALVIEW. The Carbon Dioxide Information Analysis Center (CDIAC) also provides access points to atmospheric measurement data related to methane. CDIAC's data holdings include records of the concentrations of carbon dioxide and other radioactively active gases in the atmosphere and the role of the terrestrial biosphere and the oceans in the biogeochemical cycles of greenhouse gases.

These government sources are often publicly available. In providing the public with access to these data, the government has included the community as a stakeholder. As the global warming controversy heated up after the U.S. refusal to sign the Kyoto Treaty many U.S. towns and cities adopted and ratified the principles behind the treaty, in the United States and all over the world. Global warming as a battleground expanded the environmental slogan "Think globally, act locally" to include many communities. As such, the battleground for global warming controversies often developed locally, especially around air pollution and nuclear radiation issues.

EMISSIONS INVENTORIES: DO THEY ANSWER THE QUESTION OF WHAT IS IN THE AIR?

Most U.S. environmental policy is relatively young, with the U.S. Environmental Protection Agency (EPA) forming in 1970. With many new pollution laws and a blank slate to work from, the EPA and state environmental agencies began by taking an inventory of emissions sources. Emissions inventories have evolved into policy uses of many types. They are used to help answer the question of what is in our air. With the rise in concern about global warming and scientific research into the dynamic properties of atmospheric gases proceeding, emissions inventories have increased in significance and specialization. Generally, an emissions inventory is an accounting of the amount of air pollutants and chemicals discharged into the atmosphere. It is generally characterized by the following factors.

- the chemical or physical identity of the pollutants included
- the geographic area covered
- the institutional entities covered
- the time period over which emissions are estimated
- the types of activities that cause emissions

Emission inventories are developed for a variety of purposes. Inventories of natural and human emissions are used by scientists as inputs to air quality models, by policy makers to set standards, and by facilities and regulatory agencies to establish compliance records with allowable emission rates. Because they can be used for purposes of a permit approving air emissions, industries have a vested interested in their accuracy. Many small industries remain under the radar, and large industries often complain they unfairly bear the brunt of environmental regulation. Some industries are known to keep emissions per facility just beneath the regulatory threshold and to move or build another facility that is also just under the regulatory radar rather than be regulated. If they expanded a current facility they would reach an emissions level that would require a permit. Most permits do not limit emissions substantially; some even allow them to increase. This is a persistent complaint from environmentalists about government enforcement of environmental law. The emissions inventories can often be incomplete, but they do provide valuable information in evaluating environmental impacts on the climate. Longitudinal emissions inventories can point to areas where accumulated environmental impacts can occur. Environmental impacts with human health consequences can occur when emissions form a sink.

WHAT ARE SINKS: ECOSYSTEMS AND GLOBAL WARMING

A sink is a reservoir that uptakes a chemical element or compound from another part of its cycle. For example, soil and trees tend to act as natural sinks for carbon—each year hundreds of billions of tons of carbon in the form of carbon dioxide are absorbed by oceans, soils, and trees. When sinks form in urban areas, as in air pollution, serious human health consequences can result.

POTENTIAL FOR FUTURE CONTROVERSY

Rising global temperatures are expected to raise the sea level and alter the climate. Changing regional climates alter forests, crop yields, and water supplies. The most recent reports from the United Nations Food and Agriculture Organization indicate that wheat harvests are declining because of global warming. Some experts have estimated that there will be a 3–16 percent decline in worldwide agricultural productivity by 2080. Population is expected to rise, however. In a study by environmental economist William Cline for the Centre for Global Development and Petersen Institute for International Economics, released September 12, 2007, concern was raised because the poorest nations may receive the most environmental impact from global warming. Many of these countries will not have the capacity to feed their populations or to adequately control pollution. This is a bitter point of controversy because the poor nations believe that the rich northern nations developed a high quality of life at the expense of the global environment. Some countries will be dramatically affected. India is projected to lose 29–38 percent of its agricultural productivity due to global warming. Cline states that

> Poor countries already have average temperatures near or above crop tolerance levels and, assuming no adequate reduction in greenhouse gases, seem to suffer an average 10–25 percent decline in agricultural productivity by the 2080s.

Most of the United States is expected to warm. This report predicts that U.S. agricultural productivity could decline as much as 6 percent or increase up to

8 percent. There is likely to be an overall trend toward increased precipitation and evaporation, more intense rainstorms, and drier soils. Many of the most important environmental impacts depend on whether rainfall increases or decreases, which cannot be reliably projected for specific areas.

The political controversies around global warming focus on the results of climate changes. Equity, value clashes, and uncertainty all promise to make climate change controversial. Some may gain and others will lose. There may be geopolitical shifts in world power. All efforts toward sustainability will be affected by global warming. Closer attention to environmental conditions of the climate, earth, land, air, and water is required to accommodate increasing population growth and concomitant environmental impacts. International treaties, such as Kyoto, will again seek accord. Both the effects and causes of global warming promise to be a continuing controversy.

See also Citizen Monitoring of Environmental Decisions; Climate Change; Evacuation Planning for Natural Disasters; Rain Forests

Web Resources

Global Warming International Center. Available at globalwarming.net/. Accessed January 21, 2008.

Natural Resources Defense Council. Issues: Global Warming. Available at www.nrdc.org/globalWarming/default.asp. Accessed January 21, 2008.

U.S. Environmental Protection Agency. Climate Change. Available at www.epa.gov/climatechange/. Accessed January 21, 2008.

Further Reading: Comby, Jacques, and Marcel Leroux. 2005. *Global Warming—Myth or Reality?* New York: Springer; Drake, Frances. 2000. *Global Warming: The Science of Climate Change.* New York: Oxford University Press; United Nations Environment Programme. 1997. *Global Environment Outlook.* Oxford, UK: Oxford University Press; Victor, David G. 2001. *The Collapse of the Kyoto Protocol and the Struggle to Slow Global Warming.* Princeton, NJ: Princeton University Press.

GOOD NEIGHBOR AGREEMENTS

Many communities want to make certain any industry in their neighborhood does not harm them with pollution. Some communities seek only green or sustainable industries. Environmental controversies exist in terms of the enforceability of such agreements.

Communities are now aware of the environmental impact of industries. There is little corporate accountability for decisions that affect local communities once an industry is allowed to set up operation. Industries self-report their emissions to the state environmental agency. Some of the larger industries are listed on the Toxics Release Inventory (TRI), which has mobilized community groups. The TRI, as well as increasing concerns about cumulative impacts, push communities to increase industrial environmental accountability. Community control over its health and environment includes land use and zoning, permits, and nongovernmental organization (NGO)-company contracts.

Good neighbor agreements are a form of flexible, community-based environmental protection. Agreements are formally negotiated; although some remain voluntary and without legally binding language, others are incorporated as a condition of formal permitting processes and can be legally enforced. Technically, a good neighbor agreement is an enforceable contract that details a set of commitments that the industry is required to make in order to reduce its environmental impacts on the community. There are few judicial challenges so far. Some agreements also focus on jobs that are promised and on labor conditions. Usually, an agreement contains specific commitments. Some of the commitments, for example, are

- no polluting discharge to groundwater
- oversight committee comprised of community, employee, and company representatives provided with funds for independent technical review
- some shared monitoring of emissions
- community notice of employment opportunities, youth apprenticeship, and training

CONTROVERSY AND GOOD NEIGHBOR AGREEMENTS

Good neighbor agreements usually come out of local controversy about pollution and job loss. They are relatively new, and some question their validity in the courts. It is not easy to obtain such agreements. Communities must be well organized and focused and should forge an alliance with the workers in the plant and their unions. While some industries may be willing to sign a voluntary nonbinding agreement with a well-organized and active community organization in order to defuse public agitation, an enforceable contract is much more difficult to obtain. In addition to community mobilization, some form of catalyst is usually needed to leverage corporations to move toward a meaningful agreement. This community point of intervention is usually an environmental or land-use permit requirement, which gives communities the opportunity to intervene.

For example, a change in environmental regulations requiring a permit change for an incinerator in a Rhone-Poulenc chemical plant in Manchester, Texas, followed by a sulfur dioxide release that sent 27 people to the hospital, gave the local citizens the leverage to get a good neighbor agreement. The catalyst for an agreement with several oil refineries in California was the need for a permit to modify the plant to meet new requirements. With increased citizen monitoring of environmental decisions, there are more opportunities for citizen engagement.

Leverage can also be a lawsuit. After the city of Philadelphia cited the Sun Oil Company for violations of local community health standards and violations of the federal Clean Air Act, a coalition of citizens, a labor union, and an environmental organization sued Sun and won a good neighbor agreement and a $5 million upgrade of the plant as a settlement.

THE UNOCAL AGREEMENT

Communities formed a working coalition in California to address neighborhood issues with UNOCAL. The outcome of the project was a negotiated good neighbor agreement that addressed a number of areas of concern including the following:

- There will be an independent safety and environmental audit paid for by UNOCAL. A community-based committee provides oversight for the audit, which includes internal safety management as well as emergency response and notification systems;
- The company will disclose several documents including health risk assessments and pollution monitoring results;
- UNOCAL agreed to implement a sulfur-monitoring station, develop a database of health effect information for chemicals used at the facility, and provide for appropriate landscaping on its property;
- The company agreed to provide $4.5 million to the county for transportation infrastructure and $300,000 a year to a community benefits fund for 15 years, for local citizen groups to utilize on projects that they designate to address local needs.

The community groups secured these commitments from UNOCAL by a concerted campaign in which state, local, and county officials were held accountable to withhold all approvals until a good neighbor agreement was signed. UNOCAL needed a local land-use approval to expand its plant; this and other government approvals were blocked in anticipation of a good neighbor agreement responding to community concerns. In other instances, various forms of public regulatory authority have been exercised in order to secure good neighbor–type commitments. The ability of the community organizations to assess potential environmental, health, and safety issues and to make recommendations is often a critical part of a good neighbor agreement. Numerous arrangements across the country have been reached between local industries and local community members to allow citizens to evaluate plants along with their experts and to ensure to their own satisfaction that state-of-the-art technologies and management practices are being applied.

POTENTIAL FOR FUTURE CONTROVERSY

Good neighbor agreements are relatively new. In a way they are an expression of community-based environmental planning. Communities now have a rapidly expanding knowledge base about environmental pollution and public health consequences. Technology, such as the Toxics Release Inventory, help disclose industrial emissions in the face of resistant state and federal environmental agencies. Some states protect industries from disclosure with secret audit privileges for industry. Cumulative impacts, environmental injustice, the environmental vulnerability of urban areas, and a new push for sustainability motivate communities to organize and use whatever power they have in whatever process they can. Good neighbor agreements generally arise from clusters of controversies, but do not have to.

As good neighbor agreements move to more formal contracts and heightened legal enforceability, more controversy will ensue. An industry cannot evade environmental responsibilities under the law with these agreements but is free to contract to do more. However, the question then becomes, who is the other contracting party? Often it is a community organization, or group of grassroots organizations. There are some difficult and publicly unasked questions. Industry is concerned that they may still be liable to citizen suits from environmentalists, especially once all their information is public or if the community changes its mind. Communities do want to be able to alter their priorities once all the environmental information is in, and as community leadership and environmental capacity evolve. Environmentalists do not want communities waiving statutory rights to sue. Unasked questions about industry enforcement of the contract against the community could also create controversy.

See also Citizen Monitoring of Environmental Decisions; Community-Based Environmental Planning; Public Participation/Involvement in Environmental Decisions; Toxics Release Inventory

Web Resources

Macey, Gregg P. "Seeking Good Neighbor Agreements in California." Available at http://www.epa.gov/compliance/resources/publications/ej/annual-project-reports/cbi-case-study-good-neighbor-california.pdf. Accessed January 21, 2008.

Further Reading: Boyce, James K., and Barry G. Shelley. 2003. *Natural Assets: Democratizing Ownership of Nature.* Washington, DC: Island Press; Gready, Paul. 2004. *Fighting for Human Rights.* New York: Routledge; Martello, Marybeth Long, and Sheila Jasanoff. 2004. *Earthly Politics: Local and Global in Environmental Governance.* Cambridge, MA: MIT Press; Ryder, Paul. 2006. *Good Neighbor Campaign Handbook: How to Win.* Lincoln, NE: iUniverse.

H

HEMP

Hemp is a renewable plant that it is criminally illegal to grow in the United States. Proponents tout hemp as a substitute for paper or wood and a source of clothing and a biodegradable form of plastic. Hemp is in the same general family as marijuana, which is illegal. Some fear legalizing hemp is a way to legalize marijuana.

Hemp has a long history of use all over the world. Ropes, clothes, and paper often came from hemp in the colonial United States. The recent increase in interest in sustainability has prompted an increased interest in legalizing hemp. Hemp is considered a renewable and organic crop with many uses. The environmental impacts of growing large amounts of hemp have not been fully explored, but some say these impacts can negate the environmental benefits of hemp.

ENVIRONMENTAL BENEFITS OF HEMP

Hemp is used for clothes and paper. Unlike cotton, hemp does not need pesticides. So far it is naturally resistant to pests. Hemp can grow so fast it crowds out other weeds. In many places, hemp is considered a weed. To make paper, trees must grow for many years, be logged, transported, skinned, and stored. It is a risky, labor-intensive process. All through this lengthy process the tree and then wood is subject to risks from fire, insects, and other natural disasters. It may take months and years from harvesting the tree to paper processing. Then this wood is chemically treated. Very powerful chemicals, including dioxin, are used to bleach the wood into paper. Paper mills are known for their large discharges of chemicals into waterways. Hemp, in contrast, can be harvested in a few months. Proponents claim that hemp can yield more paper with lower production costs and environmental impacts than it takes to make paper from trees. Some environmentalists note that there are other renewable sources of paper, such as banana leaves. Overall, most environmentalists would like to reduce paper usage because of its environmental impacts from creation to waste disposal. Some have principled objections to logging, or to logging old-growth timber. The creation and use of paper is an environmental battleground that facilitates hemp controversies.

HEMP IS ILLEGAL

In 1970 the U.S. Congress designated hemp, along with its relative marijuana, as a Schedule 1 drug under the Controlled Substances Act. This makes it a criminal act to grow it without a license from the U.S. Drug Enforcement Administration (DEA). Most scientists concede that industrial hemp does not contain enough psychoactive ingredients to make a user high, but it is still criminally illegal. Some do argue that legalizing hemp will allow marijuana to become legal. Marijuana is illegal because the effects these psychoactive ingredients can have on human behavior. Medical marijuana is a large controversy but not a hemp battleground.

The United States is the only developed country that does not allow hemp as an agricultural crop. The European Union has subsidized hemp production since the 1990s. Canadian farmers have been exporting huge amounts into the United States. This is not illegal. In 2005, the Canadian hemp industry tripled the amount of acreage dedicated to hemp production to meet rising U.S. demand.

HEMP: THE CONTINUING CONTROVERSY

State legislatures have become a battleground for hemp legalization. Advocates for hemp legalization come from a variety of backgrounds and often form effective coalitions in state legislatures. Farmers, and to a lesser extent agribusiness interests, have been actively involved. Many states with strong agricultural interests want the federal government to decriminalize it. State legislatures in Hawaii, Kentucky, Maine, Montana, North Dakota, and West Virginia have all passed laws that would make hemp legal if the U.S. government were to allow it. Nationally, resistance to hemp decriminalization remains dominant. A hemp farming bill introduced into Congress in 2006 by Texas Republican Ron Paul failed due to opposition from the DEA and the White House. The DEA maintains that allowing U.S. farmers to grow hemp would undermine the so-called war on drugs, as "marijuana growers could camouflage their illicit operations with similar-looking hemp plants."

Even if hemp production were legalized it would be difficult for U.S. farmers to compete in current world markets. World hemp production is dramatically down from the early 1980s and is dominated by low-cost producers. China, India, and Russia produce 70 percent of the world supply of hemp. Market risk to the U.S. farmer may be prohibitive, even if it were legal, because of these cheaper international growers.

Critics of the agricultural production of hemp point out that hemp farming is very demanding on the environment. Large areas of cultivated fields would be necessary. Irrigation would become necessary in some areas for best production. Hemp requires large amounts of nutrients.

POTENTIAL FOR FUTURE CONTROVERSY

The history of hemp use, its presence in global markets, and its potential use in sustainable programs all make its current illegality controversial. Hemp has many applications as food, clothing, paper, oils, ropes, and even soap. Each area of hemp potential will be controversial as government regulatory agencies struggle with enforcement of its illegality.

The environmental impacts of growing hemp will also be controversial. Proponents claim that it can help reduce global warming because it takes out large amounts of carbon dioxide per acre, more than most plants. Hemp can be grown in many soils, but its impact on the soil depends on the type of soil. Hemp growing does require water. To the extent it takes necessary water away from other parts of the ecosystem it could have negative environmental impacts. Cotton, one of hemp's main fiber competitors on the world market, may overcome its current pesticide issues with genetically modified crops.

See also Pesticides; Sustainability

Web Resources

Canadian Hemp Trade Alliance. Available at www.hemptrade.ca/. Accessed January 21, 2008.

North American Industrial Hemp Council. Industrial Hemp Research Index. Available at naihc.org/IndustrialFibers.html. Accessed January 21, 2008.

Vantreese, Valerie. "Industrial Hemp: Global Markets and Prices." Available at www.industrialhemp.net/pdf/abs_hemp.pdf. Accessed January 21, 2008.

Further Reading: Allen, James Lane. 2004. *Hemp: The Reign of Law: A Tale of the Kentucky Hemp Fields*. MT: Kessinger Publishing; MacAllister, William B. 1999. *Drug Diplomacy in the Twentieth Century: An International History*. UK. Routledge; Ranalli, Paolo. 1999. *Advances in Hemp Research*. New York: Haworth Press.

HORMONE DISRUPTORS: ENDOCRINE DISRUPTORS

The issue of hormone disruption is a large environmental controversy with many emerging environmental battlegrounds. Some scientists believe that a large number of the chemicals currently found in our air, water, soil, and food supply, or added to livestock to increase growth, have the ability to act as hormones when ingested by human beings and wildlife. Estrogen-mimicking growth hormones in chicken have been linked to early onset of menses and an increase in female attributes in male humans. The decrease in human sperm counts over the last 50 years may be due to hormone-disrupting chemicals. These chemicals increase profits for livestock producers and chemical manufacturers but also increase food security and provide public health protection from other environmental risks.

HUMAN REACTIONS TO CHEMICALS IN THE ENVIRONMENT

There are many potentially dangerous chemicals in the environment, both manmade and naturally occurring. There is also a range of human reactions to environmental stressors such as these chemicals. The human body has many exposure vectors, such as skin absorption, ingestion, and breathing. There is a large variation in response to chemicals in heterogeneous populations like in the United States. This makes it very difficult to predict what dose of a chemical is safe enough for use in different applications, adding fuel to the overall controversy. Children take in more of their environment as they grow than they do when they reach adulthood.

Once in the body, chemicals travel and interact with various bodily systems before they are excreted. Throughout the body there are hormone receptor sites designed specifically for a particular hormone, such as estrogen or testosterone. Hormones are produced primarily in the pituitary gland. Once attached, the hormone controls cell maturation and behavior.

The controversy about endocrine disruption begins because many common chemicals have molecular shapes that are similar to the shapes of many hormones. This means that these chemicals can fit themselves into cellular receptor sites. When this happens, the chemical either prevents real hormones from attaching to the receptor or alters the cell behavior and/or maturation. The cell growth is highly disrupted. Which chemicals have this effect, and in what dosages, is a battleground for every chemical in commerce in the United States. However, only about 2 percent of the chemicals sold in the United States are tested for public health safety. Even these tests are industry controlled, critics claim, and their results not based on vulnerability (children) or dose-response variations. Of particular concern are chemicals that contain chlorine. Chlorine-containing chemicals are themselves part of a larger class called *persistent organic pollutants* (POPs), which are also believed to be potential hormone disruptors. They pose a greater risk because they persist, or last, in the environment a long time, taking longer to break down into nondangerous components. Because they take longer to break down, they increase exposure times and vectors to humans. Because exposure is increased, the risk from these chemicals, especially endocrine disruptors, is considered greater.

ENDOCRINE DISRUPTORS: HOW DO THEY DISRUPT?

Endocrine disruptors are externally induced chemicals that interfere with the normal function of hormones. Hormones have many functions essential to human growth, development, and functioning. Endocrine disruptors can disrupt hormonal function in many ways. Here are some of them.

1. Endocrine disruptors can mimic the effects of natural hormones by binding to their receptors.

2. Endocrine disruptors may block the binding of a hormone to its receptor, or they can block the synthesis of the hormone. Finasteride, a chemical used to prevent male pattern baldness and enlargement of the prostate glands, is an antiandrogen, since it blocks the synthesis of dihydrotestosterone. Women are warned not to handle this drug if they are pregnant, since it could arrest the genital development of male fetuses.

3. Endocrine disruptors can interfere with the transport of a hormone or its elimination from the body. For instance, rats exposed to polychlorinated-biphenyl pollutants (PCBs; see following) have low levels of thyroid hormone. The PCBs compete for the binding sites of the thyroid hormone transport protein. Without being bound to this protein, the thyroid hormones are excreted from the body.

Developmental toxicology and endocrine disruption are relatively new fields of research. While traditional toxicology has pursued the environmental causes of death, cancer, and genetic damage, developmental toxicology/endocrine disruptor research has focused on the roles that environmental chemicals may have in altering development by disrupting normal endocrine function of surviving animals.

ENVIRONMENTAL ESTROGENS

There is probably no bigger controversy in the field of toxicology than whether chemical pollutants are responsible for congenital malformations in wild animals, the decline of sperm counts in men, and breast cancer in women. One of the sources of these pollutants is pesticide use. Americans use some two billion pounds of pesticides each year, and some pesticide residues stay in the food chain for decades. Although banned in the United States in 1972, DDT has an environmental half-life of about 100 years. Recent evidence has shown that DDT (dichloro-diphenyl-trichloroethane) and its chief metabolic by-product, DDE (which lacks one of the chlorine atoms), can act as estrogenic compounds, either by mimicking estrogen or by inhibiting androgen effectiveness. DDE is a more potent estrogen than DDT, and it is able to inhibit androgen-responsive transcription at doses comparable to those found in contaminated soil in the United States and other countries. DDT and DDE have been linked to such environmental problems as the decrease in the alligator populations in Florida, the feminization of fish in Lake Superior, the rise in breast cancers, and the worldwide decline in human sperm counts. Others have linked a pollutant spill in Florida's Lake Apopka (a discharge including DDT, DDE, and numerous other polychlorinated biphenyls) to a 90 percent decline in the birthrate of alligators and to the reduced penis size in the young males.

Dioxin, a by-product of the chemical processes used to make pesticides and paper products, has been linked to reproductive anomalies in male rats. The male offspring of rats exposed to this planar, lipophilic molecule when pregnant have reduced sperm counts, smaller testes, and fewer male-specific sexual behaviors. Fish embryos seem particularly susceptible to dioxin and related compounds, and it has been speculated that the amount of these compounds in the Great Lakes during the 1940s was so high that none of the lake trout hatched there during that time survived.

TIPS ON AVOIDING HORMONE DISRUPTORS

Many people, especially concerned parents, do not know where to turn to deal with fears of exposure to these chemicals. If the situation is severe enough they may be able to find some medical expertise in this area. In terms of what people themselves can do, there are some basic, commonsense steps. To help prevent hormonal disruption, here are some steps some have recommended.

1. Do not use conventional chemical cleaners or pesticides of any type because many contain chlorinated chemicals and other persistent organic pollutants.
2. Do not consume food that is processed and eat organic food.
3. Wash all conventional produce well before you prepare it to clean off any pesticide residues, preservatives, or waxes. Wash your hands afterwards because pesticide residues on skins are easily transferred from the fruit to your hands.
4. Do not heat food in any type of plastic container. Many plastics contain hormone-disrupting chemicals that affect food, especially when heated. Microwave foods in glass or ceramic containers.
5. Avoid fish and shellfish from waters suspected of being polluted. Hormone-disrupting chemicals generally accumulate in animals' fatty tissues and expose people when those animal products are ingested. Some parts of a fish may have very high concentrations of mercury, for example. Cultures that eat the whole fish and who fish for subsistence are particularly exposed. Most health risk assessments of food levels that are safe for consumption are based on eating the fillets of fish. This is a large battleground.
6. Reduce consumption of high-fat dairy products and other high-fat foods like meat (especially beef). Many hormone-disrupting chemicals including dioxins accumulate in animal fatty tissues. Some milk is tainted with a dairy growth hormone, which is not currently labeled.
7. Choose unbleached or nonchlorine-bleached paper products. Chlorine bleaching is a major source of dioxin, a particularly toxic POP. Bleached paper, including coffee filters, can pass its dioxin residues into food with which it comes into contact.
8. Women should use nonchlorine-bleached, all-cotton tampons. Most tampons are made from rayon, a wood pulp product bleached with chlorine. Scientists have detected dioxins in these products.

Some individuals have very sensitive endocrine systems and can tolerate very little chemical exposure. They claim their immune systems start breaking down and they get progressively sicker if left exposed to many common chemicals. These individuals take further precautions. They move to the areas with the cleanest air. All water is purified. All food is organic, and even some of that may be limited. Carpets, glues, curtains, drapes, and clothing are all limited. The material must be organic, without any chlorine-based processing. The items must be produced, stored, and shipped in a pesticide-free environment, which is difficult to do. Combustion motors, such as in cars, are not allowed because of the chemicals in their emissions and maintenance. Soaps and other toiletries are all organic and can be limited.

Some estrogenic compounds may be in the food we eat and in the wrapping that surrounds them, for some of the chemicals used to set plastics have been found to be estrogenic. The discovery of the estrogenic effect of plastic stabilizers was made in an unexpected way. Investigators at Tufts University Medical School had been studying estrogen-responsive tumor cells. These cells require estrogen in order to proliferate. Their studies were going well until 1987, when the experiments suddenly went awry. Then the control cells began to show high growth rates suggesting stimulation comparable to that of the estrogen-treated cells. Thus, it was if someone had contaminated the medium by adding estrogen to it. What was the source of contamination? After spending four months testing all the components of their experimental system, the researchers discovered that the source of estrogen was the plastic tubes that held their water and serum. The company that made the tubes refused to tell the investigators about its new process for stabilizing the polystyrene plastic, so the scientists had to discover it themselves. The culprit turned out to be p-nonylphenol, a chemical that is also used to harden the plastic of the plumbing tubes that bring us water and to stabilize the polystyrene plastics that hold water, milk, orange juice, and other common liquid food products. This compound is also the degradation product of detergents, household cleaners, and contraceptive creams. A related compound, 4-tert-pentylphenol, has a potent estrogenic effect on cultured human cells and can cause male carp (*Cyprinus carpis*) to develop oviducts, ovarian tissue, and oocytes.

Some other environmental estrogens are polychlorinated biphenyls (mentioned earlier). These PCBs can react with a number of different steroid receptors. PCBs were widely used as refrigerants before they were banned in the 1970s when they were shown to cause cancer in rats. They remain in the food chain, however (in both water and sediments), and have been blamed for the widespread decline in the reproductive capacities of otters, seals, mink, and fish. Some PCBs resemble diethylstilbesterol (DES) in shape, and they may affect the estrogen receptor as DES does, perhaps by binding to another site on the estrogen receptor. Another organochlorine compound (and an ingredient in many pesticides) is methoxychlor. This would severely inhibit frogs' fertility, and it may be a component of the worldwide decline in amphibian populations.

Some scientists, however, say that these claims are exaggerated. Tests on mice have shown that litter size, sperm concentration, and development were not affected by concentrations of environmental estrogens. However, recent work has shown a remarkable genetic difference in the sensitivity to estrogen among different strains of mice. The strain that had been used for testing environmental estrogens, the CD-1 strain, is at least 16 times more resistant to endocrine disruption than the most sensitive strains, such as B6. When estrogen-containing pellets were implanted beneath the skin of young male CD-1 mice, very little happened. However, when the same pellets were placed beneath the skin of B6 mice, their testes shrank, and the number of sperm seen in the seminiferous tubules dropped dramatically. This wide range of sensitivities has important consequences for determining safety limits for humans. This is sometimes known as the variance in the dose response to a given chemical.

ENVIRONMENTAL THYROID HORMONE DISRUPTORS

The structure of some PCBs resembles that of thyroid hormones, and exposure to them alters serum thyroid hormone levels in humans. Hydroxylated PCB was found to have high affinities for the thyroid hormone serum transport protein transthyretin, and it can block thyroxine from binding to this protein. This leads to the elevated excretion of the thyroid hormones. Thyroid hormones are critical for the growth of the cochlea of the inner ear, and rats whose mothers were exposed to PCBs had poorly developed cochleas and hearing defects.

DEFORMED FROGS: PESTICIDES MIMICKING RETINOIC ACID?

Throughout the United States and southern Canada there is a dramatic increase in the number of deformed frogs and salamanders in what seem to be pristine woodland ponds. These deformities include extra or missing limbs, missing or misplaced eyes, deformed jaws, and malformed hearts and guts. There is speculation that pesticides (sprayed for mosquito and tick control) might be activating or interfering with the retinoic acid pathway. The spectrum of abnormalities seen in these frogs resembles the malformations caused by exposing tadpoles to retinoic acid.

CHAINS OF CAUSATION

Whether in law or science, establishing chains of causation is a demanding and necessary task. In developmental toxicology, numerous endpoints must be checked, and many different levels of causation have to be established. For instance, one could ask if the pollutant spill in Lake Apopka was responsible for the feminization of male alligators. To establish this, one has to ask how the chemicals in the spill might contribute to reproductive anomalies in male alligators and what would be the consequences of that happening. After observing that the population level of the alligators had declined, at the organism level, unusually high levels of estrogens in the female alligators, unusually low levels of testosterone in the males, and a decrease in the number of births among the alligators were reported. On the tissue and organ level, the decline in birthrate can be explained by the elevated production of estrogens from the juvenile testes, the malformation of the testes and penis, and the changes in enzyme activity in the female gonads. On the cellular level, one sees ovarian abnormalities that correlate with unusually elevated estrogen levels. These cellular changes, in turn, can be explained at the molecular level by the finding that many of the components of the pollutant spill bind to the alligators' estrogen and progesterone receptors and that they are able to circumvent the cell's usual defenses against overproduction of steroid hormones.

HUMAN IMPACTS

It is a large battleground right now to prove the effects of environmental compounds on humans. Scientists claim that genetic variation in the human species, the lack of controlled experiments to determine the effect of any particular compound on humans, and a large range of other multiple intervening factors make any causality difficult to prove. Evidence from animal studies suggests that humans and natural animal populations are at risk from these endocrine disruptors. It may be that the damage is greater than thought because most risk assessments do not compute cumulative effects. Because of the many exposure vectors of chemicals, the cumulative impacts and risks could be far greater and have far-reaching effects on many people.

POTENTIAL FOR FUTURE CONTROVERSY

As chemical emissions accumulate in the environment, and bioaccumulate in humans, more controversy will ensue. Loss of fertility, cancer, and other health issues may not be worth the price of increased profits for industries that use or manufacture these chemicals. As more of these chemicals are used and more of their effects become known, this controversy will escalate.

See also Children and Cancer; Cumulative Emissions, Impacts, and Risks; Pesticides

Web Resources

Natural Resources Defense Council: Hormone Disruptors—Emerging Evidence of a Future Threat. Available at www.nrdc.org/health/kids/ocar/chap5e.asp. Accessed January 21, 2008.

U.S. Environmental Protection Agency. Endocrine Disruptors Research Initiative. Available at www.epa.gov/endocrine/. Accessed January 21, 2008.

Further Reading: Guillette, L. J., and D. Andrew Crain. 2000. *Environmental Endocrine Disruptors: An Evolutionary Perspective*. New York: Taylor and Francis; Hester, R. E., and Roy M. Harrison. 1999. *Endocrine Disrupting Chemicals*. London: Royal Society of Chemists; Naz, Rajesh K. 1999. *Endocrine Disruptors: Effects on Male and Female Reproductive Systems*. Boca Raton, FL: CRC Press; Wittcoff, Harold A., Jeffery S. Plotkin, and Bryan G. Reuben. 2004. *Industrial Organic Chemicals*. Hoboken, NJ: Wiley.

HUMAN HEALTH RISK ASSESSMENT

Many controversial decisions about the environment involve issues of human health risks. Reducing human health risks can affect the cost of business. Individuals seek to reduce environmental exposures that can cause cancer, asthma, and other environmentally facilitated risks. How any particular risk is actually assessed is a controversial part of many environmental decisions, and the basis for standards in environmental regulation.

BACKGROUND: EPIDEMIOLOGICAL ROOTS OF MODERN HUMAN HEALTH RISK ASSESSMENT

The basis for the current approach to health risk assessment was established in the 1800s, when cities began to industrialize and urban populations increased dramatically. Industrialization and urbanization created an awareness of public health due to the outbreak of diseases like cholera, yellow fever, dysentery, and others.

During the early decades of the twentieth century (1900–1940), qualitative understanding of health risk assessment improved as increases in labor efficiency were examined. The hazards of occupational exposure to the hundreds of chemicals that were then frequently used in the workplace impaired the strength of the labor force. Occupational exposure in this time period was much more than it is now for most people in most occupations. Work hours were long, work began in childhood, workplace health regulation did not exist, and workers were responsible for their health. Occupational exposures were large and long term. There was little to no governmental regulation and little to no knowledge about the damaging effects of chemicals on humans, except mortality.

The emergence of modern health risk assessment as an applied public policy developed about 1975, five years after the U.S. Environmental Protection Agency (EPA) was formed. The Food and Drug Administration had used these assessments to a limited extent before that time. As more and more chemicals entered into U.S. commerce, regulators wanted to make sure they were safe for people and the environment. These environmental regulators now take a lead role in developing various risk assessment methods. Risk assessment methods have changed since earlier years. They now include other stakeholders, like communities, state governments, environmentalists, and labor unions.

Risk assessment is used to predict the likelihood of many events. Health risk assessment is a different approach that uses toxicology data collected from animal studies and human epidemiology. Health risk assessment combines information about the degree of exposure to quantitatively predict the likelihood that a particular adverse response will occur in a specific human population. It focuses on a single chemical in a single vector or modality (like air or water) and its impact on a single endpoint. The assessment of toxicology data to predict health risks is an early basis for modern risk assessment. The difference between early risk assessments and those developed in the 1980s is the application of computers and massive data sets of exposure information. Computer programs like Geographic Information Systems tie human health risk assessments to environmental issues specifically. They allow large amounts of data to be developed into environmental information. This technological development allows for consideration of more factors and more complete environmental systems.

Emerging technologies like nanotechnology may increase the ability of human health risk assessments to accommodate even larger data sets, as would be necessary to study cumulative impacts.

Since 1980, environmental regulations and some occupational health standards have been based on human health risk assessments. This increased their level of controversy because many stakeholders attack regulatory standards as either being too strict or too lax. Risk-assessment methodologies were used to set some of the first standards for pesticide residues, food additives, pharmaceutical products, drinking water, soil, and ambient air. They were also used to set some of the first threshold and exposure limits for contaminants found in indoor air, consumer products, as well as all the previous categories. Environmental agencies at the state and federal level required their risk managers to use risk assessments. These are very powerful decision-making tools then and now. They are used to help decide whether a given environmental risk is significant or insignificant.

Risk assessments have been used to prevent more stringent regulation of known carcinogens. In 1980 the U.S. Supreme Court used risk assessments to conclude that more stringent regulation of benzene by the Environmental Protection Agency was unwarranted unless it lowered significant risks. Benzene was among the first chemicals regulated by the EPA in the early 1970s; it is a well-known carcinogen. With the increase in citizens' environmental literacy and continued industrial and population growth, human health risk assessment controversies will intensify.

KNOWLEDGE IN RISK PERCEPTION

An important part of environmental literacy is knowledge about the environment. Perceptions about the environment, and about risks, can vary widely depending on age, race, gender, and income. When the question becomes whose knowledge is used to make an environmental decision, a decision that could affect everyone, controversies flare.

COMPARING RISKS: THE DEVELOPMENT OF COMPARATIVE RISK ASSESSMENT (CRA)

Comparing risks is very difficult. Whenever analysts, agencies, communities, or interested parties decide to juxtapose different hazards and compare their severity, they invite controversy. Setting standards often entails a risk comparison.

When and How CRA Is Currently Undertaken

CRAs are a recent policy development that involve the comparison of categories of risks. The most prominent examples of large CRAs have come from the EPA. Reports in 1987 and 1990 both tried to show that if we would set our priorities with a more rational risk-based mindset, we could save more lives and provide greater ecological protection without increasing our total environmental budget. U.S. Supreme Court Justice Steven Breyer has published extensively on how budgets should be allocated based on risks they handle.

POTENTIAL AREAS OF FUTURE CONTROVERSY

The environmental controversies around human health risk assessment are methodological and value laden. They are also very powerful because they set the standards by which government measures pollution. The standard human health risk assessment methodology misrepresents actual exposure in several ways. Total exposure does not always equal average daily exposure times days exposed. The distribution of toxic concentrations, the different exposure vectors per individual, and the distribution of individuals' body weights all vary in uneven intervals. They may vary substantially over a lifetime. The foundational human health risk assessment formulation is appropriate only if exposure events are perfectly correlated to individual variances, and they generally are not in the real world. But this approach to human health risk assessment is the basis for almost all governmental and industry risk assessments. Environmental justice communities tend to challenge the risk assessments because they do not reflect real-world multiple exposures, at many different times and levels. Many of these risk assessments cannot measure or note chemical synergy reactions because they focus on a single chemical in a single vector with one endpoint. Industry often challenges these risk assessments, contending that risks are inflated.

Another battleground for this controversy is the use of this information to aid communities in making decisions about industry. Some communities are able to exercise political power to avoid environmentally dangerous land uses, whereas other communities are forced into the science of human health risk assessment. Both involve public hearings. Environmental justice proponents and sustainability advocates use these risk assessments to compare the environmental benefits and burdens of a given environmental decision. State and federal environmental agencies have used them for years to inform their rule making. Because of their increased use, more controversy will focus on some of the methodological flaws, especially if they are relied on in litigation.

See also Citizen Monitoring of Environmental Decisions; Cumulative Emissions, Impacts, and Risks; Ecosystem Risk Assessment; Nanotechnology

Web Resources

California Department of Pesticide Regulation. Assessing Human Health Risk. Available at www.cdpr.ca.gov/docs/risk/riskassessment.htm. Accessed January 21, 2008.

California Environmental Protection Agency. A Guide to Health Risk Assessment. Available at www.oehha.ca.gov/pdf/HRSguide2001.pdf. Accessed January 21, 2008.

U.S. Environmental Protection Agency. Human Health Risk Assessment. May 2006. Available at www.gao.gov/new.items/d06595.pdf. Accessed January 21, 2008.

Further Reading: Cutter, Susan. 1993. *Living with Risk: The Geography of Technological Hazards.* New York: Routledge, Chapman, and Hall; Finkel, Adam, and Dominic Golding, eds. 1994. *Worst Things First? The Debate over Risk-Based National Environmental*

Priorities. Washington, DC: Resources for the Future; Krimsky, Sheldon, and Dominic Golding, eds. 1992. *Social Theories of Risk.* Westport, CT: Praeger.

HURRICANES

Most hurricane disasters provoke controversies that center around emergency preparation, accurate monitoring of environmental conditions, emergency response, emergency warnings, and evacuation. Post disaster controversies occur around cleanup liability, rebuilding decisions, and insurance coverage. Hurricanes often bring wind and flood damages. Some climate change models predict that we will encounter more hurricanes of greater power in the years to come.

WHAT IS A HURRICANE?

A hurricane is more than a powerful storm. A hurricane is a severe tropical storm that forms in the North Atlantic Ocean, the Northeast Pacific Ocean east of the dateline, or the South Pacific Ocean east of 160°E. To create a hurricane, warm tropical oceans, moisture in the air, and light winds are necessary. A hurricane can produce violent winds, long series of big waves, torrential rains, and floods. In other regions of the world, these types of storms are called:

Typhoon: the Northwest Pacific Ocean west of the dateline
Severe tropical cyclone: the Southwest Pacific Ocean west of 160°E or Southeast Indian Ocean (east of 90°E)
Severe cyclonic storm: the North Indian Ocean
Tropical cyclone: the Southwest Indian Ocean

Hurricanes rotate in a counterclockwise direction around a calm center, called the *eye*. A tropical storm becomes a hurricane when winds reach 74 miles per hour (mph). There are on average six Atlantic hurricanes each year; over a three-year period, approximately five hurricanes strike the U.S. coastline from Texas to Maine. The Atlantic hurricane season begins June 1 and ends November 30. The East Pacific hurricane season runs from May 15 through November 30, with peak activity occurring during July through September. In a normal season, the East Pacific would expect 15 or 16 tropical storms. Nine of these would become hurricanes, of which four or five would be major hurricanes. The prediction of these storms is a major task of some federal and military agencies.

When hurricanes move onto land, the heavy rain, strong winds, and heavy waves can damage buildings, trees, and cars. The heavy waves are called a storm surge. Storm surge is very dangerous. It can come in a high tide that occurs with a hurricane. It can block transportation routes after the main body of a storm has passed.

Hurricanes are classified into five categories based on their wind speed, central pressure, and damage potential (see the following list). Category 3 and higher hurricanes are considered major hurricanes, although categories 1 and 2 are still extremely dangerous and warrant full attention.

Table H.1 Saffir-Simpson Hurricane Scale

Scale Number (Category)	Sustained Winds (MPH)	Damage	Storm Surge (feet)
1	74–95	Minimal: Unanchored mobile homes, vegetation, and signs.	4–5
2	96–110	Moderate: All mobile homes, roofs, small crafts, flooding.	6–8
3	111–130	Extensive: Small buildings, low-lying roads cut off.	9–12
4	131–155	Extreme: Roofs destroyed, trees down, roads cut off, mobile homes destroyed. Beach homes flooded.	13–18
5	More than 155	Catastrophic: Most buildings destroyed. Vegetation destroyed. Major roads cut off. Homes flooded.	Greater than 18

Hurricanes can produce widespread torrential rains as bands of rainstorms sweep around the eye along the path of the hurricane. Floods are the deadly and destructive result. Slow-moving storms and tropical storms moving into mountainous regions tend to produce especially heavy rain. Excessive rain can trigger landslides or mudslides, especially in mountainous regions. Flash flooding can occur due to intense rainfall. Flooding on rivers and streams may persist for several days or more after the storm.

The amount of state, federal, or local assistance before, during, and after hurricanes is currently a large controversy. Some maintain that overreliance on government is part of the problem and people need to be more self-sufficient. The majority of stakeholders that have been through a hurricane maintain that coordinated and effective monitoring, communication, emergency services, and a transportation plan are better then relying on self-sufficiency. This is what voluntary evacuations tend to do.

RACIST POLICE BLOCKED BRIDGE: HURRICANE KATRINA

Hurricane Katrina left many people homeless and in desperate need of shelter. People were fleeing the flood-ravaged city any way they could. A Louisiana police chief has admitted that he ordered his officers to block a bridge over the Mississippi River and force escaping evacuees back into the chaos and danger of New Orleans. Witnesses said the officers fired their guns above the heads of the terrified people to drive them back and protect their own suburbs. Two paramedics who were attending a conference in the city and then stayed to help those affected by the hurricane said the officers told them they did not want their community becoming another New Orleans. Many read this statement to mean that they did and want African Americans to come into their police jurisdiction.

The police blocked off the road on the Thursday and Friday after Hurricane Katrina struck on Monday, August 29. According to newspaper accounts, the evacuees left as a group of up to 800 people walking across the bridge when they heard shots and saw people running. They saw a chain of armed police officers blocking the route. When they asked about the buses, they were told there was no such arrangement and that the route was being blocked to avoid the parish becoming another New Orleans. Parishes are similar to counties in other states. They identified the police as officers from the city of Gretna, a suburb of New Orleans. The following day the evacuees tried to cross again. The police turned them away with the threat of gunfire. Police shot live ammunition over the heads of a middle-aged white couple, who were also turned back.

Arthur Lawson, chief of the Gretna police department, said he had not yet questioned his officers as to whether they fired their guns. He confirmed that his officers, along with those from Jefferson Parish and the Crescent City Connection police force, sealed the bridge and refused to let people pass. This was despite the fact that local media were informing people that the bridge was one of the few safe evacuation routes from the city. He offers little explanation for why he denied desperate people the evacuation route necessary for their survival.

Gretna is a white suburban town of around 18,000 inhabitants. In the aftermath of Katrina, three-quarters of the inhabitants still had electricity and running water. But Chief Lawson told UPI news agency: "There was no food, water or shelter in Gretna City. We did not have the wherewithal to deal with these people. If we had opened the bridge our city would have looked like New Orleans does now—looted, burned and pillaged."

Police later said they blocked the evacuees because there were no supplies or services for them on the other side of the river. The case raised widespread allegations of racism and spurred two marches across the bridge by national civil rights organizations in the months after the hurricane. Gretna Police Chief Arthur Lawson has acknowledged that his officers fired shots into the air during the blockade in an attempt to quell what he described as "unrest among the evacuees."

Denial of evacuation routes because of race harkens back to the days of slavery and Jim Crow. Many brave and unknown civil rights activists and lawyers fought for African Americans' right to ride public transportation as one of the first affirmative acts of the United States to proactively establish equality. The famous case of Rosa Parks was about transportation equality. At that time African Americans were forced to ride in the back of the bus and give up front-row seats to whites.

The racist denial of evacuation routes has sparked national controversy. National representatives want to cut off all federal monies to Gretna, and some have been arrested while marching across the bridge without a permit. Both state and federal prosecutors are examining evidence in the case, but many do not have any confidence in local investigations. The American Civil Liberties Union wants to make these investigations as public as possible, but the investigating authorities are not communicating about it with the public.

POTENTIAL FOR FUTURE CONTROVERSY

As climate change from global warming is increasing the temperature of the oceans where hurricanes develop, the forecast is for more controversy around hurricanes. For hurricane warnings to be effective with the general public they have to be accurate; otherwise, they are ignored. Issues of mandatory versus voluntary evacuation also require good public notification. Many people are not able to leave once a hurricane hits. Emergency fire, police, and medical services may not be available, increasing mortality. After the hurricane, the liability for cleanup costs is enormous. Many industries hit by a hurricane release everything into the environment as part of the devastation. This can have severe environmental impacts that can threaten public health. Should cleanup standards be lessened posthurricane to aid economic recovery? Should cities that are in the path of hurricanes be rebuilt?

Private insurance companies will not insure flood- or hurricane-prone areas. Residences are required to have government flood insurance. Should the government underwrite decisions to economically develop in areas prone to hurricanes? This is an issue of debate and is also part of the controversies around hurricanes. In February 2007 one of the largest private insurers in Mississippi, State Farm, withdrew its coverage. Stating that it will honor existing contracts, the company maintained that court judgments holding them liable for post-Katrina damage are too risky. Others criticize State Farm for underestimating the risk from natural disasters. Large insurance companies are financial centers of economic development, underwriting risk for projects that otherwise might not be able to attract enough capital. They can also make loans and influence the financial attractiveness of a business location. When they leave a state, it can impact other measures of financial soundness, such as bond ratings.

GAPING HOLE IN U.S. ENVIRONMENTAL POLICY: LACK OF INTERGOVERNMENTAL RELATIONS

In the United States, the intergovernmental relations between local, state, and federal government levels on most land-use and many environmental decisions are poor. There is very little communication, which often benefits industry and hampers citizens' access. When a disaster strikes and decisions must be made quickly, the emergency response function of government exposes the weakness of these poor intergovernmental relationships. Important life-saving decisions about deploying urgent resources and giving essential public information are delayed when the lines of intergovernmental communication and control are unclear.

Hurricanes will occur again. Some models of climate change have them occurring more frequently in the United States than now. Many controversies still remain. Should we rebuild hurricane alleys? Should the federal government continue to offer flood insurance to these areas? Can technology mitigate some of the human and environmental impacts of hurricanes? Just how much can planning and technology do in hurricane alleys?

These difficult questions remain, and so too will controversies about hurricanes.

See also Climate Change; Evacuation Planning for Natural Disasters; Floods; Global Warming

Web Resources

American Association of Insurance Services. 2005. Wind or Water? Voices Call for a Restructuring of Coverage. *Viewpoint* 30. Available at www.aaisonline.com/viewpoint/05fall2.html. Accessed January 21, 2008.

National Oceanic and Atmospheric Administration. 2005. Noteworthy Records of the 2005 Hurricane Season. Available at www.noaanews.noaa.gov/stories2005/s2540b.htm. Accessed January 21, 2008.

Further Reading: Elsner, James B., and A. Birol Kara. 1999. *Hurricanes of the North Atlantic: Climate and Society.* New York: Oxford University Press; Liu, Kam-biu, and Richard J. Murnane. 2004. *Hurricanes and Typhoons: Past, Present, and Potential.* New York: Columbia University Press; Mulcahy, Matthew. 2005. *Hurricanes and Society in the British Greater Caribbean, 1624–1783.* Baltimore: Johns Hopkins University Press; Tufty, Barbara. 1987. *1001 Questions Answered about Hurricanes, Tornadoes and Other Natural Air Disasters.* North Chelmsford, MA: Courier Dover Publications.